AF332952

NEUROSECRETION
Cellular Aspects of the
Production and Release of
Neuropeptides

NEUROSECRETION

Cellular Aspects of the Production and Release of Neuropeptides

Edited by

Brian T. Pickering
Jonathan B. Wakerley

and

Alastair J. S. Summerlee
University of Bristol
Bristol, United Kingdom

PLENUM PRESS • NEW YORK AND LONDON

Library of Congress Cataloging in Publication Data

International Symposium on Neurosecretion (10th: 1987: Bristol, Avon)
  Neurosecretion: cellular aspects of the production and release of neuropeptides /
edited by Brian T. Pickering, Jonathan B. Wakerley, and Alastair J.S. Summerlee.
      p.   cm.
  "Proceedings of the Tenth International Symposium on Neurosecretion, held August
31–September 5, 1987, in Bristol, United Kingdom" — T.p. verso.
  Includes bibliographies and index.
  ISBN 0-306-42919-5
  1. Neuropeptides — Metabolism — Congresses. 2. Neurosecretion — Congresses. I.
Pickering, Brian T. II. Wakerley, Jonathan B. III. Summerlee, Alastair, J. S. IV. Title.
  [DNLM: 1. Electrophysiology — congresses. 2. Neuropeptides — biosyntheses —
congresses. 3. Neurosecretion — congresses. W3 IN92Q 10th 1987n / WL 104 I6153
1987n]
  QP552.N39I585   1987
  599'.0188 — dc19
  DNLM/DLC   88-12440
  for Library of Congress   CIP

Proceedings of the Tenth International Symposium on Neurosecretion,
held August 31–September 5, 1987,
in Bristol, United Kingdom

© 1988 Plenum Press, New York
A Division of Plenum Publishing Corporation
233 Spring Street, New York, N.Y. 10013

All rights reserved

No part of this book may be reproduced, stored in a retrieval system, or transmitted
in any form or by any means, electronic, mechanical, photocopying, microfilming,
recording, or otherwise, without written permission from the Publisher

Printed in the United States of America

PREFACE

This volume collects together the Proceedings of the Tenth International Symposium on Neurosecretion which was held in Bristol in September 1987. This series of symposia began with a meeting in Naples in 1953 and it was particularly gratifying to welcome two members of that original gathering, Berta Scharrer and Ellen Thomsen, to the 10th Symposium. The acceptance of the invitation to meet in Bristol gave particular pleasure to the Local Organising Committee because it was the first time that a former venue was revisited - the 3rd Symposium was organised in Bristol by Hans Heller in 1961 and we were very glad that his widow, Josephine, was able to be the Guest-of-Honour at the banquet of our 10th symposium.

Neurosecretion has diversified considerably since the first meeting in 1953. From the outset, the local committee decided to focus the meeting by organising it under the title "Cellular aspects of the production and release of neuropeptides." Having sought advice from the International Committee for Symposia on Neurosecretion, we invited 28 scientists from throughout the world to present their work on aspects ranging from the organisation of the gene through to the electrical properties of peptide-secreting neurones. The progress of studies in neurosecretion owes much to the comparative approach and we attempted to reflect this in the programme of the tenth symposium by inviting the speakers, to illustrate the various aspects of neuropeptide secretion, from among scientists working with systems in many parts of the animal kingdom.

The development of Neurosecretion proceeded from the initial discovery and delineation of the concept by Ernst Scharrer. It was thus appropriate that this 10th Symposium was inaugurated with the delivery by Hal Gainer of the Scharrer Memorial Lecture of the International Society for Neuroendocrinology. This set the scientific tone of the meeting at a high level which was maintained throughout the week.

Drs. Berta Scharrer and Ellen Thomsen

Discussion was perceived to be an important function of the meeting and, to this end, there was both ample time set aside in the formal programme and encouragement of the participants to mount posters, all of which were on view throughout the meeting. Unfortunately, space does not permit either a commentary on the discussions or publication of the poster abstracts, although their titles are listed at the end of the volume..

This volume has been prepared from camera-ready copy submitted by the authors. However, to achieve a uniformity of typescript the authors submissions have been transcribed either from their word-processor disks or, sometimes, by retyping. There has been no opportunity for authors to see proofs of the final product and the editors must accept responsibility for any errors. We are confident that such errors will be few because of the careful and unstinting work of Anthea Moody and are pleased to be able to acknowledge our profound gratitude to her.

We hope that our book will give the reader some flavour of the exciting atmosphere of The 10th Symposium on Neurosecretion; although this cannot be complete without the experience of the magnificent gardens of Goldney Hall or the splendid firework display with which the City of Bristol (fortuitously) launched the meeting.

Brian Pickering
Jonathan Wakerley
Alastair Summerlee

Bristol, January, 1988.

ACKNOWLEDGMENTS

We are grateful for financial support for the Symposium from:

> The Wellcome Trust
> The Royal Society
> The British Council
> Society for Endocrinology
> International Society for Neuroendocrinology
> University of Bristol
> Bristol City Council
> Anachem
> Merck, Sharp and Dohme Ltd
> Parke Davis & Company Ltd
> Upjohn Limited

We also acknowledge that the Symposium would not have succeeded without the willing hard work of the Staff and Graduate Students of the Department of Anatomy, University of Bristol, the Permanent Officers of Bristol Students' Union and the Staff of Goldney Hall.

G. Clarke	R.W. Swann
B.K. Follett	A.J.S. Summerlee
L. Haynes	J.B. Wakerley
B.T. Pickering	R.T.S. Worley

(Local Committee)

CONTENTS

Molecular Biology and Biosynthesis of Neuropeptides

Localization of Neuropeptides

Cell Biology of Neuropeptide Secretion

OXYTOCIN AND VASOPRESSIN: AFTER THE GENES, WHAT NEXT?

Harold Gainer, Miriam Altstein, and Yoshinobu Hara

Laboratory of Neurochemistry
National Institutes of Health
NINCDS, Building 36, Room 4D-20
Bethesda, Maryland 20892

INTRODUCTION

It is now almost sixty years since Ernst Scharrer, in a paper emanating from his doctoral dissertation studies at the University of Munich (Scharrer, 1928), first described the gland-like secretory characteristics of the magnocellular neurons in the preoptic nucleus of the minnow, Phloxinus laevis. This discovery and its bold interpretation, i.e., the theory of neurosecretion, although initally received sceptically by the scientific community, was to become, largely through the subsequent heroic efforts of Ernst and Berta Scharrer, and their close colleague W. Bargmann (Bargmann and Scharrer, 1951; Sano, 1985), one of the most heuristic concepts in modern neurobiology. Not only did this concept usher in the present highly active era of neuroendocrinology, but also the even more explosive field of neuropeptide research in general (Scharrer, 1987).

The magnocellular neurons of the hypothalamo-neurohypophysial system, often referred to as the "classical neurosecretory system", have been the most intensively studied peptidergic neurons in the nervous system (Silverman and Zimmerman, 1983), and continue to serve as the principal experimental models for peptidergic (and neurosecretory) neuronal research. Excellent reviews describing recent immunohistochemical studies of this system, as well as other neuronal systems containing the biologically active peptides, oxytocin (OT) and vasopressin (VP), have been published (Castel et al., 1984; Silverman and Zimmerman, 1983; Swanson and Sawchenko, 1983; Sofroniew, 1985). As a result of these studies, three new concepts about these peptides have emerged. These include: 1) that, in addition to the "classical" magnocellular system, there are OT and VP fibers, deriving from other neuronal groups in the brain, which terminate in many rostral and caudal brain regions, 2) that in some of these systems (including the magnocellular neurons) the OT and VP coexist with other peptides, often in the same secretory vesicles, and 3) that OT and VP are produced and secreted from various peripheral tissues for local functions (Nussey et. al., 1984; Hanley et al., 1984; Wathes, 1984; Geenen et al., 1986). These observations, based largely on immunological data, indicate that the OT and VP genes are expressed in a more diverse set of cell-types than previously recognized, and raise obvious questions about the mechanisms underlying the regulation of expression of these peptide genes in these different cell-types. The recent cloning and characterizations of the OT and VP genes has transformed these questions from academic issues into experimental challenges for the next decade.

OXYTOCIN AND VASOPRESSIN GENES: STATE OF THE ART

OT and VP genes have been isolated and their nucleotide sequences have been determined for four mammalian species: the cow (Ruppert et al., 1984), rat (Ivell and Richter, 1984), human (Sausville et al., 1985), and the mouse (Hara, unpublished). The structural organization of these genes is similar in all of the species studied, and a representative example (the rat genes) is illustrated in Fig. 1. The only apparent structural distinction between the OT and VP genes is the presence of the glycoprotein sequence in the latter. The function of this highly conserved VP-associated glycoprotein remains unclear at present, although it is interesting, in this regard, that the vasotocin precursor in the toad also contains a highly homologous glycopeptide, whereas the mesotocin precursor does not (Nojiri et al., 1987). Restriction analyses of the genomic DNAs have thus far revealed only one copy each of the OT and VP genes. Studies in the human genome (Sausville et al., 1985) have shown that the OT and VP genes are physically linked within 12kb on the same chromosome with an inverted arrangement of their coding strands (i.e., they have opposite transcriptional orientations). A similar organization has been found in the mouse genome (Hara, unpublished).

While the analysis of these genes in so many species represents something of a cloning "overkill", these data do provide a unique opportunity to look for important homologies in the so-called "5' -flanking promotor" regions in the genes. The reason that this is of significance is that the DNA sequences involved in the regulation of gene expression (i.e., promoters and enhancers) are usually found in this "upstream" region of the gene, and comparing these sequences in the OT and VP genes between the species could reveal highly conserved, OT- and VP- specific sequences presumably involved in cell-specific expression and regulation. Such a comparison shows extensive interspecies homologies in the 5' -flanking sequences amongst the OT and VP genes. This is illustrated in Fig. 2 where the sequences of approximately 180 bases upstream from the cap sites in the human OT and VP genes are illustrated and compared to similar genes in other mammalian species as well to one another. Eukaryotic genes commonly have two control signals in the 5' -upstream region, a "TATA"and a "CAT" box. The TATA box is an AT rich region usually about 20-30 bases upstream from the translational initiation signal, which is believed to direct the RNA polymerase II to the correct initiation site for transcription. The CAT box is a sequence (CAAT) which is usually located upstream from the TATA box, and is believed to regulate transcription by RNA polymerase II. The VP and OT genes appear to contain several CAT sequences upstream from the TATA box, and distinct TATA boxes (also called modified Goldberg-Hogness boxes by some authors). The TATA box is between positions -20 to -30 in Fig. 2, and is highly conserved between the human OT and VP genes (i.e., 9 out of 10 of these bases are identical). The rectangles in Fig. 2 show regions of highly conserved nucleotides for the VP genes from other species as well as for the OT genes and the vertical lines depict identical nucleotides between the human VP and OT genes. It is obvious from this analysis that there are surprisingly long stretches of highly homologous sequences within the peptide gene subfamilies, whereas comparisons between the nucleotides in the human OT versus VP gene show a substantially lower degree of sequence similarity. This remarkable conservation of sequences within each peptide subfamily is suggestive that these sequences represent regulatory elements directly involved in the differential expression of the OT and VP genes in different cell-types. The point, however, is to demonstrate experimentally that this is indeed the case.

OXYTOCIN AND VASOPRESSIN GENE EXPRESSION

Although the mechanisms which are responsible for OT and VP gene regulation remain a subject for future study, the above information about gene sequences (and the mRNA sequences) has provided investigators the ability to generate either synthetic oligonucleotide or cloned cDNA probes which are essential for the descriptive analysis of OT and VP gene expression. The term gene expression is used here only to refer to the primary transcription processes leading to the production of mRNA, and does not refer to the subsequent processes which lead to the formation of the final peptide products (see later). In the past four years a large number of studies on the expression of these peptide genes under a variety of experimental conditions have been reported. Current methods for the measurement of mRNA levels are now more feasible than measurement of prohormone levels and biosynthesis rates, especially in small numbers of cells. As a result, proofs of biosynthesis of these peptides in various cells, and evaluations of their biosynthetic activities are increasingly being determined by the analysis of mRNA levels. However, it should be pointed out that mRNA levels are not only the representation of gene transcription, but also of mRNA turnover. Hence, strictly speaking, quantitative interpretations of gene expression from measurements of mRNA levels should be cautiously regarded until the individual

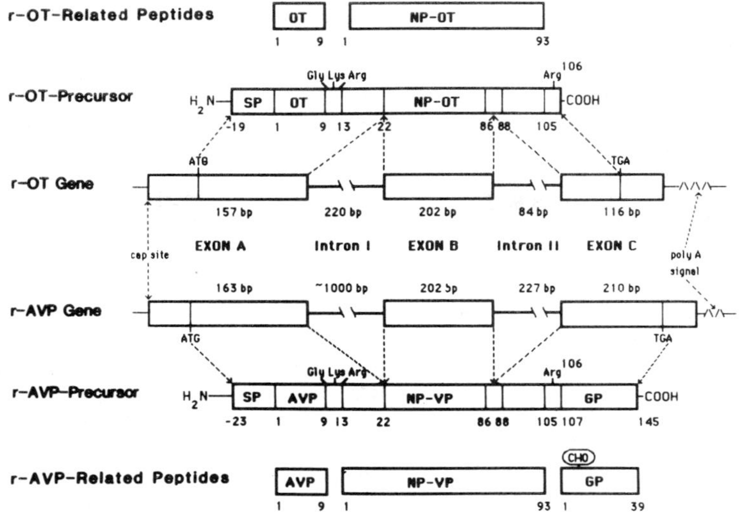

Fig. 1. Organization of the AVP (r-AVP gene) and OT(r-OT gene) genes in the rat, and the relationships of these genes to their respective precursors (pre-prohormones) and final peptide products. Both genes are composed of 3 exons (A,B. and C) shown as open rectangles, separated by 2 introns (shown as disconnected lines between exons). The "cap site" represents the site of transcription initiation, and the ATG codon the site of translation initiation. The TGA codon on the gene represents the site of translational termination. The number of nucleotides in each gene component is illustrated (bp). The amino acids in the precursors which are post-translationally modified to produce the peptide products are shown in the precursor structures. Abbreviations: SP; signal peptide; NP-OT, oxytocin associated neurophysin; NP-VP, vasopressin associated neurophysin; GP, glycopeptide; CHO, carbohydrate moiety on glycopeptide; bp, base pairs. Data from Ivell and Richter (1984).

contributions of both these processes can be accounted for. In addition, while it has been shown that mRNA levels correlate well with biosynthetic activity in some cell systems, this has rarely been performed in neurons, primarily due to difficulties in measurements of biosynthesis rates in these systems. In this regard the magnocellular neurons in the hypothalamus have again served as useful models.

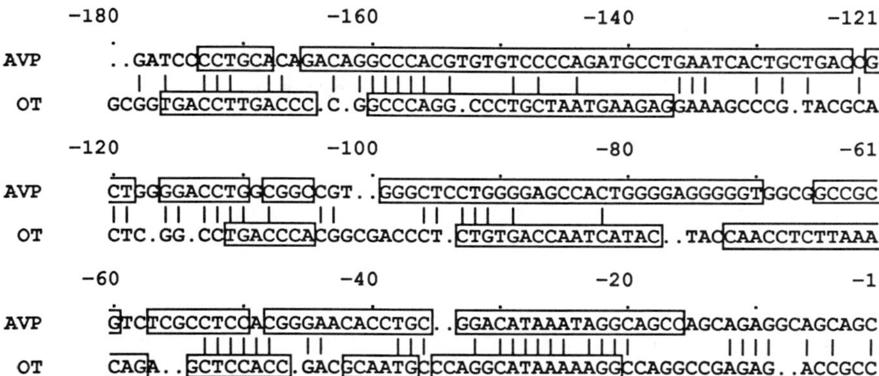

Fig. 2. Comparisons of sequence conservation in the 5' promotor regions of the VP and OT genes in various species. The nucleotide sequences shown are for the human genes (data from Sausville et al., 1985), and the nucleotide sequences enclosed by rectangles are highly conserved with respect to sequences found in comparable (VP or OT) genes in other species. The vertical lines depict identical nucleotides between the human OT and VP genes. The dots between nucleotides are gaps introduced into sequences in order to maximize the number of matches.

The magnocellular neurons in the hypothalamus are especially well suited for such studies due to their large number and compact nuclear localizations. Hence, liquid and solid phase (e.g., Northern blots) hybridization, as well as _in situ_ hybridization histochemistry (ISHH) studies are eminently feasible in this system. Indeed, several laboratories have already shown the efficacy of these methods to localize and characterize VP and OT mRNA in the hypothalamus (Burbach et al., 1986; Fuller et al., 1985a; McCabe et al., 1986a; Nojiri et al., 1985; Sherman et al., 1986a; Uhl et al., 1985; Van Tol, 1987). In addition, these methods are quite sensitive since mRNA levels have been detected in the hypothalamus at very early stages of development (Reppert and Uhl, 1986; Van Tol, 1987; Van Tol et al., 1986), comparable to the detection of peptides by immunocytochemistry (Whitnall et al., 1985).

It has been known for some time that dehydration or salt loading of rats leads to an enhanced biosynthesis of VP and OT. Recent studies using Northern blot analysis and ISHH have shown that these treatments also increase levels of the hypothalamic mRNA (Burbach et al., 1984; Sherman et al., 1986a; Van Tol, 1987; Van Tol et al., 1987 Zingg et al., 1986). Although these studies validate the concept that increased peptide mRNA levels reflect increased biosynthetic rates in these neurons , it should be pointed out that the quantitive estimates of the mRNA increases range from 2 to 7 times normal levels. This large variability in quantitation of mRNA changes between laboratories and methodologies, indicates that more methodological work is necessary to standardize the quantitative analyses. In any case, this increase in mRNA in response to osmotic stimuli has also been extended to the coexistent peptide dynorphin in the magnocellular VP neurons, whose mRNA appears to increase in a coordinate fashion with VP mRNA (Sherman et al., 1986b;c). Another correlation between increased mRNA levels and increased biosynthesis rates has been shown for oxytocin mRNA in magnocellular neurons during pregnancy and lactation (McCabe et al., 1986a; Van Tol, 1987). Of particular importance is the finding that, while osmotic perturbations increase the VP mRNA levels in the supraoptic (SON) and paraventricular (PVN) nuclei, the VP neurons in the suprachiasmatic (SCN) nucleus do not increase their mRNA content under these conditions (Burbach et al., 1984; Sherman et al., 1986a; Van Tol, 1987). Instead, the VP mRNA levels in the SCN appear to be under circadian control mechanisms, whereas the VP neurons in the SON and PVN are not (Uhl and Reppert, 1986; Reppert and Uhl, 1987).

The above demonstration of independent regulation between the SON and PVN versus SCN suggests that neuron-specific regulation of the VP gene is likely. An even more dramatic demonstration of this point is the recent finding that VP gene expression is greatly increased in rat parvocellular neurons, normally containing CRF, after adrenalectomy (Wolfson et al., 1985), and that the VP mRNA in this system can be decreased by glucocorticoids (Davis, 1986). Since in-situ hybridization histological studies show that CRF mRNA also increased in these neurons under these conditions (Young et al., 1986a), this indicates that both of these genes are being regulated by glucocorticoid feedback onto these neurons. In contrast, the magnocellular VP neurons' gene expression appears to be unaffected by glucocorticoid feedback (Young et al., 1986b). Interestingly, CRF mRNA is decreased in parvocellular neurons, but is increased in oxytocin-containing magnocellular neurons as a result of hyperosmotic stimuli (Young, 1986).

The use of molecular probes for mRNA have also been useful in evaluating the defect in the diabetes insipidus model, the Brattleboro rat. The homozygous Brattleboro rat (di/di rat) contains the normal complement of hypothalamic VP neurons, but these neurons do not contain VP, thereby leading to diabetes insipidus (Valtin et al., 1974). It is also known that these neurons do not synthesize the VP precursor (Brownstein and Gainer, 1977). Recent recombinant DNA studies by Schmale and Richter (1985) have located the genetic defect in the di/di rat. The mutant VP gene has a single nucleotide base (G) deletion in the second exon which encodes the highly conserved region of neurophysin. The consequences of this single base deletion would be profound, i.e., it would cause a shift in the reading frame of the VP precursor altering the amino acid sequence from position 64 in the VP-associated neurophysin to the C-terminus of the precursor. The result would be the loss of five of fourteen of the cysteines in the precursor, the absence of a stop codon in the mRNA, the loss of the spacer arginine (cleavage site) between the neurophysin and the glycopeptide, and the absence of the -Asn-Ala-Thr- sequence in the C-terminus region of the precursor (which acts as a signal for glycosylation). The nature of this genetic defect in the di/di VP gene, however, would not preclude transcription or normal mRNA processing. Indeed, analyses of the di/di hypothalamic mRNA have revealed that the VP gene is being expressed in the hypothalamus of the di/di rat (Fuller et al., 1984; 1985a; 1985b; Majoub et al., 1984; McCabe et al., 1986b; Schmale et al., 1984; Sherman et al., 1986b; Uhl et al., 1985; Van Tol et al., 1986; Young et al., 1986b). Therefore, as might have been predicted from the sequence data, the defect is not primarily at the transcriptional level, but at another level (see Ivell et al., 1986; Schmale et al., 1984; for discussion of this issue).

The discovery that various peripheral tissues in the di/di rat contained immunoreactive vasopressin (Hanley et al., 1984; Nussey et al., 1984; Wathes, 1984) also focused attention on the possible expression of the OT and VP genes in normal peripheral tissues. Subsequent measurements of the mRNAs in normal tissues have revealed that either VP or OT mRNAs or both are present in ovaries (Fuller et al., 1984; 1985b; Ivell and Richter, 1984b; Ivell et al., 1985;1986; Einspanier et al., 1986), in testes (Ivell et al., 1986; Rebhein et al., 1986), in adrenals (Ivell et al., 1986) and in thymus (Geenen et al., 1986; Rebhein et al., 1986). These data indicate that the VP and OT genes are being expressed selectively not only in the brain, but also selectively in peripheral tissues, presumably via different gene regulatory mechanisms, and for different functions. Interestingly, there is also an indication of post-transcriptional processing differences between brain and peripheral tissues. Normal hypothalamic VP mRNA is 720 bases in length, whereas peripheral VP mRNA is about 620 bases. However, both can be reduced to the same length (580 bases) after treatment with RNAase H and oligo (dT) (Ivell et al., 1986). This indicates that the difference in mRNA length is largely due to a differences in the poly (A) tails.

REGULATION OF POST-TRANSCRIPTIONAL AND POST-TRANSLATIONAL PROCESSING

The presence of mRNA in a cell does not necessarily mean that the peptides encoded by the mRNA will be found in the cell (see above for di/di rat). Several post-transcriptional and post-translational processing steps are also necessary in order for the final peptide forms to emerge (Douglass et al., 1984; Gainer, 1983; Loh et al., 1984). These processing steps may also be under differential regulatory control in specific cells.

The OT and VP RNA's do not appear to be alternatively spliced. However, another aspect of post-transcriptional processing may be regulated, and this is the length of the poly (A) tail. As pointed out above, hypothalamic mRNA for the VP and OT transcripts consistently have longer poly (A) tails than the mRNAs in peripheral tissues, suggesting at least a cell-specific regulation of this process. Recent studies by Zingg et al., (1986;1987) have shown that the lengths of the poly (A) tails can also be alternatively regulated in a given cell under various experimental conditions. Zingg et al., (1986) have shown that the increased VP mRNA levels during osmotic stimulation are maintained for 30 days after rehydration, raising the possibility that these newly synthesized mRNAs have an unusually long half-life. More recently, this same group (Zingg et al., 1987) has reported that the mRNA induced during osmotic stress has a longer poly(A) tail length and that this is maintained during rehydration. They suggest that this increased poly (A) tail length may increase the mRNA stability and therefore lower turnover, and could account for the prolonged half-life during rehydration.

OT and VP, in the manner of virtually all neuropeptides, are initally synthesized as larger precursor proteins, which are subsequently proteolytically cleaved and enzymatically modified to yield the final biologically active peptide products (Douglass et al., 1984; Loh et al., 1984). These post-translational processing steps generally appear to occur within the secretory vesicles (Gainer et al., 1985; Orci et al., 1985: Tooze et al., 1987). Are these post-translational processes ever regulated in the OT and VP systems? The simple structure of the OT precursor (Fig. 1) suggests that there is little room for alternative processing. However, recent studies in our laboratory on the development of OT neurons, suggest that even the simple cascade of events, from cleavage at the Lys-Arg site to amidation of the OT peptide, can be regulated.

Several laboratories have studied the ontogeny of OT and VP in the rat hypothalamo-neurohypophysial systems (HNS) and have concluded that while VP is present early in embryological development (around E16), OT is not present in the HNS until around birth (for review of this literature, see Castel et al., 1984; Whitnall et al., 1985). We have recently re-examined this issue, using radioimmunoassay (RIA) methods on the HNS of Sprague-Dawley rats ranging in age from E15 to PN 21, and have generally confirmed these conclusions (Fig.3). Taken at face value, these data might imply that the OT gene is simply being expressed later than the AVP gene in the magnocellular neurons of the HNS. However, in a more extensive set of studies, we have found that the OT gene is being expressed and the precursor is synthesized during fetal life, although at lower levels than the VP gene (see also Van Tol, 1987). We also found that during E18-21 no completely processed (amidated) OT is present in the HNS. However, significant amounts of OT-$Gly_{10}Lys_{11}Arg_{12}$ could be detected at these stages, which indicated that the Lys-Arg specific endopeptidase was present and functional in the OT neurons. The VP peptides, on the other hand, were completely processed to amidated AVP at these stages (M. Altstein and H. Gainer, in preparation). Therefore, we conclude that the expression of the OT gene and emergence of precursor are being regulated independently of the post-translational processing steps necessary to yield amidated OT. In this case, the block in processing appears to be at

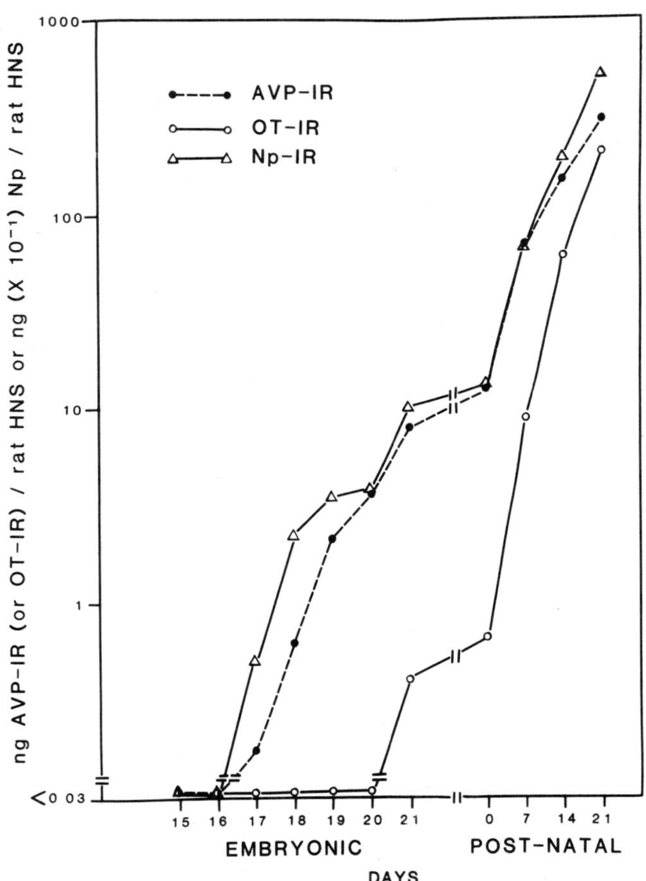

Fig. 3.   Developmental expression of the peptide products of the hypothalamo-neurohypophysial system (HNS). Data for HNS includes total hypothalamus plus neural lobe at all ages. Note that in contrast to AVP-immunoreactivity (AVP-IR) and neurophysin immunoreactivity (NP-IR), the oxytocin-immunoreactivity (OT-IR) in the HNS is virtually absent in HNS during fetal life (Embryonic days 18-21), and lags considerably behind AVP-IR in early post-natal life (M. Altstein and H. Gainer, unpublished data).

the carboxypeptidase-B-like enzymatic step, thereby resulting primarily in a OT-$Gly_{10}Lys_{11}Arg_{12}$ "final" product. Since these processing steps occur to completion in the VP cells, this "regulation" appears to be cell-specific and not due to a general developmental lag in the processing mechanisms in the hypothalamus. The biological significance of this differential development of OT and VP neurons remains to be elucidated.

CONCLUDING REMARKS

The introduction of recombinant DNA technologies into OT and VP biosynthesis research has provided a rapid accumulation of new information. Not only have the OT and VP gene structures been elucidated in four mammalian species, but molecular probes for the VP and OT specific mRNAs have been used extensively to study the expression of these genes in a wide variety of cell-types in the nervous system and in the periphery. It is apparent that with these new tools, new information about the differential regulation of these genes (and presumably the synthetic activities) in the varied cell-types which synthesize OT or VP, under many interesting physiological conditions will be soon forthcoming. The "what next?" in the title of this paper relates to the need to understand the mechanisms which underlie these differential gene expressions. Clearly there are different trans-acting-regulatory factors in the cells, and distinct cis-acting DNA sequences in the OT and VP genes, which are involved in the selective gene expressions observed as a consequence of various physiological states (e.g., glucocorticoid feedback, prolonged osmotic stress, lactation and development). The problem is that although there are emerging paradigmatic approaches to studying the mechanisms of gene expression in general, they are not easily applicable to the present cell-types that produce OT and VP. Experimental activity directed at the development of new models (e.g., continuous cell lines expressing OT and VP, and transgenic mice models) is first necessary before progress can be made in this arena.

REFERENCES

Bargmann, W., and Scharrer, E., 1951, The origin of the posterior pituitary hormones, Am. Scientist, 39:255.

Brownstein, M.J., and Gainer, H., 1977, Neurophysin biosynthesis in normal rats and in rats with hereditary diabetes insipidus, Proc. Natl. Acad. Sci. USA., 269:259.

Burbach, J.P.H., De Hoop, M.J., Schmale, H., Richter, D., De Kloet, E.R., Ten Haaf, J.A., and De Wied, D.,1984, Differential responses to osmotic stress of vasopressin-neurophysin mRNA in hypothalamic nuclei, Neuroendocrinology, 39:582.

Burbach, J.P.H., Van Tol, H.H.M., Bakkus, M.H.C., Schmale, H. and Ivell, R., 1986, Quantitation of vasopressin mRNA and oxytocin mRNA in hypothalamic nuclei by solution hybridization assays, J. Neurochem., 47:1814.

Castel, M., Dellmann, H.-D., and Gainer, H., 1984, Neuronal secretory systems, Int. Rev. Cytol., 88:303.

Davis, L.G., Arentzen, R., Reid, J.M., Manning, R.W., Wolfson, B., Lawrence, K.L., and Baldino, F., Jr., 1986, Glucocorticoid sensitivity of vasopressin mRNA levels in the paraventricular nucleus of the rat Proc. Natl. Acad. Sci.USA, 83:1145.

Douglass, J., Civelli, O., and Herbert, E., 1984, Polyprotein gene expression: generation of diversity of neuroendocrine peptides. Ann. Rev. Biochem., 53:665.

Einspanier, R., Pitzel, L., Wuttke, W., Hagendorff, G., Richter, K.D., Kardalinou, E. and Scheit, K.H., 1986, Demonstration of mRNAs for oxytocin and prolactin in porcine granulosa and luteal cells, FEBS Lett., 204:37.

Fuller, P.J., Clements, J.A., and Funder, J.W., 1985a, Localization of arginine-vasopressin-neurophysin II messenger ribonucleic acid in the hypothalamus of control and Brattleboro rats by hybridization histochemistry with a synthetic pentadecamer oligonucleotide probe, Endocrinology, 116:2366.

Fuller, P.J., Clements, J.A., Lolait, S.J., and Funder, J.W., 1984, Expression of the gene for arginine vasopressin in Brattleboro rats, J. Hypertension, 2:305.

Fuller, P.J., Clements, J.A., Tregear, G.W., Nicolaides, I., Whitfeld, P.L., and Funder, J.W., 1985b, Vasopressin-neurophysin II gene expression in the ovary: studies in Sprague-Dawley, Long-Evans, and Brattleboro rats, J. Endocrinology, 105:317.

Gainer, H., 1983, Precursors of vasopressin and oxytocin, Prog in Brain Res., 60:205.

Gainer, H., Russell, J.T., and Loh, Y.P. , 1985, The enzymology and intracellular organization of peptide precursor processing: the secretory vesicle hypothesis, Neuroendocrinology, 40:171.

Geenen, V., Legros, J.J., Franchimont, P., Baudrihaye, M., Defresne, M.P., and Boniver, J., 1986, The neuroendocrine thymus: coexistence of oxytocin and neurophysin in the human thymus, Science, 232:508.

Geenen, V., Legros, J.J., Franchimont, P., Defresne, M.P., Boniver, J., Ivell, R., and Richter, D., 1987, The thymus as a neuroendocrine organ: synthesis of vasopressin and oxytocin in human thymic epithelium, Ann. N.Y. Acad. Sci., 496:56.

Hanley, M.R., Benton, H.P., Lightman, S.L., Todd, K., Bone, E.A., Fretten, P., Palmer, S., Kirk, C.J., and Michell, R.H., 1984, A vasopressin-like peptide in the sympathetic nervous system, Nature, 309:258.

Ivell, R., and Richter, D., 1984a, Structure and comparison of the oxytocin and vasopressin genes from rat, Proc. Natl. Acad. Sci. USA, 81:2006.

Ivell, R., and Richter, D., 1984b, The gene for the hypothalamic peptide hormone oxytocin is highly expressed in the bovine corpus luteum: biosynthesis, structure, and sequence analysis, EMBO J., 3:2351.

Ivell, R., Brackett, K.H., Fields, M.H., and Richter, D., 1985, Ovulation triggers oxytocin gene expression in the bovine ovary, FEBS Lett., 190:263.

Ivell, R., Schmale, H., Krisch, B., Nahke, P. and Richter, D., 1986, Expression of a mutant vasopressin gene: differential polyadenylation and read-through of the mRNA 3'end in a frame shift mutant, EMBO J.,5:971.

Loh, Y.P., Brownstein, M.J., and Gainer, H., 1984, Proteolysis in neuropeptide processing and other neural functions, Ann., Rev. Neurosci., 7:189.

Majzoub, J.A., Pappey, A., Burg, R., and Habener, J.J., 1984, Vasopressin gene is expressed at low levels in the hypothalamus of the Brattleboro rat, Proc. Natl. Acad. Sci. USA, 81:5296.

McCabe, J.T., Morrell, J.I., and Pfaff, D.W., 1986a, Measurement of expression of the vasopressin and oxytocin genes in single neurons by in situ hybridization, in: Fink, G., Harmar, A.J., and McKerns, K.W. (eds.), Neuroendocrine Molecular Biology, Plenum Press, New York, pp. 219-229.

McCabe, J.T., Morrell, J.I., Ivell, R., Schmale, H., Richter, D., and Pfaff, D.W., 1986b, Brattleboro rat hypothalamic neurons transcribe vasopressin gene: evidence from in situ hybridization, Neuroendocrinology, 44:361.

Nojiri, H., Ishida, I., Miyashita, E., Sato, M., Urano, A. and Deguchi, T., 1987, Cloning and sequence analysis of cDNAs for neurohypophysial hormones vasotocin and mesotocin for the hypothalamus of toad, Bufo japonicus, Proc. Natl. Acad. Sci., USA, 84:3043.

Nojiri, H., Sato, M., and Urano, A., 1985, In situ hybridization of the vasopressin in mRNA in the rat hypothalamus by use of a synthetic oligonucleotide probe, Neurosci. Lett., 58:101.

Nussey, S.S., Ang, V.T.Y., Jenkins, J.S., Chowdry, H.S., and Bissett, G.W., 1984, Brattleboro rat adrenal contains vasopressin, Nature, 310:64.

Orci, L., Ravazzola, A.M., Amersherdt, M., Madsen, O., Vassali, J.D. and Perelet, A., 1985, Direct identification of prohormone conversion site in insulin-secreting cells, Cell. 42:671.

Rebhein, M., Hillers, M., Mohr, E., Ivell, R., Morley, S., Schmale, H. and Richter D., 1986, The neurohypophysial hormones vasopressin and oxytocin, Biol. Chem. Hoppe-Seyler, 367:695.

Reppert, S.M. and Uhl, G.R., 1987, Vasopressin messenger ribonucleic acid in supraoptic and suprachiasmatic nuclei: appearance and circadian regulation during development, Endocrinology, 120:2483.

Ruppert, S.D., Scherer, G., and Schutz, G., 1984, Recent gene conversion involving bovine vasopressin and oxytocin precursor genes suggested by nucleotide sequence, Nature, 308:554.

Sano, Y., 1985, History of Neurosecretion, in: Neurosecretion and the Biology of Neuropeptides (Kobayashi, H., Bern, H.A., and Urano, A., Eds.) pp 1-7, Japan Sci. Soc. Press, Tokyo/Springer-Verlag, Berlin.

Sausville, E., Carney, D., and Battey, J., 1985, The human vasopressin gene is linked to the oxytocin gene and is selectively expressed in a cultured lung cancer cell line, J. Biol. Chem., 260:10236.

Scharrer, B., 1987, Neurosecretion: Beginnings and new directions in neuropeptide research, Ann. Rev. Neurosci., 10:1.

Scharrer, E., 1928, Die lichtempfindlichkeit blinder elritzen (Untersuchungen uber das zwischenhirn der fische), Z. Vergl. Physiol., 7:1.

Schmale, H., and Richter, D., 1984, Single base deletion in the vasopressin gene is the cause of diabetes insipidus in Brattleboro rats, Nature, 308:705

Schmale, H., Heinsohn, S., and Richter, D., 1983, Structural organization of the rat gene for the arginine-vasopressin-neurophysin precursor, EMBO J., 2:763.

Schmale, H., Ivell, R., Breindel, M., Darmer, D., and Richter, D., 1984, The mutant vasopressin gene from diabetes insipidus (Brattleboro) rats is transcribed but the message is not efficiently translated, EMBO J., 3:3289.

Sherman, T.G., Civelli, O., Douglass, J., Herbert, E., Burke, S., and Watson, S.J., 1986b, Hypothalamic dynorphin and vasopressin mRNA expression in normal and Brattleboro rats, Federation Proc., 45:2323.

Sherman, T.G., Civelli, O., Douglass, J., Herbert, E., and Watson, S.J., 1986c, Coordinate expression of hypothalamic pro-dynorphin and pro-vasopressin mRNAs with osmotic stimulation, Neuroendocrinology, 44:222.

Sherman, T.G., McKelvy, J., and Watson, S.J., 1986a, Vasopressin mRNA regulation in individual hypothalamic nuclei: a northern and in situ hybridization analysis, J. Neurosci., 6:1685.

Silverman, A., and Zimmerman, E.A., 1983, Magnocellular neurosecretory system, Ann. Rev. Neurosci., 6:357.

Sofroniew, M.V., 1985, Vasopressin, oxytocin and their related neurophysins, in: Handbook of Chemical Neuroanatomy, vol. 4, Part I, edited by A. Bjorklund, and T. Hokfelt. pp. 93-165. Elsevier Science Publishers, Amsterdam.

Swanson, L.W., and Sawchenko, P.E., 1983, Hypothalamic integration: organization of the paraventricular and supraoptic nuclei, Ann. Rev. Neurosci., 6:269.

Tooze, J., Hollinshead, M., Frank, R., and Burke, B., 1987, An antibody specific for an endoproteolytic cleavage site provides evidence that pro-opiomelanocortin is packaged into secretory granules in AtT20 cells before its cleavage, J. Cell Biol., 105:155.

Uhl, G.R., and Reppert, S.M., 1986, Suprachiasmatic nucleus vasopressin messenger RNA: Circadian variation in normal and Brattleboro rats, Science, 232:390.

Uhl, G.R., Zingg, H.H., and Habener, J.F., 1985, Vasopressin mRNA in situ hybridization: Localization and regulation studied with oligonucleotide cDNA probes in normal and Brattleboro rat hypothalamus, Proc. Natl. Acad. Sci. USA, 82:5555.

Valtin, H., Stewart, J., and Sokol, H.W., 1974, Genetic control of the production of posterior pituitary principles.. in: Handbook of Physiology, Section 7, Endocrinology, Vol. IV, Part I, edited by R.O. Greep, and E.B. Astwood, pp. 131-171. Amer. Soc. Physiol., Wash., D.C.

Van Tol, H.H.M., 1987, Regulation of vasopressin and oxytocin gene expression in the hypothalamo-neurohypophyseal system of the rat, Ph.D. thesis, University of Utrecht, The Netherlands, 177 pp.

Van Tol, H.H.M., Voorhuis, D.T.A.M., and Burbach, J.P.H., 1987, Oxytocin gene expression in discrete hypothalamic magnocellular cell groups is stimulated by prolonged salt loading, Endocrinology, 120:71.

Van Tol, H.H.M., Voorhuis, T.A.M., Snijdewint, F.G.M., Boer, G.J., and Burbach, J.P.H., 1986, Vasopressin gene expression is attenuated in the fetal Brattleboro rat, FEBS Lett., 204:101.

Wathes, D.C., 1984, Possible actions of gonadal oxytocin and vasopressin, J. Reprod. Fertil., 71:315.

Whitnall, M.H., Key, S., Ben-Barak, Y., Ozato, K., and Gainer, H., 1985, Neurophysin in the hypothalamo-neurohypophysial system. II. Immunocytochemical studies of the ontogeny of oxytocinergic and vasopressinergic neurons, J. Neurosci., 5:98.

Wolfson,B., Manning, R.W., Davis, L.G., Arentzen, R., and Baldino, F. Jr., 1985, Co-localization of corticotropin releasing factor and vasopressin mRNA in neurons after adrenalectomy, Nature, 315:59.

Young, W.S. III, 1986, Corticotropin-releasing factor mRNA in the hypothalmus is affected differently by drinking saline and by dehydration, FEBS Lett., 208:158.

Young, W.S., III, Mezey, E., and Siegel, R.E., 1986a, Quantitative _in situ_ hybridization histochemistry reveals increased levels of corticotropin-releasing factor mRNA after adrenalectomy in rats, _Neurosci. Lett._, 20:198.

Young, W.S., III, Mezey, E., and Siegel, R.E., 1986b, Vasopressin and oxytocin mRNAs in adrenalectomized and Brattleboro rats: analysis by quantitative _in situ_ hybridization histochemistry, _Mol. Brain Res._, 1:231.

Zingg, H.H., Almazan, G., and Lefebvre, D.L., 1987, Osmotic stress induces an important increase in vasopressin mRNA poly (A) tail length: A novel mechanism of vasopressin gene control? _69th Ann. Meet. Endocrine Society_, pp. 135 (Abstract 456).

Zingg, H.H., Lefebvre, D., and Almazan, G., 1986, Regulation of vasopressin gene expression in rat hypothalamic neurons. Response to osmotic stimulation. _J. Biol. Chem._ 261:12956.

THE EVOLUTION OF FMRFamide-LIKE NEUROPEPTIDE GENES

John R. Nambu and Richard H. Scheller

Department of Biological Sciences
Stanford University
Stanford, CA 94305, U.S.A.

Biologically active peptides constitute an important class of extracellular molecular messengers. Consisting of short chains of amino acids, they mediate a range of behavioral, developmental, and physiological processes and are major components of both neuronal and endocrine communication networks (Krieger, 1983). In general, peptides are initially contained on larger precursor proteins which undergo post-translational proteolytic processing (Loh and Gainer, 1983). These precursors often serve as polyproteins to liberate multiple bioactive peptide products, a property that greatly increases their potential information capacity.

In the nervous system, peptides can function as transmitters, modulators or hormones, depending on their site of release, duration of action and identity of target tissue (Scheller et al., 1984). Neuropeptides have evolved to assume more diverse functional roles than classical neurotransmitters such as biogenic amines, amino acids, or acetylcholine, and studies on coelenterates suggest that peptides may actually be the archetypal neuronal messenger, as in these animals in which primitive nervous systems first arose, peptides have been identified while classical small molecule transmitters have not (Grimmelikhuijzen and Graff, 1985).

One of the most widely studied peptides in invertebrates is the amidated tetrapeptide Phe-Met-Arg-Phe-NH$_2$ (FMRFamide). FMRFamide was originally purified from the venus clam <u>Macrocallista</u> <u>nimbosa</u> on the basis of its potent cardioexcitatory properties (Price and Greenberg, 1977). The distribution of authentic FMRFamide appears to be limited to molluscs. However, FMRFamide immunoreactivity has been detected throughout the animal kingdom and a large number of peptides bearing structural resemblance have been isolated from invertebrate as well as vertebrate species (Boer et al., 1980; Dockray et al., 1983; Grimmelikhuijzen and Graff, 1986; Nambu et al., 1988). The sequences of several of these peptides, as determined either via biochemical characterization or inferred from cloned gene sequences, are presented in Fig. 1. The identified peptides vary in size from 4 to 9 amino acids and display a wide range of structural homologies. In addition, many organisms appear to express several distinct FMRFamide-like peptides (Ebberink and Joosse, 1985; Price, 1986; Shyamala et al., 1986; Taussig and Scheller, 1986; Nambu et al., 1988; E. Kravitz, personal communication).

The most highly related peptide is FLRFamide, which also appears to be specific to molluscs and is identical to FMRFamide except for a leucine substitution at the methionine position (Price, 1986; Taussig and Scheller; 1986). There are a number of less related peptides which also possess this leucine substitution and are extended amino-terminally, including peptides isolated from the land snail, lobster, and chicken (Dockray et al., 1983; Price et al., 1985; E.Kravitz, personal communication). This conservative substitution is reminiscent of the mammalian enkephalin precursor where there exists both Met- (YGGFM) and Leu-enkephalin (YGGFL) structures (Noda et al., 1982), and may confer enhanced stability, as leucine is less susceptible to inactivation by oxidation than methionine. This property could be especially important to the long-range hormonal actions of these peptides

(Price, 1986). Another closely related structure is that of a peptide isolated from the fruit fly (Drosophila) which contains a FMRFamide core sequence and is extended amino-terminally (Nambu et al., 1988). The vertebrate peptide Met-enkephalin-Arg-Phe (YGGFMRF) is also an amino-terminal extension of the same 4 amino acid core sequence, although this peptide does not possess a carboxy-terminal amide (Noda et al., 1982).

In addition, there exist numerous peptides which display lower levels of homology to FMRFamide. In particular, several peptides have been identified which share the Arg-Phe-amide structure, including a sea anemone and jellyfish peptide, the sea hare (Aplysia) L5 peptide, cockroach leucosulfakinin, and vertebrate gamma-MSH (Nakanishi et al., 1979; Grimmelikhuijzen and Fraffi, 1986; Nacham et al., 1986; Shyamala et al., 1986; McFarlane et al., 1987). Although these peptides could be ancestrally related to FMRFamide, they may also possess Arg-Phe-amide moieties as the result of convergent evolution. Carboxy-terminal amide groups are thought to confer enhanced stability, and amidated peptides are in general quite common (Tatemoto and Mutt, 1978; Loh and Gainer, 1983). The Arg-Phe-amide moiety, consisting of both a charged and hydrophobic amino acid residue, appears to be a valuable motif in peptide structure and may be particularly well suited for physiologically active peptides due to its amphipathic character. Both immunological and structure/function studies on FMRFamide-like peptides have demonstrated that it is this carboxy-terminal region that is the main antigenic determinant and the critical region for biological actions (Weber et al., 1981; Painter et al., 1982).

FMRFamide and related peptides exhibit a number of distinct physiological activities. In molluscs, FMRFamide, FLRFamide, and the land snail peptide pQDPFLRFamide, generally have cardioexcitatory actions, although FMRFamide is cardioinhibitory in some species (Painter and Greenberg, 1982; Price et al., 1985; Price, 1986). These peptides have similar actions on non-cardiac muscles as FMRFamide and pQDPFLRFamide enhance contractions of the radula protractor muscle in Busycon, while FMRFamide inhibits gut contractions and enhances gill contractions in Aplysia (Austin et al., 1982; Weiss et al., 1984; Price et al., 1985). FMRFamide also has defined neuronal activities, including inhibitory, excitatory, or modulatory effects on identified neurons in Aplysia and Helix (Cottrell et al., 1984; Sossin et al., 1987). These peptides may thus have a variety of distinct actions within the same organism. In vertebrates, authentic FMRFamide has not been identified, however FMRFamide does excite rat medullary neurons and inhibit cultured mouse spinal neurons (Gayton, 1982; McCarthy and Cottrell, 1984). In addition, FMRFamide has systemic actions to inhibit the glucose stimulated release of insulin and somatostatin from the rat pancreas, and to elevate arterial blood pressure and respiration (Barnard and Dockray, 1984; Sorensen et al., 1984). The chicken peptide, LPLRFamide, has closely related effects, exciting rat brain stem neurons and increasing blood pressure (Dockray et al., 1983; Barnard and Dockray, 1984). The overall similarity of their actions suggests that different FMRFamide-like peptides are likely to function in related physiological capacities in their respective source organisms.

In Aplysia, immunocytochemical studies have indicated that FMRFamide immunoreactive material is expressed in 100-200 cells in the central nervous system (Schaefer et al., 1985; Shyamala et al., 1986). However, as the specificity of FMRFamide antisera is generally limited to the Arg-Phe-amide carboxy-terminus (Weber et al., 1981), it is not clear precisely how many distinct FMRFamide-like peptides are synthesized. Molecular biology investigations in Aplysia have utilized differential cDNA screening procedures to isolate two genes which encode different precursors that give rise to several such peptides. Thus the FMRFamide immunoreactive abdominal ganglion neuron L5 was shown to express a prominent mRNA which encodes a 112 amino acid precursor; consistent with the immunoreactivity, the precursor contains a predicted peptide product which ends in Arg-Phe-amide (Shyamala et al., 1986). Interestingly, this peptide is homologous to Antho-RFamide (pQGRFamide), isolated from sea anenome (Grimmelikhuijzen and Graff, 1986). A comparison of these two sequences is presented in Fig. 2. The single difference between these peptides is the pyro-glutamic acid moiety of Antho-RFamide, which is formed by

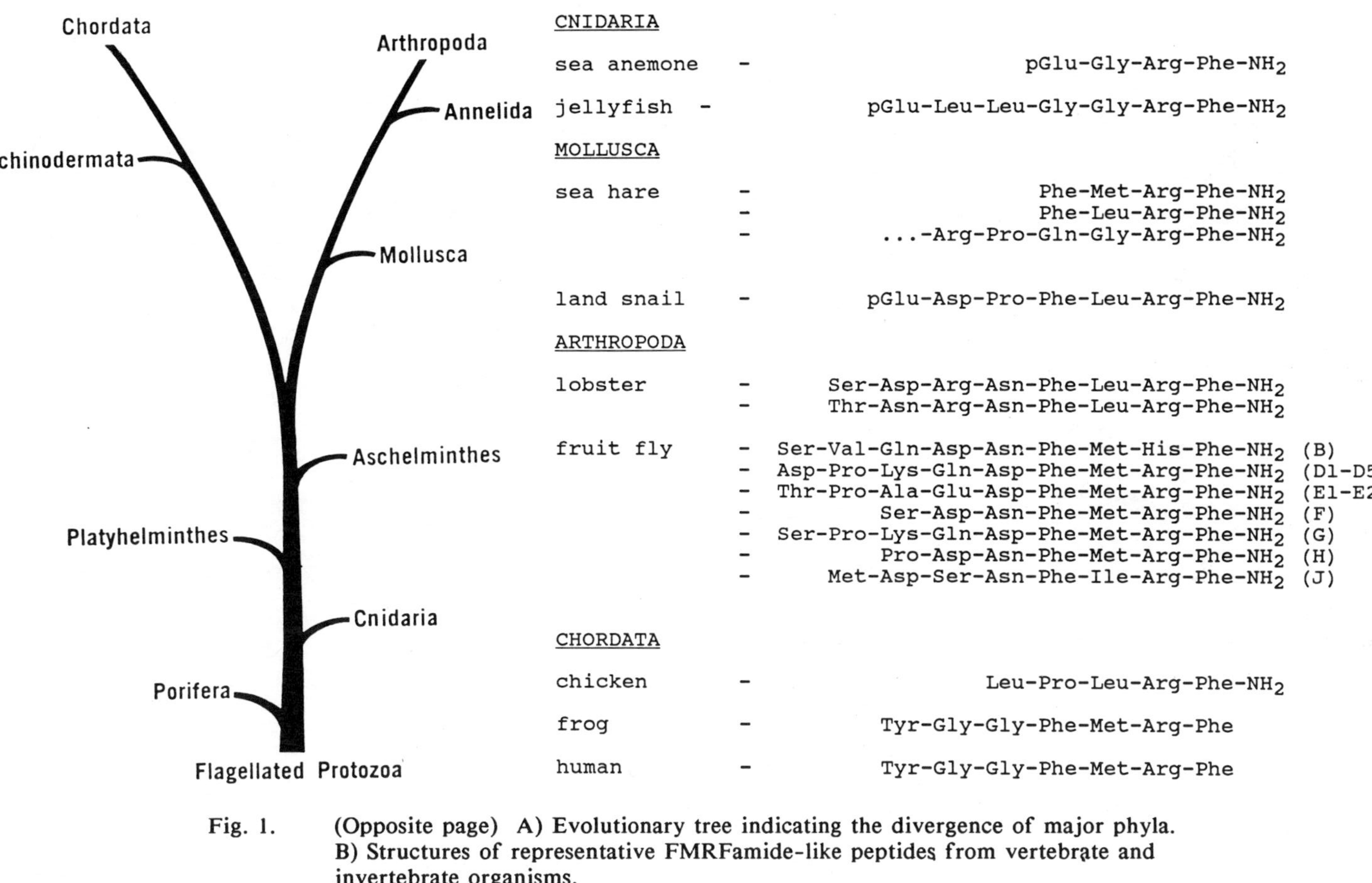

Fig. 1.   (Opposite page) A) Evolutionary tree indicating the divergence of major phyla. B) Structures of representative FMRFamide-like peptides from vertebrate and invertebrate organisms.

13

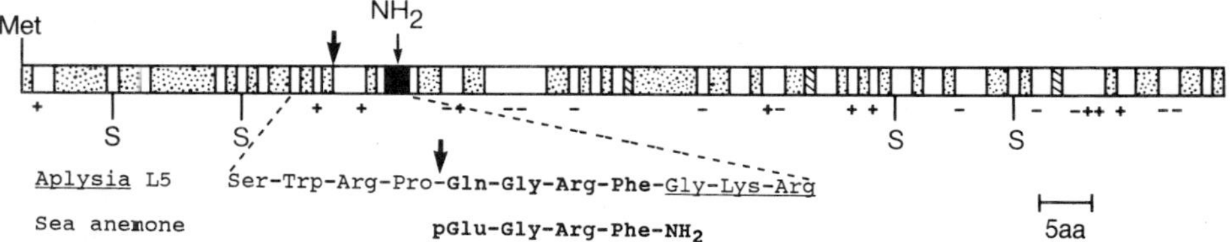

Fig. 2.   Schematic representation of the Aplysia L5 neuropeptide precursor protein, including sequence of the region containing predicted FMRFamide immunoreactive peptide. The putative signal sequence cleavage site is indicated by the large arrow (↓). The predicted carboxy-terminal dibasic residue cleavage site (↓), acidic residues (−), basic residues (+), cysteine residues (S), and the putative amidation site (NH2 and small arrow (↓)) are also noted. Presented below is the sequence of sea anenome Antho-RFamide for comparison. Identical residues between the two peptides are indicated by boldface. Size bar = 5 amino acids.

residue at precisely the corresponding position. However, because the signal sequence cleavage site is unknown, the amino-terminus of this peptide is not defined and may extend beyond the glutamine. Nonetheless, this homology suggests an ancestral relationship between the Aplysia and sea anenome peptides, and it will be interesting to compare the structures of the two precursor proteins as well as the physiological actions of the peptide products. In the sea anenome, Antho-RFamide is localized in neurons and the peptide is active on both muscles and neuronal systems (McFarlane et al., 1987).

The gene encoding authentic FMRFamide has also been cloned in Aplysia (Schaefer et al., 1985; Taussig and Scheller, 1986). The FMRFamide gene is present in a single copy per haploid genome and is expressed in all the major ganglia as well as in an exocrine reproductive organ, the atrial gland. The predicted precursor protein, depicted in Fig. 3, is 597 amino acids in length and contains 28 identical copies of FMRFamide, 1 copy of FLRFamide, and 1 copy of an undefined peptide ending in LRFamide. In addition, there are numerous acidic spacer regions separating many of the peptide copies. These spacers exhibit variability in both sequence and length.

In Drosophila, FMRFamide immunoreactivity is observed in a small but stereotypic subset of 30-40 central neurons in both larval and adult stages (White and Valles, 1985; White et al., 1986). In addition, immunoreactivity in the larva was also detected in fibers innervating the neurohemal ring gland, in the ganglion innervating the gut, and in the gastric caeca. A 9 amino acid FMRFamide related peptide, DPKQDFMRFamide, has been biochemically purified from Drosophila (Nambu et al., 1988). Use of a synthetic oligonucleotide probe led to the cloning of the corresponding gene, which is present in a single copy per haploid genome. The predicted precursor protein, shown in Fig. 3, is 347 amino acids in length and contains 5 tandemly repeated copies of DPKQDFMRFamide as well as 10 additional predicted peptides having amidated carboxy-termini. These additional peptides exhibit varying degrees of structural relatedness; many are also amino-terminal extensions of a FMRFamide core sequence (Figs. 1 and 3).

The homology of the peptide products and the organization of the precursors indicate that the Aplysia FMRFamide and Drosophila DPKQDFMRFamide genes are ancestrally related. Comparison of the two precursors indicates many structural similarities. Characteristic of neuropeptide precursor proteins, both have hydrophobic amino-terminal

14

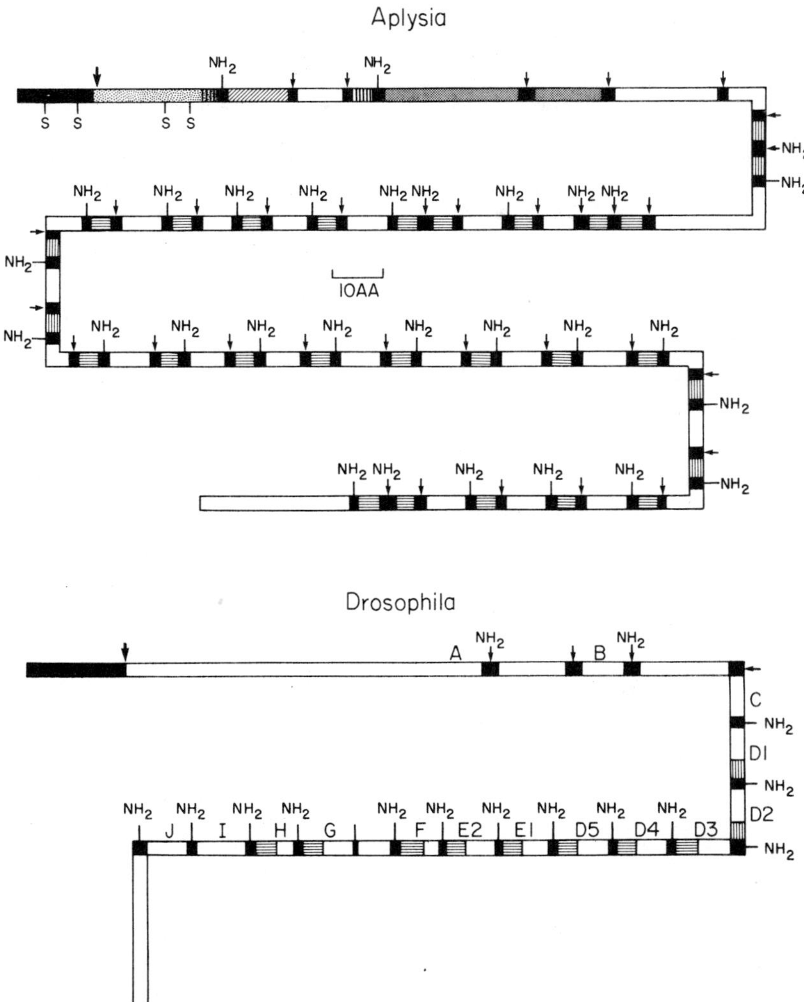

Fig. 3.   (Opposite page) Schematic representations of the *Aplysia* FMRFamide precursor protein and the *Drosophila* DPKQDFMRFamide precursor protein. The putative signal sequences are indicated by (■) and the large arrows (↓) denote the predicted cleavage sites. Cysteine residues are indicated by (s). The positions of single or multiple basic residue cleavage sites are depicted by lines (|) or small arrows (↓) respectively. FMRF sequences are presented as (≡) while NH2 denotes potential amidation sites. Letter/number designations above the DPKQDFMRFamide precursor indicate the representation of distinct peptide structures. Size bar = 10 amino acids.

signal sequences. In addition, there are numerous copies of FMRFamide related peptides flanked by single or multiple basic residue cleavage sites, and acidic spacer regions (Fig. 3). While the FMRFamide precursor contains 28 exact copies of FMRFamide and two FLRFamide derived peptides, the DPKQDFMRFamide precursor exhibits much greater heterogeneity as it contains 15 potentially amidated peptides which have 10 distinct structures. These observations raise two intriguing questions: 1) Why are there so many peptide copies on the two precursors, and 2) given that the common ancestor of these precursors almost certainly already possessed multiple copies of a FMRFamide related peptide, how is it that the degree of homology exhibited by peptides on the same precursor is so much higher than between peptides on the two different precursors?

Some insight is provided by an examination of the precursor structures. It is clear that several intragenic duplication events must have occurred in the evolution of both polyprotein precursors. The numerical disparity of related peptides on the _Aplysia_ and _Drosophila_ precursors is then the result of additional independent duplication or deletion events on the arthropod and molluscan lines. At least two explanations can be offered as to why this multi-peptide arrangement has been selected for. One is that the elevated peptide copy number facilitates high levels of peptide biosynthesis which are required for specific physiological processes. The second is that multiple copies of these small peptides may shield useful peptide structures from deleterious mutations, while still allowing beneficial alterations to become incorporated. In all likelihood, both these and other factors have probably been important in generating and maintaining the multiple peptide copies observed in these precursors. With respect to higher peptide homologies within a precursor, it seems that either: 1) as the individual peptide copies have diverged, in both precursors a series of multiple, independent mutation events have yielded similar or identical structures; or, 2) that the individual peptide units on these precursors are in fact not diverging independently of one another, but rather, are evolving in a concerted fashion.

It is unlikely that 28 identical copies of FMRFamide could be maintained in _Aplysia_ soley by positive selection, especially given that FMRFamide or related peptides are present in other molluscs which diverged from _Aplysia_ hundreds of millions of years ago (Ebberink and Joosse, 1985; Price, 1986). Furthermore, for both the FMRFamide and the DPKQDFMRFamide precursor, if the individual peptide units have been evolving independently, one would expect the homology of the peptides on one precursor to be about the same as between peptides on the two precursors. Yet this is precisely the opposite of what is observed; the intraspecific amino acid and DNA sequence homologies of the peptides are much higher than the interspecific homologies. Finally, nearby acidic spacer regions in the FMRFamide precursor are more homologous than those which are further apart, and in the DPKQDFMRFamide precursor, identical or very similar peptides are more closely linked. Taken together these observations suggest that during the evolution of these genes, intragenic interactions have served to homogenize the peptide units. Such interactions between distinct genes have long been thought to be responsible for the higher than expected intraspecific homology observed between members of multi-gene families, where both gene conversion and unequal crossing-over mechanisms can act to alter gene structures and copy numbers (Baltimore, 1981; Arnheim, 1983). It is likely that related processes have been important in the evolution of these two peptide precursors, and have acted to facilitate the rapid fixation of advantageous mutations, and also, to alter the number of peptide units present. Similar conclusions were reached with regards to the homology of the polypeptide repeats within the polyubiquitin protein (Sharp and Li, 1987). Both the FMRFamide and DPKQDFMRFamide precursors have probably experienced recent internal duplication and/or correction events as there is very high DNA sequence homology among several of their peptide units. Presumably the repetitive nature of the genes facilitates such events. Consistent with these notions is the occurrence of distinct peptide sequences and copy numbers on the FMRFamide and DPKQDFMRFamide precursor proteins.

It is possible that all the _Drosophila_ FMRFamide-extended peptides act in a similar fashion, with their amino-terminal regions merely altering stability or potency. However, it is quite unlikely that each of the remaining peptides on the precursor could act at the same target receptors (Nambu et al., 1988) Thus, in addition to supplying high levels of one peptide, DPKQDFMRFamide, this precursor probably liberates several related peptides which may act at different targets and elicit distinct responses. This is in contrast to the

Aplysia FMRFamide precursor where the peptides probably all share common target receptors. In Drosophila, peptide J, MDSNFIRFamide, has a 5 out of 8 residue identity with the lobster peptide, SDRNFLRFamide, suggesting a common ancestry. This, and the existence of several related yet distinct peptides on the Drosophila precursor implies that many of the different FMRFamide-like peptides from other organisms may in fact be ancestrally related.

One particularly intriguing possible relationship is that of the FMRFamide and DPKQDFMRFamide precursors to the mammalian enkephalin precursor, which contains multiple copies of met-enkephalin derived peptides, including YGGFMRF, and a related leu-enkephalin peptide. This has prompted speculation on the ancestral relationship between these genes (Greenberg and Price, 1983; Schaefer et al., 1985; Taussig and Scheller, 1986). Although this relationship is uncertain, there are a number of Arg-Phe-amide peptides which have been isolated from vertebrates, including the chicken peptide LPLRFamide, suggesting a common ancestry for at least some of the vertebrate and invertebrate FMRFamide-like peptides.

The phyletic distribution of FMRFamide related peptides is diverse and attests to an ancient origin for this peptide family. It will be of great interest to determine the identities of additional family members. Examination of the corresponding gene structures may aid in the analysis of ancestral relationships; thus FMRFamide and the L5 peptide probably exhibit limited homology due to convergent evolution, while the L5 peptide and Antho-RFamide are probably ancestrally related, as are FMRFamide, FLRFamide, and peptides on the DPKQDFMRFamide precursor. Studies of these, and other FMRFamide related peptides have greatly enhanced our understanding of neuropeptide expression, evolution, and action. Perhaps through manipulation of the cloned DPKQDFMRFamide gene it will now be possible to utilize the powerful molecular and genetic techniques available in Drosophila to gain further insights into the physiological significance of the primary structures and copy numbers of these peptides.

REFERENCES

Arnheim, N., 1983, in: "Evolution of Genes and Proteins," M. Nei and R.K. Koehn, eds., Sinauer, Sunderland.

Austin, T., Weiss, S., and Lukowiak, K., 1982, FMRFamide effects on spontaneous and induced contractions of the anterior gizzard in Aplysia, Can. J. Physiol. Pharmacol., 61:949.

Baltimore, D., 1981, Gene conversion: some implications for immunoglobulin genes, Cell, 24:592.

Barnard C.S., and Dockray, G.J., 1984, Increasing arterial blood pressure in the rat in response to a new vertebrate neuropeptide, LPLRFamide, and a related molluscan peptide, FMRFamide, Reg. Peptides, 8:209.

Boer, H.H., Schot, L.P.C., Veenstra, J.A., and Reichelt, D., 1980, Immunocytochemical identification of neural elements in the central nervous systems of a snail, some insects, a fish, and a mammal with antiserum to the molluscan cardio-excitatory tetrapeptide FMRFamide, Cell and Tissue Res., 213:21.

Cottrell, G.A., Davies, N.W., and Green, K.A., 1984, Multiple actions of a molluscan cardioexcitatory neuropeptide and related peptides on identified Helix neurons, J. Physiol., 365:315.

Dockray, G.J., Reeve Jr., J.R., Shively, J., Gayton, R.J., and Barnard, C.S., 1983, A novel active pentapeptide from chicken brain identified by antibodies to FMRFamide, Nature, 305:328.

Ebberink R.H.M., and Joosse, J., 1985, Molecular properties of various snail peptides from brain and gut, Peptides, 6(3):451.

Gayton, R.J., 1982, Mammalian neuronal actions of FMRFamide and the structurally related opiod Met-enkephalin-Arg^6-Phe^7, Nature, 298:275.

Greenberg M.J., and Price, D.A., 1983, Invertebrate neuropeptides: native and naturalized, Ann. Rev. Physiol., 45:271.

Grimmelikhuijzen C.J.P., and Graff, D., 1985, Arg-Phe-amide-like peptides in the primitive nervous systems of coelenterates, Peptides, 6(3):477.

Grimmelikhuijzen C.J.P., and Graff, D., 1986, Isolation of <Glu-Gly-Arg-Phe-NH2 (Antho-RFamide), a neuropeptide from sea anenomes, Proc. Natl. Acad. Sci. USA, 83:9817.

Kreiger, D.T., 1983,Brain peptides: what, where, and why?, Science, 222:975..

Loh Y.P. and Gainer, H., 1983, Biosynthesis and processing of neuropeptides, in: "Brain Peptides", D.T. Krieger, M.J. Brownstein, and J.B. Martin, eds., Wiley, New York.

McCarthy P.W., and Cottrell, G.A., 1984, Responses of mouse spinal neurones in culture to locally applied Phe-Met-Arg-Phe-NH2, Comp. Biochem. Physiol., 79C:383.

McFarlane, I.D., Graff, D., and Grimmelikhuijzen, C.J.P., 1987, J. Exp. Biol., in press.

Nacham, R.J., Holman, G.M., Haddon, W.F., and Ling, N., 1986, Leucosulfakinin, a sulfated insect neuropeptide with homology to gastrin and cholecystokinin, Science, 234:71.

Nakanishi, S., Inoue, A., Kita, T., Nakamura, M., Chang, A.C.Y. Cohen, S.N., and Numa, S., 1979, Nucleotide sequence of cloned cDNA for bovine corticotropin-beta-lipotropin precursor, Nature, 278:423.

Nambu, J.R., Murphy-Erdosh, C., Andrews, P.C., Feistner, G.J., and Scheller, R.H., 1988, Isolation and characterization of a Drosophila neuropeptide gene, submitted.

Noda, M., Furutani, Y., Takahashi, H., Toyosato, M. Hirose, T., Inayama, S., Nakanishi, S., and Numa, S., 1982, Cloning and sequence analysis of cDNA for bovine adrenal preproenkephalin, Nature, 295:202.

Painter S.D.,and Greenberg, M.J., 1982, A survey of the responses of bivalve hearts to the molluscan neuropeptide FMRFamide and to 5-hydroxytryptamine, Biol. Bull., 162:311.

Painter, S.D., Morley, J.S., and Price, D.A., 1982, Structure-activity relations of the molluscan neuropeptide FMRFamide on some molluscan muscles, Life Sci., 31:2471.

Price D.A., and Greenberg, M.J., 1977, The structure of a molluscan cardioexcitatory peptide, Science, 197:670.

Price, D.A., 1986, Evolution of a molluscan cardioregulatory neuropeptide, American Zoologist, 26:1007.

Price, D.A., Cottrell, G.A., Doble, K.E., Greenberg, M.J. Jorenby, W., Lehman, H.K., and Riehm, J.P., 1985, A novel FMRF-related peptide in Helix: pQDPFLRFamide, Biol. Bull., 169:256.

Schaefer, M., Picciotto, M.R., Kreiner, T., Kaldany, R.R., Taussig, R., and Scheller, R.H., 1985, Aplysia neurons express a gene encoding multiple FMRFamide neuropeptides, Cell, 41:457.

Scheller, R.H., Kaldany, R.R., Kreiner, T., Mahon, A.C., Nambu, J.R., Schaefer, M., and Taussig, R., 1984, Neuropeptides: mediators of behavior in Aplysia, Science, 225:1300.

Sharp P.M., and Li, W.H., 1987, Ubiquitin genes as a paradigm of concerted evolution in tandem repeats, J. Molec. Evol., 25:58.

Shyamala, M. Fisher, J.,and Scheller, R.H., 1986, A neuropeptide precursor expressed in Aplysia neuron L5, DNA, 5:203.

Sorensen, R.L., Sasek, C.A., and Elde, R., 1984, Phe-Met-Arg-Phe-amide (FMRF-NH2) inhibits insulin and somatostatin secretion and anti-FMRF-NH2 sera detects pancreatic polypeptide cells in the rat islets, Peptides, 5:777.

Sossin, W.S., Kirk, M.D., and Scheller, R.H., 1987, Peptidergic modulation of neuronal circuitry controlling feeding in Aplysia, J. Neurosci., 7(3):671.

Tatemoto K., and Mutt, V., 1978, Chemical determination of polypeptide hormones, Proc. Natl. Acad. Sci. USA., 75:4115.

Taussig R., and Scheller, R.H., 1986, The Aplysia FMRFamide gene encodes sequences related to mammalian brain peptides, DNA, 5:453.

Weber, E., Evans, C.J., Samuelsson, S.J., and Barchas, J.D., 1981, A novel peptide neuronal system in rat brain and pituitary, Science, 214:1248.

Weiss, S., Goldberg, J.I., Chohan, K.S., Stell, W.K., Drummond, G.I., and Lukowiak, K., 1984, Evidence for FMRFamide as a neurotransmitter in the gill of Aplysia californica, J. Neurosci., 4:1994.

White K., and Valles, A.M., 1985, Immunohistochemical and genetic studies of serotonin and neuropeptides in Drosophila, in: "Molecular Basis of Neuronal Development," G.M. Edelman, W. Gall, W.M. Cowan, eds., Wiley, New York.

White, K., Hurteau, T., and Punsal, P., 1986, Neuropeptide FMRFamide-like immunoreactivity in Drosophila: development and distribution, J. Comp. Neurol., 247(4):430.

DEVELOPMENTAL AND MOLECULAR STUDIES OF NEURONS THAT EXPRESS

FMRFamide-RELATED GENES IN INSECTS

Paul H. Taghert

Department of Anatomy and Neurobiology
Washington University School of Medicine
660 South Euclid Avenue
Saint Louis, Missouri 63110

Ernst and Berta Scharrer first demonstrated the glandular nature of magnocellular neurons more than 40 years ago. Since that time considerable effort has been made to prove the neuronal nature of neurosecretory cells (NSC) in the mammalian hypothamalmus and elsewhere: it has been shown for example, that these cells have active membrane properties (Copenhaver and Truman, 1986), that they have complex neuronal geometries (Taghert and Truman, 1982), and that they receive varied neuronal synaptic inputs (Jansen and Bos, 1984). Perhaps the sole distinction that can still be drawn between NSCs and conventional neurons is that, while the latter contact and synapse upon discrete cellular elements, NSCs instead form "blind", synaptoid endings within structures termed Neurohaemal Organs (NHO). It is from such NSC terminations that various neuropeptides are released into the circulation and eventually act on cellular targets that are microns to meters away.

It may be presumed therefore, that NSC axons are not obliged to find precise cellular targets during development as say motoneurons and interneurons must do. Nevertheless, NSCs examined to date all have reproducible patterns of axonal projections that form sterotypical NHOs (Carrow et al., 1984). What are the cellular and molecular rules that govern the development of this special neuronal cell type? How precisely do the growth cones of NSC navigate in the embryonic terrain? Furthermore, how is morphological development coupled with the expression of the neurosecretory products to help form a functional neurosecretory effector system? In this chapter, I review our recent developmental and molecular studies of insect NSCs that express neuropeptides related to FMRFamide (Phe-Met-Arg-Phe-amide). The developmental experiments have revealed a precise and stereotyped program of axonal outgrowth that includes specific interactions of NSCs with a cellular element that may act as a fictive target for the NSC axons. Molecular and immunological studies reveal the expression of a single gene very early in the development of these NSCs, just as they are contacting their presumptive NHO and their "fictive target".

The insect nervous system is a favorable preparation for examining a variety of physiological and developmental issues by virtue of its numerical simplicity ($\sim 10^5$ neurons) and the large size and identifiability of the individual neurons (Goodman and Heitler, 1979; Taghert and Goodman, 1984). The segmental nerve cord of insects contains a classically defined NHO - the Transverse Nerve (TN); a single TN is associated with each segmental neuromere. This nerve has also been called variously the Peri-Visceral Organ, the Peri-Sympathetic Organ and the Median Nerve (see Raabe, 1982 for a review). We have identified each of the 16 NSCs that in the larvae of the moth <u>Manduca</u> <u>sexta</u> project to each TN with respect to pattern of arborization, axonal projection, peptide immunoreactivity and/or neuropeptide content (reviewed in Taghert et al., 1986). The cell body positions and axonal pathways of the typical complement of neurons are schematized in Figure 1. Also

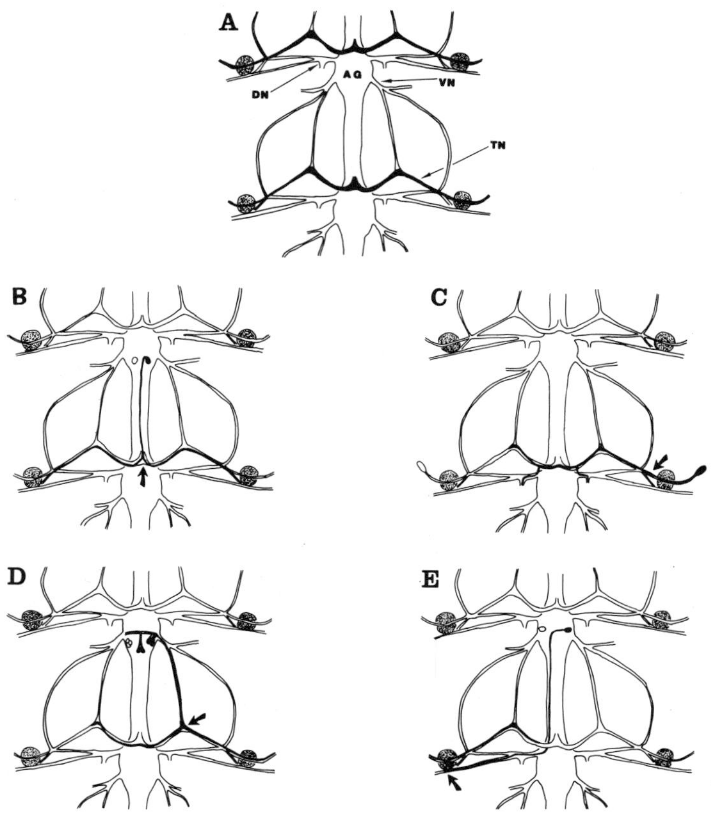

Fig. 1.   Schematic diagrams indicating the anatomy of the neurohaemal Transverse Nerve
and its constituent neurons. (A) The arrangement of ganglia and nerves in two
segments of the mature <u>Manduca</u> larva. AG: abdominal ganglion; DN: dorsal
segmental nerve; VN: ventral segmental nerve; TN: transverse nerve. TN is
highlighted by shading; proportions not correct. (B) The cell body position and
axonal projection of the identified spiracular closer muscle motorneurons;
contralateral homologue is shown in outline. (C) The peripheral NSC L1; (D)
the diverse NSC that project to the periphery via the ventral nerve; these contain
3 of the 4 identified bursicon-containing neurons. (E). The fourth bursicon-
containing neuron. Arrows indicate the site at which the axons of the various
neurons join the TN.

included in this set are 2 motoneurons that project to the TN and ultimately synapse upon a
specific skeletal muscle.

L1 - An Identified Neurosecretory Neuron

Peripheral neurosecretory neurons in insects were first described in stick insects and
flies (Finlayson and Osborne, 1969), and later in the larva of a moth (Hinks, 1976). We
discovered a similar neuron in the periphery of each segment of moth embryos that projects
to the TN. We call this neuron L1, for Link neuron 1 (c.f. Finlayson and Osborne, 1969).
In the mature larva, the L1 cell on each side projects its main axon centrally over each
segmental ganglion (Figure 1C); the cell is bipolar (and in some segments, multipolar) with a
distal process that in abdominal segments projects to the heart. The processes of L1 never
enter the CNS but are exclusively found in the periphery. They ramify extensively along
various nerve branches and show a surprising degree of stereotyped morphology in that L1
processes and varicosities are always found at specific locations along the TN and other
nerve branches.

L1 Expresses FMRFamide-like Neuropeptides in a Segment-Specific Pattern

Immunocytochemical staining with an anti-FMRFamide antiserum reveals a precise and segment-specific pattern of expression. Among peripheral neurons in the body wall, only the L1 is stained; further, only L1 cells in the anterior 4 segments (the three thoracic and first abdominal). In the remaining abdominal segments, L1 survives but shows little and usually no specific immunoreactivity. FMRFamide-like immunoreactivity has previously been noted in specific neurosecretory cells of the brain of <u>Manduca</u> (Carrol et al., 1986); immunoreactive peptides are also released by generalized depolarization of these neurosecretory terminals. In locusts, FMRFamide-like neuropeptides modulate neuromuscular transmission in the leg (Evans and Myers, 1986). It is reasonable to propose therefore, that FMRFamide-like peptides are released by the L1 cell from along the length of the neurohaemal TN to act as hormonal modulators. As yet, no FMRFamide-like neuropeptides have been purified in insects, but, as discussed later in this chapter, molecular genetic techniques have recently provided their structure in <u>Drosophila</u> (Schneider and Taghert, submitted). Because the L1 cell is easily identified at all stages of embryogenesis, we next turned to these earlier stages in order to describe the schedule of morphological and biochemical development of an identified neurosecretory cell.

Morphological Differentiation of the L1 Neuron in Embryogenesis

Using the technique of Lucifer Yellow (LY) dye injection followed by staining with the anti-LY antiserum (Taghert et al., 1982), we traced the entire period of axonal outgrowth of the L1 cell. We can first find the L1 cell at 37% of embryonic development: at this point it has not yet elaborated a major process. Once begun, the elongation of the L1 axon is precise and smooth. At no time do we find evidence of error in projection or branches that are retracted. The timeline of growth is shown in Figure 2. Starting at ~39% of embryonic development, L1 sends its growth cone in the direction of the prospective nerve. The growth cone contacts the prospective nerve at ~41%, and at this time it flattens to extend filopodial processes in many directions and can first be visualized with the anti-FMRFamide antibody. It is the first cell in <u>Manduca</u> to begin expression of FMRFamide-like neuropeptides. Growth continues as the axon is elongated through the developing TN over the segmental ganglion.

This cell is flat and lies at the segment boundary atop the large intersegmental muscles; it is syncytial, with 4-5 nuclei present. It is dye coupled to the developing nerve (data not shown) and hence intimately involved in its development. The specific adhesion of L1 filopodia to the "Syncytium" is shown in Figure 3 where it is clearly seen that the L1 specialization is distinct from the L1 growth cone and hence is likely not to be a

Interestingly, while the L1 growth cone morphology remains simple, a morphological specialization can be seen at a particular point along the L1 axon at its entry point to the TN. This specialization takes the form of numerous filopodia and often a large lamellopodium that are first elaborated just as the growth cone reaches this position and are retained there even as the growth cone has gone past it. It is a transient structure, lasting 10-15%; very little if any remnant of it is evident in mature larvae. We searched to find some aspect of the local microenvironment around this specialization and were surprised to find a specific cell that appeared to be focus for this L1 specialization (Figure 3).

specialization that subserves growth cone navigation. What is this function of this transient, cell-specific embryonic interaction?

All TN Neurosecretory Cells Interact with the Syncytium

As mentioned above the TN contains the axons of a number of distinct NSCs besides the L1 cell and it also contains the axons of a pair of motorneurons. We examined the growth of each of these neurons as their growth cones approached and joined the presumptive nerve during embryogenesis (Taghert et al., 1986; Carr and Taghert, submitted). At no time do processes of the motorneurons demonstrate any recognition or adhesion to the specific syncytial cell despite the fact that they pass within 10 μm of it (as close as does the growth cone of L1). In contrast, the growth cones of the bursicon-containing neurons

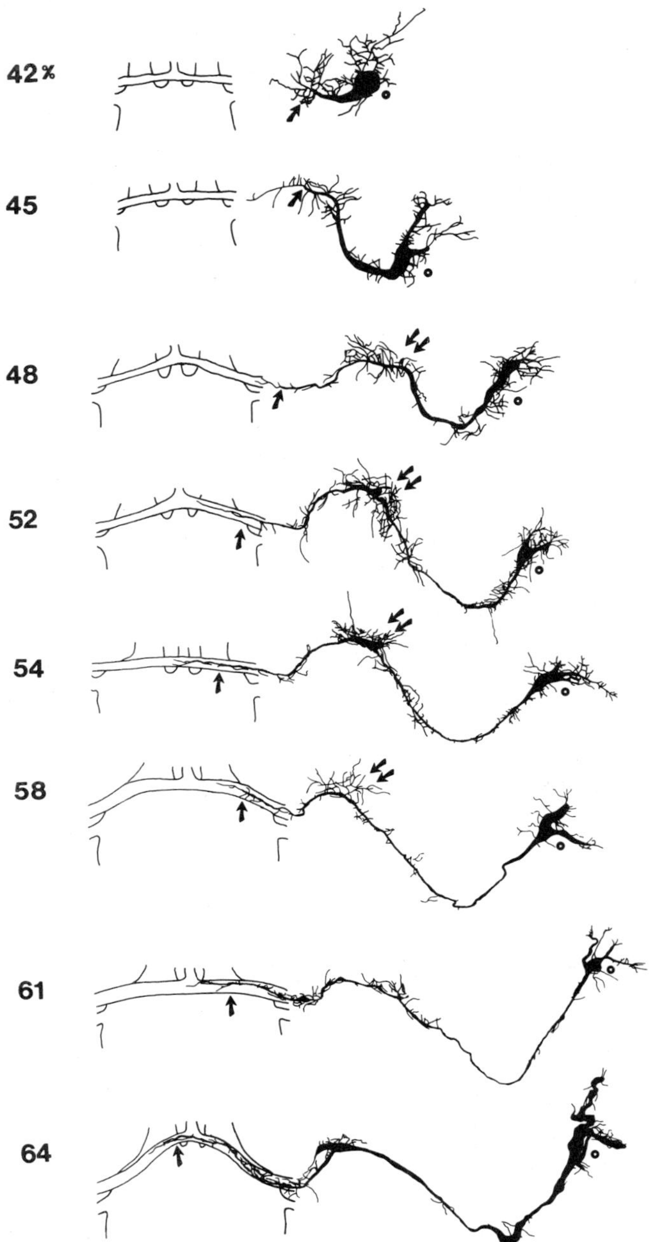

Fig. 2.   (Opposite page). Morphological differentiation of the identified neurosecretory neuron L1. This period of growth is characterized by the smooth and error-free elaboration of the axon (42-58%) followed by the extensive ramification with the neurohaemal organ the Transverse Nerve (at 61-64%). Developmental time is rated as percent of total embryonic development (numbers to the left). Drawings are <u>camera</u> <u>lucida</u> images of individual neurons that in sequentially older embryos were filled with the dye Lucifer Yellow and secondarily processed with an anti-LY antiserum. A simplified segmental ganglion and Transerve Nerve is indicated in the left half of each panel. Asterisks indicate the positions of the L1 cell body; the single arrow indicates the position of the L1 growth cone - the leading edge of the developing axon; the double arrow indicates a morphological specialization that is cell-specific and transient (see text).

behave as does that of L1: as soon as they are within filopodial grasp of the Syncytium, they each adhere to its surface. These NSC growth cones remain attached to it for approximately 10-15% of development before they disattach and proceed to grow within the developing TN. Hence, during the normal development of the neurohaemal TN, NSC growth cones transiently interact with a specific syncytial cell near the site of their terminal differentation, whereas motorneuron growth cones do not. Perhaps a clue as to the functional nature of this interaction lies in this observation of adhesion by specific (neurosecretory) cell type.

How does this interaction fit into the general schedule of NSC differentiation? Ultimately, each individual NSC ramifies extensively within the TN; many of these terminal branches become varicose. In the case of these NS neurons in moth embryos, the process of terminal differentiation occurs following the period of interaction between NSCs and the specific syncytial cell. Whether these observations indicate a causal relationship can only be determined by experimental analysis in embryo culture (exp'ts. in progress). Whatever its nature, our observations of normal development clearly suggest the involvement of specific non-neuronal cells during the differentiation of neurosecretory cells. Electron microscopic studies of developing hypothalamic NSCs (Dellman et al., 1981) indicate similar interactions with non-neuronal cells (pituicytes). The utility of model systems such as insect embryos lies in the fact that we can study and manipulate unique identifiable NSCs and non-neuronal cells.

Insect FMRFamide-Encoding Genes

We wish to understand the integration of morphological and biochemical differentiation in these neurons. In order to study the regulation of neuropeptide expression in specific NSCs it is necessary to know the nature of the secretory molecules, their primary structures and that of the gene that encodes them. We have recently cloned the Drosophila gene that encodes FMRFamide-related peptides (Schneider and Taghert, submitted) by its homology with a molluscan cDNA (Schaeffer et al., 1985).

The gene is present in single copy in the haploid genome and is transcribed into a ~1650 bp RNA that in turn encodes a precursor protein of ~40 kd. We sequenced a cDNA of 1352 bp from which we deduced the primary structure of the Drosophila preproFMRFamide protein (Figure 4). Like numerous other neuropeptide precursors, this polyprotein encodes a large number of putative biologically active peptides. The sequence -FMRFG- (Gly is the likely site of destructive amidation) occurs 10 times, in each case it is extended towards the amino terminus by 3 to 5 residues. Further, the peptides are separated by single or di-basic residues which are likely sites of endoproteolytic cleavage. In all, 13 FMRFamide-like peptides are present - the 3 additional peptides in this category include: a -FMHFamide, a -FVRSamide and a -FIRFamide. Other, novel insect peptides are also predicted including a CRF-like neuropeptide: homology is seen at 9 of 27 residues with either CRF or related peptides. These homologies have been noted previously for the case of the Aplysia FMRFamide gene and the results presented here further support the contention that the FMRFamide gene is evolutionarily conserved across numerous animal phyla.

The significance of cloning a neuropeptide gene in Drosophila lies not just in the knowledge it provides with regard to the primary structure of the neuropeptides, but also in the potential for novel approaches to studying neuropeptide expression and function. For example, as indicated in the description above, insects are excellent systems for developmental studies. Individual neurons are large, identifiable and accessible at all stages

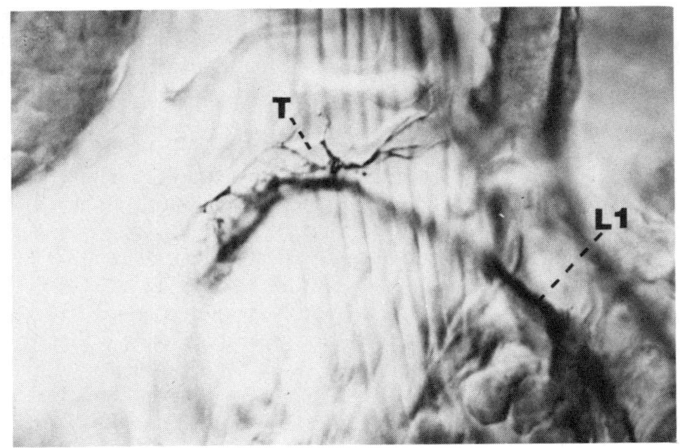

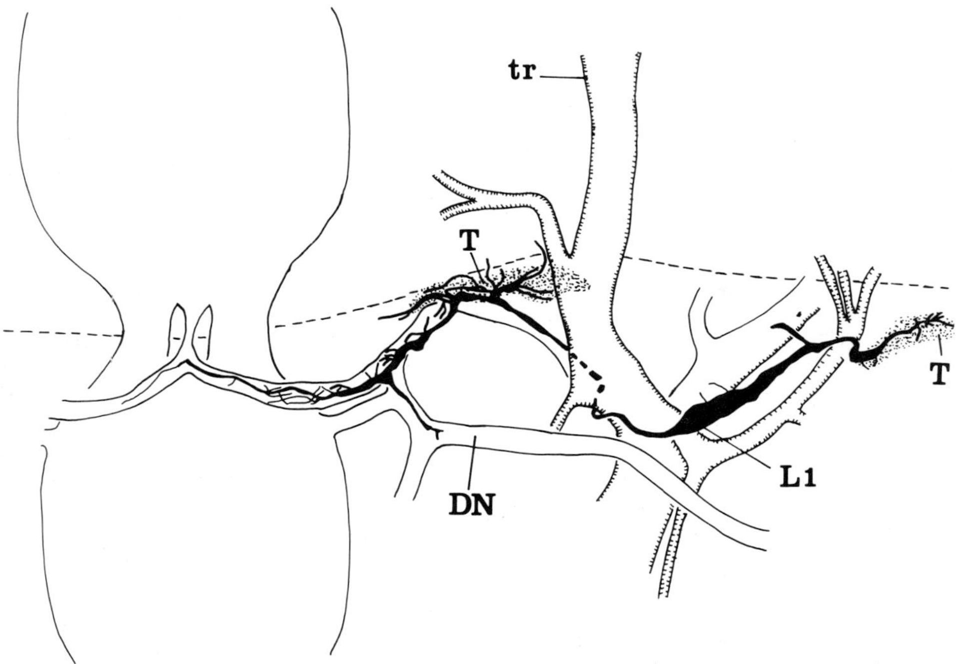

Fig. 3.   A transient embryonic interaction between the identified NSC L1 and a specific syncytial cell (T). (Top) L1 was filled with LY dye at ~55% of embryonic development and processed as above. Note the extensive and long filopodial processes of the L1 cell that are in intimate contact with the syncytial cell (T). This stereotyped interaction takes place just as the L1 growth cone has entered the presumptive nerve (the neurohaemal organ) and just before the elaboration of L1 terminal branches within the nerve. L1: the L1 neuron; tr: trachea; DN: the dorsal segmental nerve. Note also that a second syncytial cell (also marked T) is contacted by the distal (to the right) process of the L1 cell. Image is of the right hand side of the first abdominal segment; segmental ganglia are outlined to the left side. The bottom panel is a <u>camera lucida</u> tracing of the same preparation.

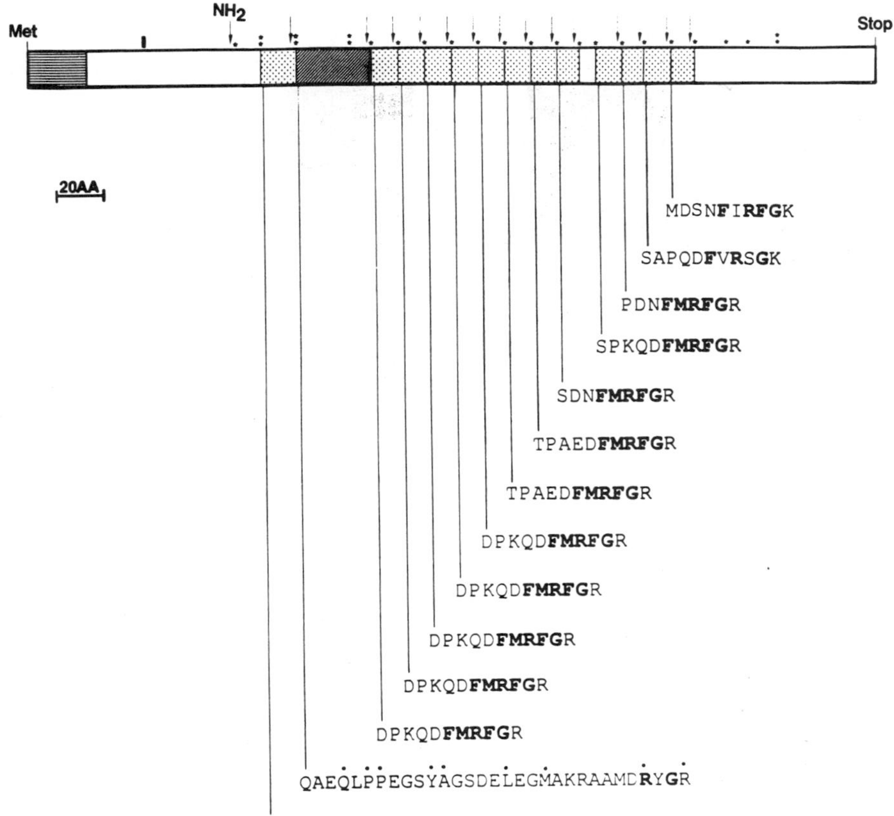

Fig. 4. Schematic diagram that indicates the structure of the deduced protein encoded by the _Drosophila_ FMRFamide gene. Thirteen separate FMRFamide-like peptides are encoded. Arrows indicate positions of Gly residues that are likely to serve as sites of amidation; asterisks indicate positions of single or dibasic residues where post-translational cleavage may occur. Filled areas are highlighted as follows: horizontal lines - putative signal sequence; dots - FMRFamide (or related sequence) -containing peptides; diagonal lines - a CRF-related peptide. Relevant peptide sequences are provided below the precursor; FMRFG or related sequences are in boldface. Dots over the CRF-like peptide indicate residues that are homologous with CRF or related peptides.

of development from neuroblasts to fully differentiated neurons (Goodman and Spitzer, 1979). Furthermore, we can now culture _Manduca_ embryos throughout most of embryonic development, from the time of initial neurogenesis through the complete differentiation of the L1 neuron (Wall and Taghert, unpublished results). Finally, the advanced genetics and molecular techniques that are available in the study of _Drosophila_ offer the potential of a detailed genetic analysis of neuropeptide expression and function. The identification of neuropeptide gene mutations could lead to an investigation of putative developmental roles for neuropeptides. The identification of mutations that affect neuropeptide gene expression without affecting the neuropeptide gene itself could begin a genetic analysis of neuropeptide gene regulation.

Naturally, we would like to clone the homologous gene in <u>Manduca</u>, so as to analyse the development of individual peptidergic neurons like L1 at the molecular level. We believe a homologous gene exists in the moth because the pattern of FMRFamide-immunoreactive neurons is nearly equivalent between <u>Manduca</u> and <u>Drosophila</u> (Copenhaver and Taghert, unpublished results). In addition, we have recently isolated homolgous <u>Manduca</u> cDNAs but, at this time, do not yet know if they represent the true FMRFamide gene homologue. The following two questions are examples of the types of analyses to be done with a <u>Manduca</u> FMRFamide gene. (i) We know when FMRFamide-like immunoreactivity first appears in the L1 neuron relative to its morphological differentiation but when is neuropeptide gene transcription first initiated? This question is not trivial for, as shown in the case of oxytocinergic neurons of the hypothalamus, neuropeptide gene transcription and translation may be quite separate temporally from the onset of post-translational processing to produce biologically active molecules (Whitnall et al., 1985). Secondly, what is the molecular basis that underlies the transient expression of FMRFamide-like peptides in the L1 neuron in posterior segments of <u>Manduca</u> embryos (see above)? This transience could represent changes in gene transcription, translation or post-translational processing. Having nucleic acid probes to the specific gene should allow us to address this process as well as the cellular and molecular mechanisms that control it during development.

SUMMARY

The studies reviewed here indicate the manner by which a simple nervous system can be utilized to analyze fundamental problems of neuronal development and cellular differentiation. Our focus on neurosecretory neurons has led us to search and find cellular candidates that may act as "fictive targets" for developing neurosecretory cells. Such targets may trigger the arrest of axon outgrowth, the elaboration of terminal differentiation and/or may regulate aspects of neuropeptide gene expression. In addition, our molecular studies have succeeded in the first isolation of a neuropeptide gene in insects - the FMRFamide gene of <u>Drosophila</u>. The ability in insects to study neuropeptide genes and their regulation genetically and in single identified neurons represent effective strategies that will contribute to our general understanding of peptidergic neurosecretory neurons.

REFERENCES

Carrol, L.S., Carrow, G.M. and Calabrese, R.L., 1986, Localization and release of FMRFamide-like immunoreactivity in the cerebral neuroendocrine system of <u>Manduca sexta</u>., <u>J. Exp. Biol.</u>, 126:1.

Carrow, G.M., Calabrese, R.L. and Williams, C.M., 1984, Architecture and physiology of insect cerebral neurosecretory cells, <u>J. Neurosci.</u>, 4:1034.

Copenhaver, P. and Truman, J.W., 1986, Control of neurosecretion in the moth <u>Manduca sexta</u>: physiological regulation of the eclosion hormone cells, <u>J. Comp. Physiol.</u>, 158:445.

Dellman, H.C., Sikora, K. and Castel, M., 1981, Fine structure of the rat supraoptic nucleus and neural lobe during pre- and post-natal development, <u>in</u>: "Neurosecretion: Molecules, Cells and Systems", D.Farner and K. Lederis, eds., Plenum Press: New York, pp. 177-186.

Evans, P.D. and Myers, C.M., 1986, Physiology of FMRFamide-related peptides in the neuromuscular junction of the locust, <u>J. Exp. Biol.</u>, 126:403.

Finlayson, L. and Osborne, M.P., 1969, Peripheral neurosecretory cells in the stick insect (<u>Carausius</u> <u>morosus</u>) and the blow-fly (<u>Phormia terra novae</u>), <u>J. Insect Physiol.</u>, 14:1793.

Goodman, C.S. and Heitler, W., J., 1979, Electrical properties of insect neurones with spiking and non-spiking somata: normal, axotomized and colchicine-treated neurones, <u>J. Exp. Biol.</u>, 83:95.

Goodman, C.S. and Spitzer, N.C., 1979, Embryonic development of identified neurones: differentiation from neuroblast to neurone, <u>Nature</u>, 280:208.

Hinks, C.F., 1976, Peripheral neurosecretory cells in some Lepidoptera, <u>Cand. J. Zool.</u>, 53:1035.

Jansen, R.F. and Bos, N.P.A., 1984, An identified neuron modulating the activity of the ovulation hormone producing caudo-dorsal cells of the pond snail <u>Lymnae stagnalis</u>.,<u>J. Neurobiol.</u>, 15:161.

Raabe, M., 1982, "Insect Neurohormones", Plenum Press: New York.

Schaeffer, M., Piciotto, M.R., Kreiner, T., Kaldany, R.R., Taussig, R. and Scheller, R.H., 1985, <u>Aplysia</u> neurons express a gene encoding multiple FMRFamide neuropeptides, <u>Cell</u>, 41:457.

Taghert, P.H. and Truman,J.W., 1982, Identification of the bursicon-containing neurones in abdominal ganglia of the tobacco hornworm, <u>Manduca sexta.</u>, <u>J. Exp. Biol.</u>, 98:385.

Taghert, P.H., and Goodman, C.S., 1984, Cell determination and differentiation of identified serotonin-immunoreactive neurons in grasshopper embryos, <u>J. Neurosci.</u>, 4:989.

Taghert, P.H., Bastiani, M.J., Ho, R.K. and Goodman, C.S., 1982, Guidance of pioneer growth cones: filopodial contacts and coupling revealed with an antibody to Lucifer Yellow, <u>Dev. Biol.</u>, 94:391.

Taghert, P.H., Carr, J.N., Wall, J.B. and Copenhaver, P.F., 1986, Embryonic formation of a simple neurosecretory nerve in the moth <u>Manduca sexta</u>. <u>in</u>: "Insect Neurochemistry and Neurophysiology", A.B. Borkovec and D.B. Gelman, eds., The Humana Press: Clifton, N.J., pp. 143-172.

Whitnall, M.H., Key, S., Ben-Barak, Y., Ozato, K. and Gainer, H., 1985, Neurophysin in the hypothalamo-neurohypophysial system. II. Immunocytochemical studies of the ontogeny of the oxytocinergic and vasopressinergic neurons, <u>J. Neurosci.</u>, 5:98.

THE ROLE OF RNA SPLICING AND POST-TRANSLATIONAL PROTEOLYTIC PROCESSING IN THE BIOSYNTHESIS OF NEUROPEPTIDES

Anthony J. Harmar and *Lindsay Sawyer

MRC Brain Metabolism Unit
Royal Edinburgh Hospital
Morningside Park
Edinburgh EH10 5HF

*Department of Biochemistry
University of Edinburgh
Hugh Robson Building
George Square
Edinburgh EH8 9XD

INTRODUCTION

The neuropeptides are synthesised by mechanisms which closely resemble those for other secreted proteins. Transcription of a neuropeptide gene results in the synthesis of precursor RNA which is processed in the nucleus into mature messenger RNA (mRNA). Only certain segments of the precursor RNA (exons) are present in mRNA and the remaining segments of RNA (introns) are excised during RNA maturation. In some cases, more than one species of mRNA may be generated from a single gene as the result of alternative patterns of RNA splicing.

Translation of mRNA in the neuronal cell body leads to the synthesis of a precursor polypeptide containing at its N-terminus a signal sequence which directs the nascent polypeptide into the intracisternal space of the endoplasmic reticulum, where it is cleaved from the precursor. As the precursor passes into the Golgi apparatus and is packaged into secretory vesicles, further proteolytic cleavages take place, often at sites flanked by pairs of basic amino acids (arginine and lysine), liberating the mature neuropeptide from its precursor. These events, together with other enzymic modifications of individual amino acids within the peptide sequence, are known collectively as post-translational processing. In some cases, a single neuropeptide precursor polypeptide may give rise to different spectra of biologically active products as the result of tissue-specific pathways of post-translational processing.

There are thus two mechanisms by which the expression of a single neuropeptide gene can give rise to alternative patterns of biologically active peptides: by variations in the splicing of the primary gene transcript and by differential processing. Both of these two mechanisms may operate in a tissue-specific manner. In this chapter, we summarise our knowledge of the contributions of these two mechanisms to the biosynthesis of the neuropeptides and demonstrate that both RNA splicing and post-translational processing contribute to the diversity of peptides derived from the gene coding for the substance P precursor (the preprotachykinin gene).

ALTERNATIVE RNA SPLICING

The best understood example of a neuropeptide gene susceptible to tissue-specific RNA splicing is that coding for calcitonin and calcitonin gene-related peptide (CGRP: Rosenfeld et al., 1984). The primary transcript of this gene contains six exons. In the C cells of the thyroid, exons 1-4 are spliced together to generate an mRNA coding for the precursor to calcitonin, a hormone which plays an essential role in calcium homeostasis. The same gene is also expressed in some parts of the nervous system: here, exons 1-3 are spliced to exons 5 and 6 to generate a mRNA encoding the precursor to CGRP, a potent vasoactive neuropeptide. The tissue-specific splicing of calcitonin/CGRP gene transcripts appears to result from the presence in certain neurones of specific regulatory machinery - presumably a protein or proteins - which is required for the generation of CGRP transcripts (Crenshaw et al., 1987). In non-neuronal cells where this machinery is absent, the predominant mRNA derived from the gene is that encoding the calcitonin precursor.

Two further genes coding for peptides of importance in the nervous system are known to produce primary transcripts which are subjected to alternative RNA processing: the gene encoding the two plasma proteins which are precursors to bradykinin (the kininogens) and that encoding the precursor to gastrin-releasing peptide (GRP). In man, two bradykinin precursors are synthesised in the liver and circulate in plasma (Tagaki et al., 1985; Muller-Esterl et al., 1986): a high molecular weight (HMW) kininogen of 626 amino acids and a low molecular weight (LMW) kininogen of 409 amino acids. The N-terminal 383 amino acids of the two kininogens, including the sequence encoding the bradykinin moiety, are identical; they are encoded by exons 1-9 and the 5' part of exon 10 of the kininogen gene. HMW kininogen mRNA is generated by the splicing together of exons 1-9 and the whole of exon 10 (the 3' portion of which encodes the region specific to HMW kininogens). In the synthesis of LMW kininogen mRNA, an alternative donor site within exon 10 is spliced to an 11th exon which specifies the LMW-specific domain of the precursor. Kallikrein and related proteases can act on either kininogen to release bradykinin, but the C-terminal polypeptide unique to HMW kininogen plays a specific role in initiating blood coagulation following blood vessel trauma.

GRP may function as a transmitter in the central and peripheral nervous systems; the peptide is also present in neuroendocrine cells in the lung, where it may act as a pulmonary growth factor (see Polak, this volume and Spindel, 1986). The GRP precursor is encoded by a gene containing three exons. Since two alternative splice-donor sites are present in exon 2 and two possible splice-acceptor sites in exon 3, at least three distinct mRNAs are generated from a single gene (Spindel et al., 1987). Translation of these RNAs should generate a family of prohormones, which share a common N-terminal region containing the GRP sequence, but which differ in the C-terminal region. The functions of these alternative forms of the GRP precursor have not yet been established.

TISSUE-SPECIFIC POST-TRANSLATIONAL PROCESSING

Pro-opiomelanocortin (POMC) is the common precursor to adrenocorticotrophic hormone (ACTH), the melanocyte-stimulating hormones and a family of opioid peptides (beta-lipotropin and the endorphins). In the anterior pituitary gland, the predominant products of processing are ACTH and beta-lipotropin, whereas in the pars intermedia, further processing steps lead to the generation of the melanocyte-stimulating hormones and acetylated, biologically inactive forms of the endorphins (Roberts et al., 1985). In this case, tissue-specific, post-translational processing of a single polypeptide generates alternative patterns of peptides with different biological functions and targets.

In contrast, variations in the post-translational processing of the precursors to somatostatin (SS) and cholecystokinin (CCK) lead to the production of families of peptides each containing a single biologically active sequence with an N-terminal extension of variable size. The sequence of SS-14 is located at the C-terminus of its precursor, preprosomatostatin. Several N-terminally extended forms of SS, including a peptide of 28 amino acids (SS-28) can be generated from the polypeptide. In most tissues (for example pancreas, spinal cord and extrahypothalamic brain) SS-14 predominates, though significant amounts of SS-28 are present. However, in the median eminence (Pierotti and Harmar, 1985; Pierotti et al., 1985) and the neurohypophysis (Gomez et al., 1983), SS-14 and SS-28 are present in approximately equal amounts and in intestinal D cells (Baskin and Ensinck, 1984), SS-28 predominates. Both forms of the peptide appear to act at a single class of receptors, but SS-28 has a longer half-life in plasma than SS-14 (Polonsky et al., 1982). This may account for the production of SS-28 by intestinal D cells (which may regulate pancreatic

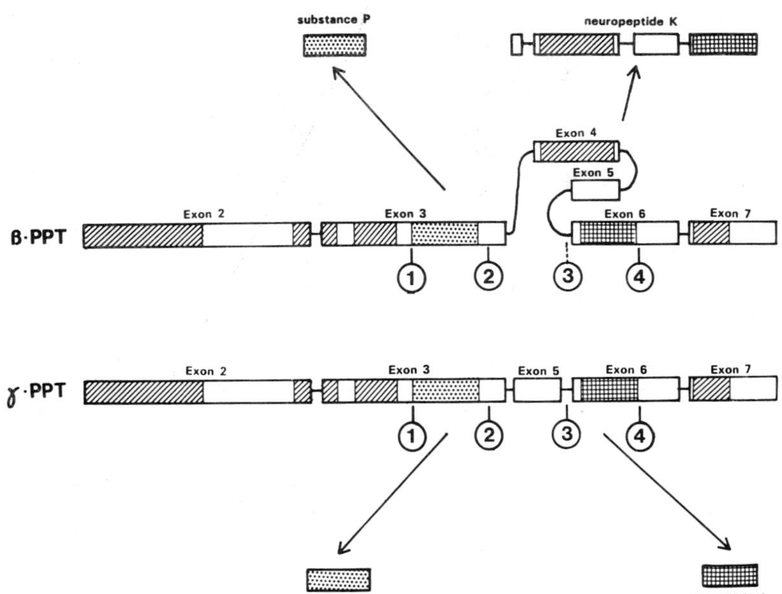

Fig 1.   Post-translational processing of beta- and gamma-preprotachykinin. The beta-PPT polypeptide includes sequences encoded by exons 2 - 7 of the PPT gene; the sequences of substance P and neurokinin A are flanked by potential sites of post-translational processing (numbered circles). Secondary structure analysis (Sawyer et al., 1987) predicts the presence of 4 alpha-helical regions within the precursor polypeptide (diagonal shading). The predominant products of processing of this precursor are substance P and neuropeptide K: the presence of an alpha-helical domain encoded by exon 4 may influence the conformation of the precursor to prevent proteolysis at the potential processing site 3. This domain is absent from the sequence of gamma-PPT: in this polypeptide, site 3 may be accessible to processing enzymes and gamma-PPT may function as the physiological precursor of neurokinin A.

hormone secretion by releasing SS into the bloodstream) and by hypothalamic neurosecretory neurones, which modulate the secretion of pituitary hormones by releasing SS into hypophysial portal blood (Millar et al., 1983).

The cholecystokinin family of peptides are synthesised from a precursor of 115 amino acids (preprocholecystokinin). At least four forms of CCK can be produced in tissues (Dockray et al., 1987), including an octapeptide (CCK-8: the main form of CCK synthesised in the central nervous system) and larger forms (including CCK-33 and CCK-39) which predominate in the gastrointestinal tract. As in the case of somatostatin, it is thought that the larger forms of CCK, which have a longer plasma half-life than CCK-8 and are resistant to degradation by the liver, are produced by the gut to fulfil a hormonal role.

SUBSTANCE P PRECURSORS (PREPROTACHYKININS)

Substance P (SP), neurokinin A (NkA) and neurokinin B (Kimura et al., 1983) are the mammalian representatives of the tachykinins, a family of peptides which share the common C-terminal amino acid sequence Phe-X-Gly-Leu-Met-NH$_2$. Nakanishi and his colleagues (Nawa et al., 1982, 1983) have shown that SP and NkA are encoded by a single gene, the preprotachykinin (PPT) gene, consisting of 7 exons. Exons 3 and 6 contain, respectively, the sequences of SP and NkA, each flanked by pairs of basic amino acids which are the sites of post-translational proteolysis. Three mRNAs may be generated as the result of alternative splicing (Kawaguchi et al., 1986; Krause et al., 1987):

1)   alpha-PPT (exons 1-5 and 7) lacks exon 6 and thus encodes only a single tachykinin sequence, that of SP.

2)   beta-PPT (exons 1-7) contains the sequences of three tachykinins: SP, NkA and neuropeptide K (NPK), an N-terminally extended form of NkA (Tatemoto et al., 1985).

3)   gamma-PPT (exons 1-3 and 5-7) differs from beta-PPT in lacking a pentadecapeptide sequence in the region separating SP and NKA (corresponding to residues 3-17 of NPK).

We are currently attempting to characterise the patterns of biologically active peptides generated by the post-translational processing of each of the three substance P precursors. Unfortunately, in most tissues, all three precursors are present together with precursors to the related peptide neurokinin B: the complex spectrum of peptides generated is difficult to interpret. However, we have recently cloned and sequenced cDNA encoding human beta-PPT using tissue from a carcinoid tumour as the source of mRNA (Harmar et al., 1986). Tissue from this tumour proved to be a relatively homogeneous source of the peptides derived from beta-PPT; chromatographic analysis indicated that the predominant tachykinins produced were SP and NPK (Harmar et al., 1987). Since only small amounts of NkA were detected, it appears that a putative processing site (Lys-Arg) at the N-terminus of the NkA sequence is not cleaved. If this site were similarly resistant to proteolysis when present within the sequence of gamma-PPT, a characteristic peptide (des[3-17]NPK) should be produced. Since there is no evidence for the presence of this peptide in tissues, it seems likely that this cleavage site is not utilised when present within the beta-PPT sequence.

We propose that beta- and gamma-PPT may function, respectively, as the physiological precursors of NPK and NkA. Support for this hypothesis has been obtained from predictions of the secondary structures of the two precursors (Figure 1). The presence in the beta-PPT sequence of an alpha-helical segment encoded by exon 4 may influence the secondary structure of the polypeptide so that a potential processing site is protected from proteolysis and NPK is generated. Gamma-PPT may adopt a different conformation in which the processing site is accessible and NkA may be produced.

The PPT gene may constitute the first example of a neuropeptide gene in which tissue-specific RNA splicing leads to the inclusion or exclusion in the encoded precursor polypeptide of domains devoid of intrinsic biological activity but which influence the secondary structure of the precursor to determine its post-translational processing.

REFERENCES

Baskin, D.G. and Ensinck, J.W., 1984, Somatostatin in epithelial cells of intestinal mucosa is present primarily as somatostatin 28, Peptides (Fayetteville), 5:615.

Crenshaw, E.B. III, Russo, A.F., Swanson, L.W. and Rosenfeld, M.G., 1987, Neuron-specific alternative RNA processing in transgenic mice expressing a metallothionein - calcitonin fusion gene, Cell, 49:389.

Dockray, G.J., Dimaline, R., Pauwels, S. and Varro, A., Gastrin and CCK-related peptides, in: "Prohormones, hormones and their fragments", J. Martinez, ed., Ellis Horwood. In press.

Gomez, S., Morel, A., Nicolas, P.and Cohen, P., 1983, Regional distribution of the Mr15000 somatostatin precursor, somatostatin-28 and somatostatin-14 in the rat brain suggests a differential intracellular processing of the high molecular weight species, Biochem. Biophys. Res. Commun., 112:297.

Harmar, A.J., Armstrong, A., Pascall, J., Chapman, K., Rosie, R., Curtis, A., Going, J.J., Edwards,C.R.W. and Fink, G., 1986, cDNA sequence of human beta-preprotachykinin, the common precursor to substance P and neurokinin A, FEBS Lett., 208:67.

Harmar, A.J., Pierotti, A.R., Sanchez-Watts, G., Going, J.J. and Edwards, C.R.W., 1987, Production by a metastatic laryngeal carcinoid of peptides derived from the substance P precursor, beta-preprotachykinin, J. Endocrinol., 112:277.

Kawaguchi, Y., Hoshimaru, M., Nawa, H. and Nakanishi, S., 1986, Sequence analysis of cloned cDNA for rat substance P precursor: existence of a third substance P precursor, Biochem. Biophys. Res. Commun., 139:1040.

Kimura, S., Oada, M., Sugita, Y., Kanazawa, I. and Munekata, E, 1983, Novel neuropeptides, neurokinin alpha and beta, isolated from porcine spinal cord, Proc. Jap. Acad. Ser.B, 59:101.

Krause, J.E., Chirgwin, J.M., Carter, M.S., Xu, Z.S. and Hershey, A.D., 1987, Three rat preprotachykinin mRNAs encode the neuropeptides substance P and neurokinin A, Proc. Natl. Acad. Sci. USA, 84:881.

Millar, R.P., Sheward, W.J., Wegener, I and Fink, G., 1983, Somatostatin-28 is a hormonally active peptide released into hypophysial portal blood, Brain Res., 260:334.

Muller-Esterl, W., Iwanaga, S.and Nakanishi, S., 1986, Kininogens revisited, Trends Biochem. Sci., 2:336.

Nawa, H., Hirose, T., Takashima, H., Inayama, S. and Nakanishi, S., 1983, Nucleotide sequence of cloned cDNAs for two types of bovine brain substance P precursor, Nature, 306:32.

Nawa, H., Kotani, H. and Nakanishi, S., 1984, Tissue-specific generation of two preprotachykinin mRNAs from one gene by alternative RNA splicing, Nature, 312:729.

Pierotti, A.R. and Harmar, A.J., 1985, Multiple forms of somatostatin-like immunoreactivity in the hypothalamus and amygdala of the rat: selective localization of somatostatin-28 in the median eminence, J. Endocr., 105:383.

Pierotti, A.R., Harmar, A.J., Tannahill, L. and Arbuthnott, G.W., 1985, Different patterns of molecular forms of somatostatin are released by the rat median eminence and hypothalamus, Neurosci. Lett., 57:215.

Polonsky, K.S., Jaspan, J., Berelowitz, M., Pugh, W., Moossa, A . and Ling,N., 1982, The in vivo metabolism of somatostatin 28: Possible relationship between diminished metabolism and enhanced biological action, J. Endocr., 111:1698.

Roberts, J.L., Chen, C-L.C., Dionne, F.T. and Gee, C.E., 1985, Peptide hormone gene expression in heterogeneous tissues: the pro-opiomelanocortin system, in: "Neurotransmitters in Action", D. Bousfield, ed., Elsevier Biomedical Press, Amsterdam, pp 226.

Rosenfeld, M.G., Amara, S.G. and Evans, R.M., 1984, Alternative RNA processing: determining neuronal phenotype, Science, 225:1315.

Sawyer, L., Tollin, P. and Wilson, H.R., 1987, A comparison between the predicted secondary structures of potato virus X and papaya mosaic virus coat proteins, J. Gen. Virol., 68:1229.

Spindel, E.R., 1986, Mammalian bombesin-like peptides, Trends Neurosci., 9:130.

Spindel, E.R., Zilberberg, M.D. and Chin, W.W., 1987, Analysis of the gene and multiple messenger ribonucleic acids (mRNAs) encoding human gastrin-releasing peptide: alternate RNA splicing occurs in neural and endocrine tissue, Mol. Endocrinol., 1:224.

Takagaki, Y., Kitamura, N. and Nakanishi, S., 1985, Cloning and sequence analysis of cDNAs for human high molecular weight and low molecular weight prekininogens, J. Biol. Chem., 260:8601.

Tatemoto, K., Lundberg, J.M, Jornvall, H. and Mutt, V., 1985, Neuropeptide K: Isolation, structure and biological activities of a novel brain tachykinin, Biochem. Biophys. Res. Commun., 128:947.

PRECURSORS OF UROTENSINS AND THEIR CO-EXPRESSION IN THE CAUDAL NEUROSECRETORY SYSTEM

Tomoyuki Ichikawa[1], Isao Ishida[2], Shunji Ohsako[2] and Takeo Deguchi[2]

Departments of [1]Anatomy and Embryology and [2]Molecular
Neurobiology, Tokyo Metropolitan Institute for
Neurosciences, Tokyo, Japan 183

INTRODUCTION

An array of neurosecretory neurons in the posterior spinal cord and their associated neurohemal organ, the urophysis, comprise the caudal neurosecretory system (CNS) of fishes. The CNS is credited with the synthesis and release of at least 2 neurohormones, urotensins I and II (UI and UII). There is no clear evidence for the existence of urotensin III as a separate chemical entity. Urotensin IV appears to be chemically and pharmacologically indistinguishable from arginine vasotocin (Lacanilao, 1972a,b), which occurs in small amounts in the urophysis of only some teleost species (Lacanilao and Bern, 1972). Also present in the teleost urophysis are large amounts of acetylcholine of unknown significance (Kobayashi et al., 1963; Ichikawa, 1978; Ichikawa and Kobayashi, 1978) and putative carrier proteins analogous to the neurophysins, the urophysins (Berlind et al., 1972; Moore et al., 1975; Lederis et al., 1981).

The primary structures of UI peptides from the carp, _Cyprinus carpio_ (Ichikawa et al., 1982) and sucker, _Catostomus commersoni_ (Lederis et al., 1982) have been determined. Urotensin I is a 41-residue amidated peptide homologous to mammalian corticotropin-releasing factor (CRF) (see Ichikawa, 1985; Ichikawa et al., 1986). A single form of UII from the goby, _Gillich-thys mirabilis_, was the first of peptides of the CNS to be structurally characterized (Pearson et al., 1980), followed by the identification of 2 forms of UII (UII_A and UII_B) from the sucker (McMaster and Lederis, 1983) and 3 forms of UII ($UII\alpha$, $UII\beta$ and $UII\gamma$) from the carp (Ichikawa et al., 1984). All forms of UII peptides so far isolated are cyclic dodecapeptides that share an identical 6-residue disulfide-bridged ring, while multiple substitutions occur in amino terminal sequences (see Ichikawa 1985; Ichikawa et al., 1986). A partial sequence homology between UII and mammalian somatostatin has been indicated (Pearson et al., 1980; see Ichikawa et al., 1986).

The structural characterization of urotensins has led to a dramatic increase in our knowledge of the CNS at many levels. Sequencing of UI and UII peptides, for example, has broadened our concepts of their possible functions implicated from the homologies to other known neuropeptide families. There is increasing evidence for hypophysiotropic effects of urotensins in addition to their osmoregulatory roles (see Bern et al., 1985; Fryer and Lederis, 1986; Ichikawa et al., 1986). The availability of synthetic urotensins has also allowed the production of specific antibodies, and recent immunohistochemical studies have presented evidence for the co-localization of UI and UII in the same caudal neurosecretory neurons (see Bern et al., 1985; Ichikawa et al., 1986; Kobayashi et al., 1986). Further, knowledge of sequences of UI and UII has provided information necessary to embark on molecular cloning studies, which have elucidated the nucleotide sequences of mRNAs for the precursors of both peptides. This, in turn, makes it possible to carry out _in situ_ hybridization by which a particular mRNA can be visualized in individual cells on tissue sections with a labelled cDNA probe.

In this chapter, we describe our recent molecular biological studies which revealed the structures of the precursors of urotensins, and demonstrated the co-expression of them in the caudal neurosecretory neurons by an _in situ_ hybridization technique.

PRECURSORS OF UROTENSINS

The primary structures of the precursors of UI (Ishida et al., 1986), and UIIα and UIIγ (Ohsako et al., 1986) of the carp, and UII of the goby (Pearson et al., 1986) have been inferred by analyzing the nucleotide sequences of cloned DNAs complementary to mRNAs encoding them. The precursor of carp UI consists of 145 amino acid residues including a 22-residue signal peptide (Fig. 1A). The carboxyterminus represents the 41-amino acid sequence of UI, preceded by Lys-Arg, a processing signal, and followed by Gly-Lys, a processing and carboxyterminal amidation signal (Fig. 1A). The precursors of carp UIIα and UIIγ are composed of 125 amino acid residues including a 21-residue signal peptide, and active UII peptides exist at the carboxytermini preceded by a processing signal, Arg-Lys-Arg (Fig. 1B). An additional pair of basic amino acids, Arg-Arg, is located at positions 107 and 108. Only 24 nucleotide and 12 amino acid substitutions exist between the precursors of UIIα and UIIγ. The precursor of goby UII consists of 117 amino acid residues including a 19-residue signal peptide, and shows a similar organization to the precursors of carp UII peptides.

In spite of the similar organization of the precursors of UI and UII, there is no homology in amino acid nor nucleotide sequences between them. A high homology in both amino acid and nucleotide sequences is observed between the precursors of carp UI and ovine CRF (Furutani et al., 1983) (see Ishida et al., 1986); 55% of the nucleotides in the hormonal portion of CRF mRNA occur in an equivalent region in that of UI mRNA, and a 51% homology in amino acid sequence is present between the 2 peptides. The nonhormonal portion excluding a signal peptide (amino acid residues 23-100) of UI precursor exhibits a 21% homology with residues 21-100 of ovine CRF. These data suggest that the 2 precursor proteins may be evolutionarily related. Evidence for a possible role of UI as an effector of

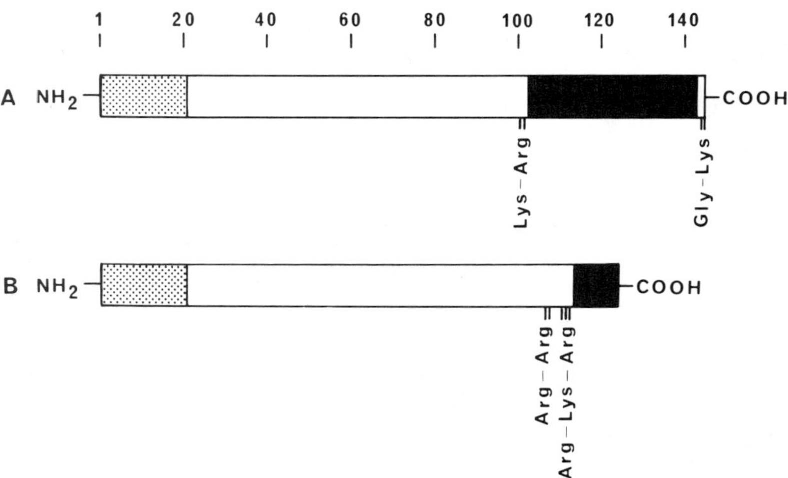

Fig. 1.   Schematic representation of organization of precursors of UI (A) and either UIIα or UIIγ (B). Only 12 amino acid substitutions are observed between precursors of UII peptides. The UI and UII sequences are indicated by black boxes and signal peptides by stippled boxes. Processing signals, Lys-Arg in UI precursor, and Arg-Arg and Arg-Lys-Arg in UII precursor are shown. A processing and carboxyl terminal amidation signal, Gly-Lys in UI precursor is also shown. Amino acid numbers are given at the top.

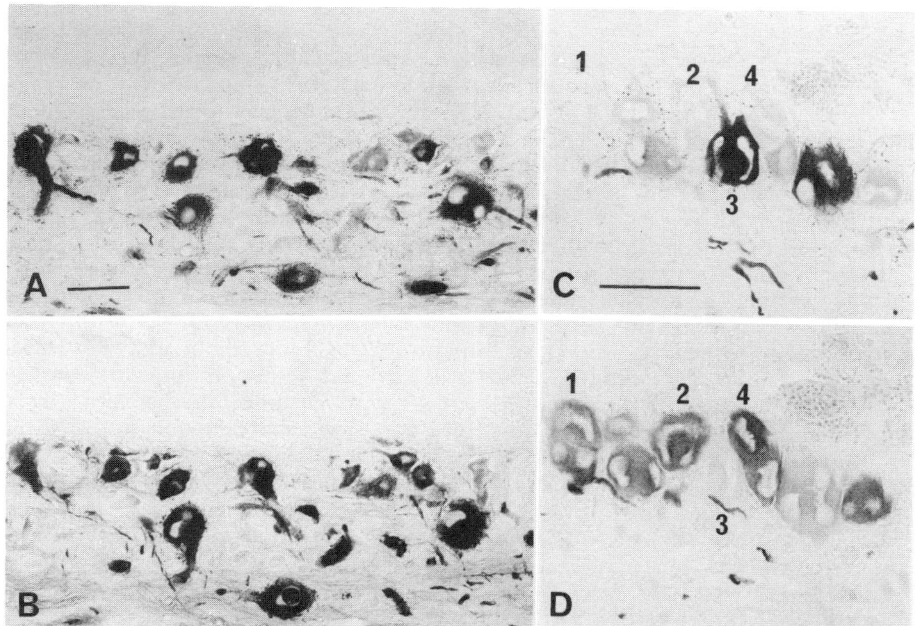

Fig. 2.　Adjacent parasagittal 10-μm paraffin sections (A and B, and C and D) of the caudal spinal cord of the carp fixed in Bouin's solution without acetic acid and immunostained for UI (A and C) and UII (B and D).　Note that all the neurons immunoreactive to UI in A are also immunoreactive to UII in B.　In C and D, neurons 1, 2 and 4 are UI-negative but UII-positive, while neuron 3 is UI-positive but UII-negative.　Other neurons are both UI- and UII-positive although the staining intensity is varied.　Bar indicates 100 μm.

ACTH release in fishes has been accumulated (see Fryer and Lederis, 1986; Ichikawa et al., 1986).　A more than 90% homology in both amino acid and nucleotide sequences between UIIα and UIIτ suggests that the genes for them were generated from a common ancestral gene by duplication.　There is little similarity between the precursors of UII peptides and somatostatin (Montminy et al., 1984).

After cleavage of the signal peptide and proteolytic processing, carp UI precursor may yield a peptide consisting of 78 amino acid residues, and either carp UII precursor may yield an 85-residue peptide and a dipeptide, Gln-Phe, in addition to the active UI and UII peptides.　These larger peptides may represent putative urotensin carrier proteins, the urophysins (Berlind et al., 1972; Moore et al., 1975; Lederis et al., 1981).　One of the peptides recently isolated from the carp urophysis is indistinguishable in amino acid composition from the 85-residue peptide (amino acid residues 22-106 of UIIτ precursor; Ichikawa, unpublished).　Its specific binding with UII peptides is under investigation to clarify whether or not this peptide is the specific carrier protein analogous to the neurophysin.　The dipeptide may also have a physiological role, as yet unknown.

RNA blot transfer analyses indicate that mRNAs encoding the precursors of UI, UIIα and UIIτ are present in the carp spinal cord, but not in the brain, intestine, liver or kidney (Ishida et al., 1986; Ohsako et al.,1986).　In the brain of the carp, although a complex network of UI-positive fibers is present, there is no clear evidence for the occurrence of UI-immunoreactive cell bodies (Ichikawa, unpublished), and no structure immunoreactive to anti-UII serum could be detected (Owada et al., 1985).　In contrast to these observations, the presence of UI-positive cells and a fiber network has been reported in the sucker brain (Yulis et al., 1986).

The presence of at least 2 populations of the caudal neurosecretory neurons has long been postulated, based on the morphology, electrophysiological responses and neural inputs (see Chan and Bern, 1976; Kriebel et al., 1985), and it was assumed that one cell type might produce UI and the other UII. There is, however, strong immunohistochemical evidence that a single neuron can contain both UI and UII. Lederis and colleagues observed that all the caudal neurosecretory neurons of the sucker were UI-immunoreactive, and cast doubt on the two-cell hypothesis (Fisher et al., 1984). Later, they demonstrated that most of the UI-positive neurons were also UII-positive (Yulis and Lederis, 1986). Similarly, in the goby, all the neurons in the CNS were shown to contain UI, but only a small portion of them were UII-immunoreactive (Bern et al., 1985; Larson et al., 1987). In the carp,3 populations of neurons have been identified; neurons immunoreactive to either UI or UII, and those immunoreactive to both UI and UII (Ichikawa et al., 1986; Kobayashi et al., 1986; Yamada et al., 1986) (Fig. 2). It should be noted that different fixatives result in considerable variability in the proportions of the 3 populations of neurons in the carp (Ichikawa, unpublished). The presence of 3 types of neurons was also reported in the CNS of the midshipman, Porichthys notatus (Onstott and Elde, 1986). Thus, while the wide-spread co-localization of UI and UII in the same caudal neurosecretory neurons is apparent, there is a considerable inter-species variability in the existence of neurons exclusively immunoreactive for either UI or UII.

In situ hybridization offers another technique to examine the co-expression of urotensins in the caudal neurosecretory neurons. As shown in Fig. 3, using a ^{32}P-labelled cDNA probe coding the entire sequence of carp UI precursor, we have succeeded to detect UI mRNA in individual caudal neurosecretory neurons of the carp (Ichikawa, Ishida and Deguchi, unpublished). Almost all the morphologically identifiable neurons were heavily labelled with the probe. However, a remarkable sequence homology between mRNAs for UIIα and UIIγ required the use of specific synthetic oligonucleotide probes to discriminate them.

We have synthesized the specific oligonucleotide probes for mRNAs for UI, UIIα and UIIγ, and carried out in situ hybridization in the caudal neurosecretory neurons of the carp (Ichikawa, Ishida, Ohsako and Deguchi, unpublished) with the procedures as described (Ohsako et al., 1986). The specificity of the method rested on 3 criteria; (1) the same neurons were labelled with 2 different probes for each mRNA; (2) a synthetic control probe having an identical sequence to UI mRNA did not label any caudal neurosecretory neurons; and (3) a combination of in situ hybridization with immunohistochemistry revealed the co-localization of mRNA and its corresponding peptide.

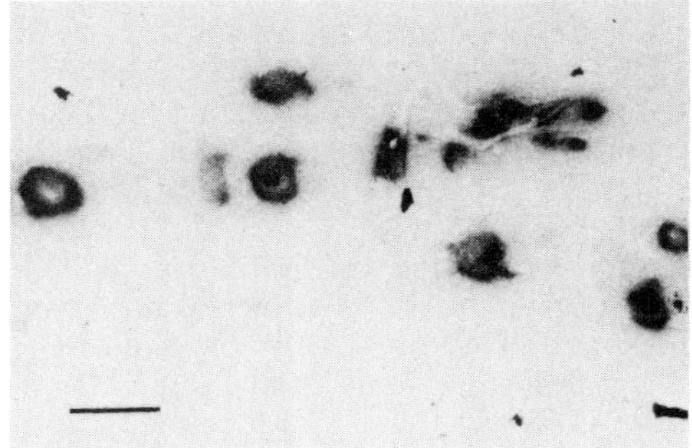

Fig. 3.   In situ hybridization of UI mRNA in the caudal neurosecretory neurons of the carp using ^{32}P-labelled UI cDNA probe. Note that autoradiographic grains were accumulated densely in the cytoplasm of the neuron. Bar indicates 100 μm.

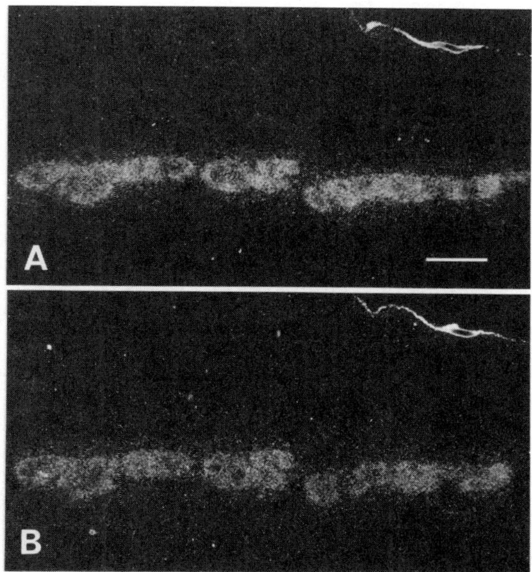

Fig. 4. <u>In situ</u> hybridization of mRNAs for UI (A) and UIIα (B) in the caudal neurosecretory neurons of the carp in serial sections using ^{32}P-labelled synthetic oligonucleotide probes. Note that the labelled neurons in A correspond well to those in B. Bar indicates 100 μm.

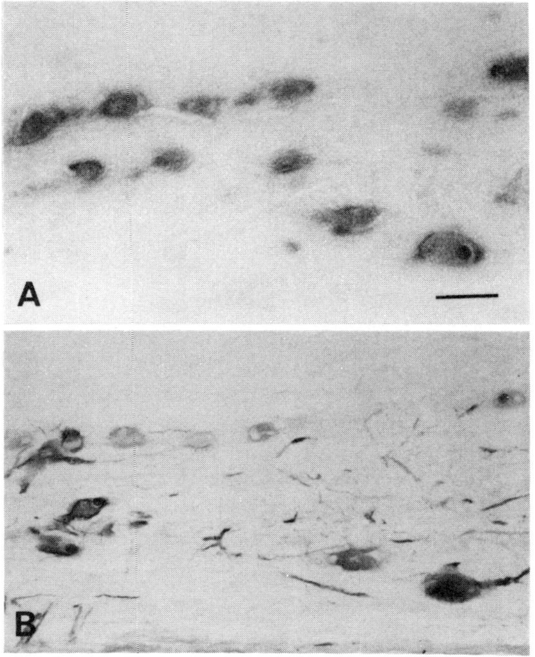

Fig. 5. <u>In situ</u> hybridization of mRNA for UI using a ^{32}P-labelled synthetic oligonucleotide probe (A) and immunohistochemistry for UI (B) in the caudal spinal cord of the carp in serial section. Note that the labelled neurons in A do not necessarily exhibit immunoreactivity in B. Bar indicates 100 μm.

The synthetic oligonucleotide probe for UI labelled almost all the caudal neurosecretory neurons. Distributions of neurons labelled with either UI and UIIα (Fig. 4), UI and UIIγ or UIIα and UIIγ were identical in paired adjacent sections. These results suggest that essentially all the caudal neurosecretory neurons of the carp co-express UI, UIIα and UIIγ. The combination of in situ hybridization with immunohistochemistry revealed the co-localization of mRNA and its corresponding peptide in the same neurons (Fig. 5). Although there was no conspicuous difference in the density of grains in the hybridized neurons, the intensity of immunoreactivity varied among neurons; a portion of neurons showed strong positivity, and others showed moderate to weak positivity or no reaction. These observations suggest that most neurons synthesize actively all 3 hormones, but turnover and/or release of them may differ from neuron to neuron.

CONCLUSION

We have elucidated the structures of precursors of UI, UIIα and UIIγ of the carp and revealed evolutionary relationships between UI and CRF and between UIIα and UIIγ. In situ hybridization using specific synthetic oligonucleotide probes demonstrated the co-expression of UI, UIIα and UIIγ in the caudal neurosecretory neurons of the carp. This strongly suggests that these 3 hormones have coordinated functions in fishes. In situ hybridization combined with immunohistochemistry gives information of the active state of synthesis and release of hormones in individual neurons and it may provide a powerful technique to study a coherent function of the CNS.

REFERENCES

Berlind, A., Lacanilao, F., and Bern, H.A., 1972, Teleost caudal neurosecretory system: Effects of osmotic stress on urophysial proteins and active factors, Comp. Biochem. Physiol., 42A:345.

Bern, H.A., Pearson, D., Larson, B.A., and Nishioka, R.S., 1985, Neurohormones from fish tails: The caudal neurosecretory system. I. "Urophysiology" and the caudal neurosecretory system of fishes, Recent Prog. Horm. Res., 41:533.

Chan, D.K.O., and Bern, H.A., 1976, The caudal neurosecretory system. A critical evaluation of the two-hormone hypothesis, Cell Tissue Res., 174:339.

Fisher, A.W.F., Wong, K., Gill, V., and Lederis, K., 1984, Immunocytochemical localization of urotensin I neurons in the caudal neurosecretory system of the white sucker (Catostomus commersoni), Cell Tissue Res., 235:19.

Fryer, J.N., and Lederis, K., 1986, Control of corticotropin secretion in teleost fishes, Amer. Zool., 26:1017.

Furutani, Y., Morimoto, Y., Shibahara, S., Noda, M., Takahashi, H., Hirose, T., Asai, M., Inayama, S., Hayashida, H., Miyata, T., and Numa, S., 1983, Cloning and sequence analysis of cDNA for ovine corticotropin- releasing factor precursor, Nature, 301:537.

Ichikawa, T., 1978, Acetylcholine in the urophysis of several species of teleosts, Gen. Comp. Endocrinol., 35:226.

Ichikawa, T., 1985, The chemistry of urotensins: Present status, in: "Neurosecretion and the Biology of Neuropeptides," H. Kobayashi, H.A. Bern and A. Urano, eds., Japan Sci. Soc. Press, Tokyo/Springer-Verlag, Berlin.

Ichikawa, T., and Kobayashi, H., 1978, Acetylcholine in the urophysis and release of urophysial hormones by neurotransmitters in vitro, in: "Neurosecretion and Neuroendocrine Activity. Evolution, Structure and Function," W. Bargmann, A. Oksche, A. Polenov and B. Scharrer, eds., Springer-Verlag, Berlin.

Ichikawa, T., Lederis, K., and kobayashi, H., 1984, Primary structures of multiple forms of urotensin II in the urophysis of the carp, Cyprinus carpio, Gen. Comp. Endocrinol., 55:133.

Ichikawa, T., McMaster, D., Lederis, K., and Kobayashi, H., 1982, Isolation and amino acid sequence of urotensin I, a vasoactive and ACTH- releasing neuropeptide, from the carp (Cyprinus carpio) urophysis, Peptides, 3:859.

Ichikawa, T., Pearson, D., Yamada, C., and Kobayashi, H., 1986, The caudal neurosecretory system of fishes, Zool. Sci., 3:585.

Ishida, I., Ichikawa, T., and Deguchi, T., 1986, Cloning and sequence analysis of cDNA encoding urotensin I precursor, Proc. Natl. Acad. Sci. USA, 83:308.

Kobayashi, H., Owada, K., Yamada, C., and Okawara, Y., 1986, The caudal neurosecretory system in fishes, in: "Vertebrate Endocrinology: Fundamentals and Biomedical Implications," P.K.T. Pang and M.P. Schreibman, eds., Vol. I. "Morphological Considerations," A. Gorbman, ed., Academic Press, New York.

Kobayashi, H., Uemura, H., Oota, Y., and Ishii, S., 1963, Cholinergic substance in the caudal neurosecretory storage organ of fish, Science, 141:714.

Kriebel, R.M., Parsons, R.L., and Miller, K.E., 1985, Innervation of caudal neurosecretory cells, in: "Neurosecretion and the Biology of Neuropeptides," H. Kobayashi, H.A. Bern and A. Urano, eds., Japan Sci. Soc. Press, Tokyo/Springer-Verlag, Berlin.

Lacanilao, F., 1972a, The urophysial hydrosmotic factor of fishes. I. Characteristics and similarity to neurohypophysial hormones, Gen. Comp. Endocrinol., 19:405.

Lacanilao, F., 1972b, The urophysial hydrosmotic factor of fishes. II. Chromatogaphic and pharmacologic indications of similarity to arginine vasotocin, Gen. Comp. Endocrinol., 19:413.

Lacanilao, F., and Bern, H.A., 1972, The urophysial hydrosmotic factor of fishes. III. Survey of fish caudal spinal cord regions for hydrosmotic activity, Proc. Soc. Exp. Biol. Med., 140: 1252.

Larson, B.A., Bern, H.A., Lin, R.J., and Nishioka, R.S., 1987, A double sequential immunofluorescence method demonstrating the co-localization of urotensins I and II in the caudal neurosecretory system of the teleost, Gillichthys mirabilis, Cell Tissue Res., 247:233.

Lederis, K., Ichikawa, T., and McMaster, D., 1981, Urophyseal peptides and proteins, in: "Neurosecretion: Molecules, Cells, Systems," D.S. Farner and K. Lederis, eds., Plenum Press, New York.

Lederis, K., Letter, A., McMaster, D., Moore, G., and Schlesinger, D., 1982, Complete amino acid sequence of urotensin I, a hypotensive and corticotropin-releasing neuropeptide from Catostomus, Science, 218:162.

McMaster, D., and Lederis, K., 1983, Isolation and amino acid sequence of two urotensin II peptides from Catostomus commersoni urophyses, Peptides, 4:367.

Montminy, M.R., Goodman, R.H., Horovitch, S.J., and Habener, J.F., 1984, Primary structure of the gene encoding rat preprosomatostatin, Proc. Natl. Acad. Sci. USA, 81:3337.

Moore, G., Burford, G., and Lederis, K., 1975, Properties of urophysial proteins (urophysins) from the white sucker, Catostomus commersoni, Mol. Cell. Endocrinol., 3:297.

Ohsako, S., Ishida, I., Ichikawa, T., and Deguchi, T., 1986, Cloning and sequence analysis of cDNAs encoding precursors of urotensin II-α and -τ, J. Neurosci., 6:2730.

Onstott, D., and Elde, R., 1986, Coexistence of urotensin I/corticotropin- releasing factor and urotensin II immunoreactivities in cells of the caudal neurosecretory system of a teleost and an elasmobranch fish, Gen. Comp. Endocrinol., 63:295.

Owada, K., Kawata, M., Akaji, K., Takagi, A., Moriga, M., and Kobayashi, H., 1985, Urotensin II-immunoreactive neurons in the caudal neurosecretory system of freshwater and seawater fish, Cell Tissue Res., 239:349.

Pearson, D., Cheers, Y., Atmani, D., Ishida, I., Deguchi, T., Nishioka, R. S., and Bern, H.A., 1986, The prohormone for urotensin II from the caudal neurosecretory complex of the goby, Abstr. 1st Cong. Neuroendocrinol., 198.

Pearson, D., Shively, J.E., Clark, B.R., Geschwind, I.I., Barkley, M., Nishioka, R.S., and Bern, H.A., 1980, Urotensin II: A somatostatin-like peptide in the caudal neurosecretory system of fishes, Proc. Natl. Acad. Sci. USA, 77:5021.

Yamada, C., Owada, K., Ichikawa, T., Iwanaga, T., and Kobayashi, H., 1986, Immunohistochemical localization of urotensins I and II in the caudal neurosecretory neurons of the carp Cyprinus carpio and the sharks Heterodontus japonicus and Cephaloscyllium umbratile, Arch. Histol. Japan., 49:39.

Yulis, C.R., and Lederis, K., 1986, Extraurophyseal distribution of urotensin II immunoreactive neuronal perikarya and their processes, Proc. Natl. Acad. Sci. USA, 83:7079.

Yulis, C.R., Lederis, K., Wong, K., and Fisher, A.W.F., 1986, Localization of urotensin I- and corticotropin-releasing factor-like immunoreactivity in the central nervous system of Catostomus commersoni, Peptides, 7:79.

IN SITU HYBRIDIZATION STUDY OF NEUROHYPOPHYSIAL HORMONE mRNAS

Akihisa Urano, Susumu Hyodo and Moriyuki Sato[*]

Department of Regulation Biology, Saitama University
Urawa, Saitama 338, Japan
*Tokyo Research Laboratories, Kyowa Hakko Kogyo Co.
Machida, Tokyo 194, Japan

INTRODUCTION

Neurohypophysial hormones, nonapeptides synthesized in hypothalamic magnocellular neurosecretory neurons, have been considered to function not only as neurohormones, but also as neurotransmitters or neuromodulators (Buijs, 1984; Jokura and Urano, 1987). In mammals, vasopressin is involved in osmotic and cardiovascular regulation, and oxytocin stimulates uterine contraction and milk ejection. However, there is no known role for oxytocin in males. Further, physiological roles of many other neurohypophysial hormones in lower vertebrates are not clear yet, mainly because of the lack of knowledge concerning secretory and synthetic activities of neurosecretory neurons in response to physiological stimuli. It is therefore probable that identification of stimuli which regulate synthetic activity of neurohypophysial hormones will help understanding of a biological significance for these hormones.

Investigation of regulatory mechanisms for gene expression of neurohypophysial hormones will be the most straightforward approach to the problem mentioned above. We considered the application of an in situ hybridization (ISH) method to this study, since an immunohistochemical technique has a serious limitation that it cannot demonstrate neuropeptides which are actively synthesized and secreted. A problem arising here is that in the conventional cDNA-mRNA ISH method, by which a particular mRNA in individual cells is visualized with a labelled cDNA probe, the probe may hybridize with mRNAs for which the homology is about 65% (Bloch et al., 1985). As has been reported by Ivell and Richter (1984), the structures of rat arginine vasopressin (AVP) and oxytocin (OXT) genes share a 95% homologous region, the exon B which encodes the conservative region of neurophysin (NP). Moreover, the use of the ISH technique depends entirely on availability of cDNAs, including preparation of cloned ones. We have overcome these problems by using synthetic deoxyoligonucleotides as probes for the ISH (Nojiri et al., 1985 and 1986; Fujiwara et al., 1985).

In this paper, we describe: an oligonucleotide mRNA ISH method with which AVP and OXT mRNAs are discriminable on paraffin sections; specificity of the present method, including a result of cross species hybridization with the mRNAs of toad neurohypophysial hormones, nucleotide sequences of which were recently determined by Nojiri et al., (1987); and then effects of salt loading and water deprivation on the AVP and OXT mRNA levels in the rat hypothalamus.

OLIGONUCLEOTIDE-mRNA IN SITU HYBRIDIZATION

The use of oligonucleotides as probes for the ISH method can have several advantages over the use of cDNAs: high specificity (Lathe, 1985; Majzoub et al., 1983; Wallace et al., 1979); easy design of a probe with a desirable sequence; easy preparation and labelling of a probe without facilities for the recombinant DNA technique; easy accessibility of probes to cellular mRNAs (Brahic and Haase, 1978; Lawrence and Singer, 1985; Moench et al., 1985). Disadvantages are: the number of labels in each probe molecule is limited to one; probes are so specific that homologous mRNAs in other species cannot be detected.

<u>Synthetic Probes</u>

The lengths and sequences of oligonucleotides are crucial to obtain specific hybridization signals on tissue sections. In our study, 4 types of 22mer oligonucleotides (Fig. 1) were synthesized to examine specificity of oligonucleotide-mRNA hybridization. They are complementary to the loci in the rat AVP mRNA encoding AVP (2-9) and AVP-NP (1-8), to that in the rat OXT mRNA encoding OXT-NP (1-8), and to the region in the toad vasotocin (AVT) mRNA encoding AVT ((-1)-7). Here, they are referred to as AVP, AVP-NP, OXT-NP and AVT/OXT probes, respectively. The nucleotide sequence of AVT/OXT probe is exactly complementary to the corresponding region of OXT mRNA, while the nucleotide at position 9 of this probe is mismatched with the counterpart of AVP mRNA. The AVP probe has 2 mismatching positions with the AVT mRNA, and 6 mismatching positions with the OXT mRNA. The AVP-NP and OXT-NP probes differ at 10 positions.

The probes were labelled at the 5' ends with T4 polynucleotide kinase using [gamma-^{32}P]ATP to a final specific activity of 3-4 x 10^5 cpm/pmol by the method of Maxam and Gilbert, (1980). The radiolabelled probes were diluted to 1 x 10^4 cpm/microliter in hybridization buffer (0.9 M NaCl, 6 mM EDTA, 0.2% bovine serum albumin, 0.2% Ficoll, 0.2% polyvinylpyrrolidone and 100 microgram/ml denatured salmon sperm DNA in 90 mM Tris buffer, pH 7.5), and were applied to tissue sections.

<u>Hybridization Procedure</u>

A detailed procedure for the ISH method has appeared elsewhere (Nojiri et al., 1985, 1986) and is described only briefly here.

<u>Preparation of tissue sections</u>. Male Wistar-Imamichi rats (6-8 weeks old) and adult Japanese toads (<u>Bufo japonicus</u>) of both sexes were mainly used as experimental animals. They were killed by decapitation. The hypothalami were rapidly taken out, and were immersed in fixative solutions (Table 1) at 4°C for 2 days. The fixative routinely used contains 2% paraformaldehyde, 1% glutaraldehyde and 1% picric acid in 0.05 M phosphate buffer (pH 7.3). After fixation, the hypothalami were washed twice in 70% ethanol at 4°C for 24 h, and then embedded in paraplast. Sections (8 or 10 μm) were separated into several groups and mounted on gelatinized slides.

```
<rat pre-Vasopressin>
-1   1                                 9      Np1
Ala Cys Tyr Phe Gln Asn Cys Pro Arg Gly --- Ala Thr Ser Asp Met Glu Leu Arg
GCC TGC TAC TTC CAG AAC TGC CCA AGA GGA --- GCC ACA TCC GAC ATG GAG CTG AGA
        [G AAG GTC TTG ACG GGT TCT CCT]     [GG TGT AGG CTG TAC CTC GAC TC]
          ▲                     ▲

<rat pre-Oxytocin>
Ala Cys Tyr Ile Gln Asn Cys Pro Leu Gly --- Ala Ala Leu Asp Leu Asp Met Arg
GCC TGC TAC ATC CAG AAC TGC CCC CTG GGC --- GCT GCG CTA GAC CTG GAT ATG CGC
 ** *** *** T** *** *** *** **              [GA CGC GAT CTG GAC CTA TAC GC]

<toad pre-Vasotocin>
Ala Cys Tyr Ile Gln Asn Cys Pro Arg Gly --- Ser Tyr Pro Asp Thr Ala Val Arg
GCC TGC TAC ATC CAG AAC TGC CCC AGA GGA --- TCC TAC CCC GAC ACA GCA GTC CGA
[GG ACG ATG TAG GTC TTG ACG GG]
          ▲

<toad pre-Mesotocin>
Ala Cys Tyr Ile Gln Asn Cys Pro Ile Gly --- Ser Val Ile Asp Phe Met Asp Val
GCG TGT TAT ATA CAA AAC TGC CCC ATA GGA --- TCG GTG ATA GAC TTC ATG GAC GTA
```

Fig. 1.   Synthetic 22mer oligonucleotide probes for <u>in situ</u> hybridization study of neurohypophysial hormone mRNAs. Sequences of probes are shown in rectangles below the sequences of cDNAs. Small filled triangles indicate mismatching positions between the AVP probe and the AVT mRNA, and that between the AVT/OXT probe and the AVP mRNA.

<u>Hybridization</u>. After rehydration, tissue sections were treated with proteinase K (1 µg/ml; Sigma, type XI) in 0.1 M Tris buffer (pH 8.0) containing 50 mM EDTA at 37°C for 30 min and washed briefly in doubly deionized water. This procedure is essential for tissues fixed with glutaraldehyde. After the treatment with proteinase, sections were rinsed in 2 x SSC (1 x SSC contains 0.15 M NaCl and 0.015 M sodium citrate), preincubated in hybridization buffer, and then incubated with a radiolabelled probe at 30°C overnight. They were then washed and processed for emulsion autoradiography.

<u>Immunohistochemistry</u>. Adjacent or mirror image tissue sections were immunohistochemically stained to compare the distribution of hybridization signals with that of neurohypophysial hormone immunoreactivity. We have adopted the avidin-biotin-peroxidase complex (ABC) method, including specificity tests, as described by Nojiri et al., (1986). When tissue sections were first stained immunofluorescently with fluorescein-labelled avidin D which was replaced with peroxidase complex, the sections could be doubly stained with immunohistochemical and ISH methods without a marked loss of autoradiographic silver grains.

Table 1.   Appropriate fixatives for ISH of the AVP mRNA with immunohistochemical staining (IHC) of adjacent tissue sections in the rat hypothalamus. PFA, paraformaldehyde; GLA, glutaraldehyde; PA, picric acid; AA, acetic acid.

Fixatives	Treatment with Proteinase	ISH	[IHC] Batch # of anti AVP 001	1285
Bouin's solution	Yes	+	++	++
Bouin's without AA	Yes	++	not tested	++
4% PFA	No	++++	-	++
2% PFA + 1% GLA	Yes	++	not tested	
2% PFA + 1% GLA + 1% PA	Yes	+++	-	+
2% PFA + 1% GLA + 1% PA	No	+	+++	+++

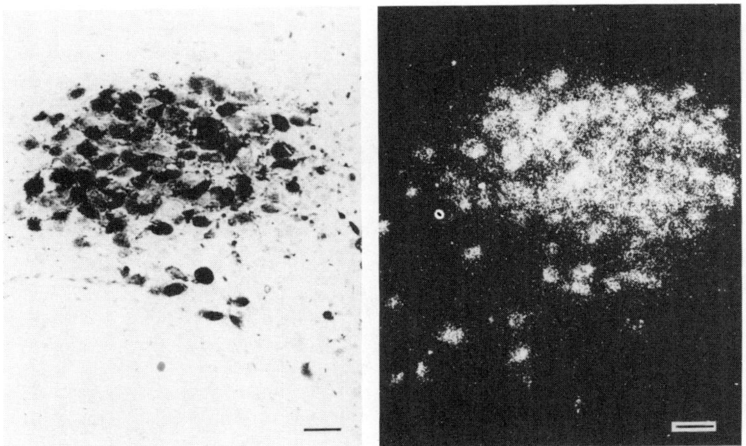

Fig. 2.   Parallel distribution of AVP-immunoreactive neurons (left side) and autoradiographic signals of a mixture of radiolabelled AVP and AVP-NP probes (right side) in the magnocellular part of the paraventricular nucleus in mirror image sections of the rat hypothalamus. Scale bar, 50 µm.

In the rat hypothalamus, autoradiographic silver grains which are considered to show locations of oligonucleotide-mRNA hybrids were densely localized over the magnocellular neurons in the supraoptic (SON) and the paraventricular (PVN) nuclei, the circular nucleus, the anterior commissural nucleus and other accessory magnocellular nuclei. The localization of silver grains, i.e., hybridization signals, was consistent with the immunohistochemical distribution of the corresponding neurohypophysial hormone. The ventral part of the SON and the rather dorsolateral part of the PVN, where AVP-immunoreactive neurons are predominant, contained most of the neurons densely labelled with radioactive AVP and/or AVP-NP probes (Fig. 2), while the dorsal part of the SON and the ventromedial part of the PVN, where OXT-immunoreactive neurons prevail, included many neurons labelled with the OXT-NP probe. Noticeable hybridization signals were not found, however, in the suprachiasmatic nucleus and the parvocellular part of the PVN in which AVP-immunoreactive parvocellular neurons have been localized, probably because the transcription rate of the AVP mRNA, i.e., the synthetic rate of AVP, in these loci is lower than that in the SON and the PVN. Significant autoradiographic signals above back ground were rarely observed in other hypothalamic nuclei, the median eminence or the pars nervosa, the terminal region of neurosecretory fibers.

A similar result was obtained in the toad hypothalamus. The distribution pattern of the AVT/OXT probe was almost the same as the localization of AVT-immunoreactive neurons in the magnocellular part of the preoptic nucleus (Fig. 3-A,B). The ISH of toad brain sections with the AVP probe gave very weak autoradiographic signals (slightly above background) in this locus (Fig. 3-C). The AVP probe has 2 mismatching positions with the corresponding region of AVT mRNA, however, the AVT/OXT probe has 5 mismatching positions with the toad mesotocin (MT) mRNA. This fact strongly supports the conclusion that the AVT/OXT probe formed highly specific pairs with the toad AVT mRNA, but not with the mesotocin (MT) mRNA. Further,it warrants that the AVP probe, the sequence of which differs from the OXT mRNA at 6 positions was not hybridized with OXT mRNA.

At present, there is no standardized method to test the specificity of staining results by in situ hybridization. We therefore have checked it by various tests including: utilization of naturally occurring homologues of neurohypophysial hormone mRNAs; cross species hybridization results of which are mentioned partially above; comparison with immunohistochemical staining. Details of these specificity tests will appear elsewhere (Hyodo et al., submitted to Zool. Sci.).

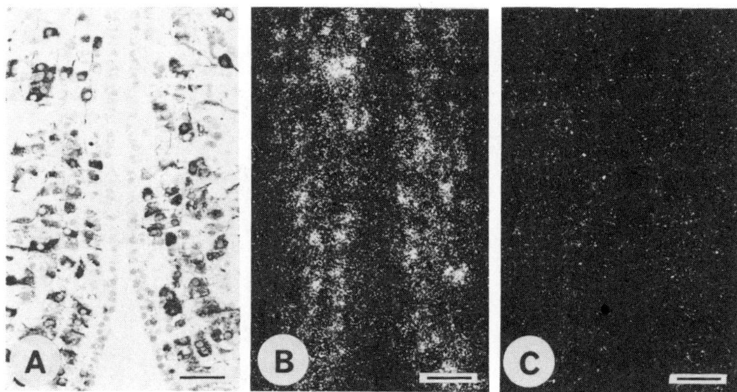

Fig. 3.   Immunoreactive AVT neurons (A) and in situ hybridization of the AVT mRNA in the toad preoptic nucleus with the AVT/OXT probe (B) and the AVP probe (C). Note that the antisense AVT/OXT probe gave much stronger autoradiographic signals than the AVP probe which has 2 mismatches.

RNase treatment. Following rehydration and a proteinase treatment, tissue sections from the rat hypothalamus were incubated with ribonuclease (100 μg/ml) prior to hybridization. This pretreatment diminished hybridization signals in the SON and the PVN to the background level, showing that the oligonucleotide probes were actually hybridized with tissue RNAs.

Estimation of melting temperature (Tm). When a probe molecule is paired with the complementary mRNA region by hydrogen bonds, a value of Tm experimentally determined should coincide with that empirically estimated or theoretically calculated. In a filter hybridization, Tm values of oligonucleotides of around 20 bases in length were empirically 50-60°C. Thus an experimental Tm value was determined by modifying washing temperature of hybridized sections. In this experiment, the AVT/OXT probe was used to show the effect of the presence of a single mismatch (see Fig. 1). Hybridization signals were localized in both OXT and AVP predominant loci of the rat with similar intensity. Autoradiographic silver grains in 0.1 mm x 0.1 mm squares were counted in the OXT and AVT immunoreactive regions in the PVN, and were plotted to estimate the Tm value graphically. When washing temperature was raised to 50°C, autoradiographic signals were reduced. Signals decreased further with elevation of washing temperature, and at ca. 65°C almost all silver grains were re moved. The Tm values estimated from the plots (Fig. 4) were about 51°C for the hybrids of the AVT/OXT probe and OXT mRNA, and 47°C for the pairing of the probe and AVP mRNA. A similar experiment in the toad, showed that the Tm is about 45°C for pairing with toad AVT mRNA. These Tm values indicate that the oligonucleotide probes were actually paired with tissue mRNAs by hydrogen bonds.

Absorption and competition tests. Addition of excess amounts of unlabelled sense and antisense oligonucleotides of the probes to the hybridization media markedly reduced specific hybridization signals. On the other hand, an excess amount of unlabelled mismatching probe, e.g., the unlabelled AVP-NP probe to the labelled OXT-NP probe, did not change the localization pattern or intensity of signals.

Use of different probes to the same mRNA. The AVP and AVP-NP probes recognize different regions in the same mRNA. This implies that both AVP and AVP-NP probes should be localized in the same loci in the rat hypothalamus. As expected, hybridization signals with these probes were localized in the same loci exactly. Application of a mixture of AVP and AVP-NP probes resulted in a conspicuous increase in hybridization signals the localization of which was consistent with that of AVP-immunoreactivity (see Fig. 2). The AVP and the AVP-NP probes do not seem to interact each other.

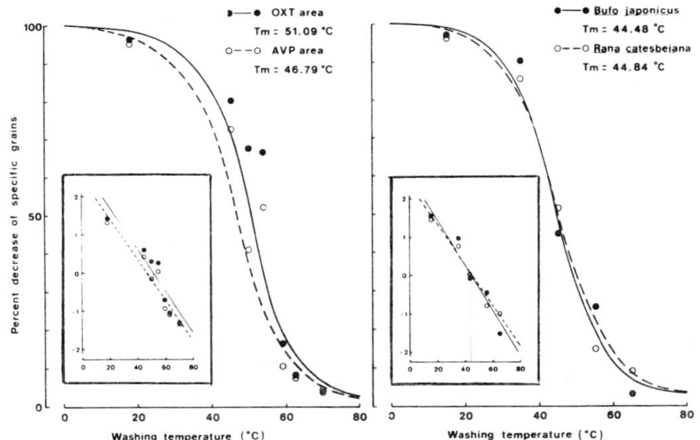

Fig. 4.   Analyses of melting temperature (Tm) compairing of the AVT/OXT probe and the neurohypophysial hormone mRNAs in tissue sections from the hypothalami of the rat (left), the Japanese toad and the bullfrog (right). Inlets show the logit plots for thermal denaturation of probe-mRNA duplexes. (Kozono and Hyodo, unpublished data)

CROSS SPECIES HYBRIDIZATION STUDY

<u>Recognizability of Mismatches</u>

As described above, the AVP probe hardly hybridized with the AVT mRNA, probably because of the presence of 2 mismatches. In contrast, in the rat hypothalamus, hybridization signals given by the AVT/OXT probe were localized not only in the region where OXT neurons are predominant, but also with a similar intensity in the region occupied by AVP neurons. However, the Tm value of the pair of AVT/OXT probe with AVP mRNA was slightly lower than that of the pair with OXT mRNA (Fig. 4, left). Since the AVT/OXT probe has one mismatch with AVP mRNA, the above fact suggests that the present 22mer oligonucleotide probes can detect one mismatch, i.e., a point mutation, when the design of probe is appropriate.

<u>Hybridization in Other Vertebrate Species</u>

Precursors of AVT and MT are probably ancestral to those of AVP and OXT, respectively (Nojiri et al., 1987). Their mRNAs showed about 70% homology, except for extraordinarily high homology between the AVP and OXT mRNAs. It is thus surprising that the nucleotide sequence encoding the cyclic region of AVT which differs from the MT one in 4 positions is just the same as that of OXT, and has only 1 base pair mismatch with the AVP mRNA (Fig. 1). A survey for the presence of this nucleotide sequence in other vertebrate species may offer valuable information for considering the molecular evolution of neurohypophysial hormones, and some results are shown in Table 2.

Table 2.   Cross species hybridization of the AVT/OXT probe on hypothalamic sections from several vertebrate species. (Kozono, unpublished data)

Species	Hybridization Signals	Species	Hybridization Signals
<u>Anguilla japonica</u>	–	<u>Coturnix c. japonica</u>	–
<u>Xenopus laevis</u>	+	<u>Poephila guttata</u>	–
<u>Bufo japonicus</u>	+++	Mouse (ICR strain)	+++
<u>Rana catesbeiana</u>	+++	Rat (Wistar-Imamichi)	+++

Considerable hybridization signals probably indicating interaction of the AVT/OXT probe and neurohypophysial hormone mRNA were found in the hypothalamic magnocellular nuclei of the Japanese toad, the bullfrog (<u>Rana cates beiana</u>), the mouse (ICR strain) and the rat. As is shown in Fig. 4, thermal denaturation curves of the probe-mRNA hybrids between the toad and the bullfrog are nearly identical suggesting that the nucleotide sequences of toad AVT and bullfrog AVT are almost identical.

On the other hand, only weak signals were seen in the preoptic nucleus of <u>Xenopus laevis</u> and no signals above background were observed in the hypothalami of the eel (<u>Anguilla japonica</u>), the Japanese quail (<u>Coturnix coturnix japonica</u>), and the zebra finch (<u>Poephila guttata</u>). Since AVP and AVT immunoreactivities in the magnocellular neurons of the species studied were comparable, it is probable that the teleostean and avian AVT mRNAs., and MT mRNAs, do not show sequence matching with the AVT/OXT probe. The reason for the conservation of the anuran AVT mRNA sequence in the rat OXT and AVP mRNAs is not known at present, although it may reflect a conversion between the AVT gene and the OXT gene as suggested for the AVP and OXT genes (Ruppert et al., 1984).

EFFECTS OF SALT LOADING AND WATER DEPRIVATION

Both AVP and OXT are considered to be involved in osmotic regulation. Hyperosmotic stimulation depleted pituitary neurohypophysial hormones and raised their

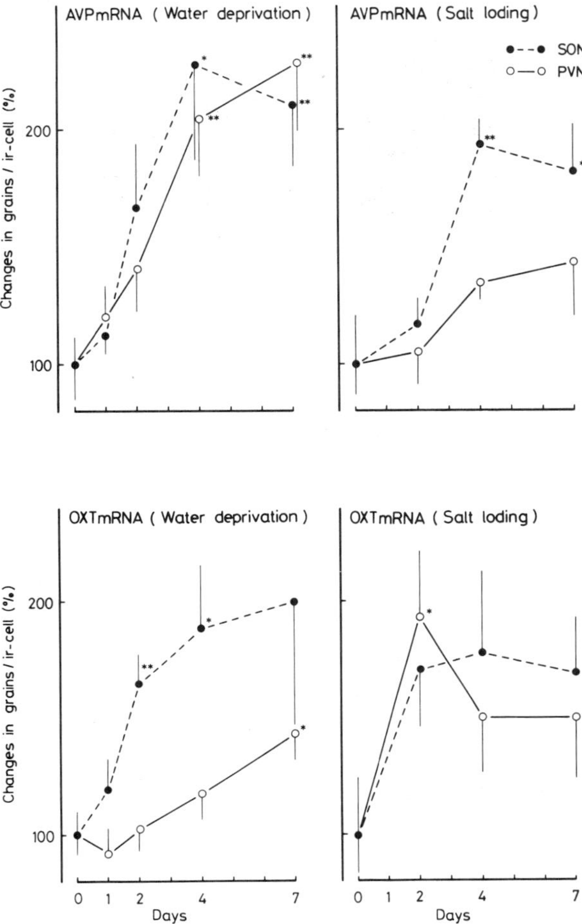

Fig. 5.   Percent changes in the levels of AVP and OXT mRNAs in the rat SON and PVN
following oral 2% saline and water deprivation. Autoradiographic silver grains in
a 0.1 mm x 0.1mm square were counted, and were divided by the number of
immunoreactive neurons in the corresponding area in the adjacent or mirror
image section. (Mean ± SE, n=5) saline and water deprivation on the levels of
AVP and OXT levels in the magnocellular neurosecretory neurons of the rat SON
and PVN (Fig. 5).

plasma levels (see Van Tol et al., 1987). Chronic hypertonic stimuli also increased the
amounts of AVP and OXT mRNAs in the rat hypothalamus. Meanwhile, both light- and
electron-microscopic studies showed various cytological changes in the rat magnocellular
neurons within a few days after the start of osmotic stimuli (see Nojiri et al., 1986). These
facts suggest that the levels of AVP and OXT mRNAs, i.e. expression of AVP and OXT
genes, are promptly elevated within a few days of the start of the stimuli. We therefore
have examined the effects of oral hypertonic saline and water deprivation on the levels of
AVP and OXT levels in the magnocellular neurosecretory neurons of the rat SON and PVN
(Fig. 5).

<u>Effects of Oral Hypertonic Saline</u>

The rats given 2% NaCl, orally, showed polydipsia and polyurea with a rapid and
significant increase in the plasma Na^+ level. The level of AVP mRNA was gradually
elevated by this salt-loading stimulus compared to the phasic increase in the level of OXT
mRNA. The magnitude of increase in the SON was more conspicuous than that in the PVN,

as was reported by Sherman et al., (1986). This result supports the idea that AVP neurons in the SON is more responsible for osmotic regulation than PVN neurons. The rapid increase of the OXT mRNA may imply that OXT neurons have some unknown physiological role in salt and water metabolism.

Effects of Water Deprivation

Water-deprived rats showed severe antidiuresis, although the plasma osmolarity and Na^+ level were rather stable. In contrast to salt loading, water deprivation induced a rapid and conspicuous increase in the level of AVP mRNA in the PVN, suggesting a difference of the PVN and the SON in osmotic and cardiovascular control. It is noteworthy that the increase in the OXT mRNA in the SON was much more significant than that in the PVN, although the physiological meaning of this is not known.

The SON and the PVN thus showed different patterns of change in the levels of AVP and OXT mRNAs following salt-loading and water-deprivation. This result suggests differential roles for the SON and the PVN in homeostatic regulation of body fluids that remain to be clarified.

CONCLUSION

The oligonucleotide-mRNA *in situ* hybridization method, a procedure which is much simpler than the cDNA-mRNA one, is highly specific and sufficiently sensitive to examine the localization and the relative quantity of neurohypophysial hormone mRNAs. A probe, 22mer in length, of which the Tm is around $50^{\circ}C$ may discriminate a single mismatch, when the design of probe is appropriate.

A cross species hybridization study showed that the particular region in the toad AVT mRNA encoding the AVT N-terminus is localized in the mouse and the rat, but not found in the eel or in birds. This result may indicate that a cloning and sequencing study of teleostean and avian neurohypophysial hormone mRNAs is required for understanding the molecular evolution of neurohypophysial hormones.

REFERENCES

Bloch, B., Le Guellec, D., and De Keyzer, Y., 1985, Detection of the messenger RNAs coding for the opioid peptide precursors in pituitary and adrenal by *in situ* hybridization: study in several mammal species, <u>Neurosci. Lett.</u>, 53:141.

Brahic, M., and Haase, A. T.,1978, Detection of viral sequences of low reiteration frequency by *in situ* hybridization, <u>Proc. Natl. Acad. Sci. USA</u>, 75:6125.

Buijs, R. M., 1984, Extrahypothalamic pathways of a neurosecretory system: their role in neurotransmission, <u>in</u>: "Neurosecretion and the Biology of Neuropeptides," H. Kobayashi, H. A. Bern, and A. Urano, ed., Jap. Sci. Soc. Press, Tokyo.

Fujiwara, M., Hyodo, S., Sato, M., and Urano, A., 1985, Changes in vasopressin and oxytocin mRNA levels in the rat hypothalamus by oral hypertonic saline, <u>Zool. Sci.</u>, 2:990.

Ivell, R., and Richter, D., 1984, Structure and comparison of the oxytocin and vasopressin genes from rat, <u>Proc. Natl. Acad. Sci. USA</u>, 81:2006.

Jokura, Y., and Urano, A., 1987, Extrahypothalamic projection of immunoreactive vasotocin fibers in the brain of the toad, <u>Bufo japonicus</u>, <u>Zool. Sci.</u>, 4:675.

Lawrence, J. B., and Singer, R. H., 1985, Quantitative analysis of *in situ* hybridization methods for the detection of actin gene expression, <u>Nucleic Acids Res.</u>, 13:1777.

Maxam, A., and Gilbert, W., 1980, Sequencing end-labelled DNA with base-specific chemical cleavages, <u>Methods in Enzymology</u>, 65:499.

Moench, T. R., Gendelman, H. E., Clements, J. E., Narayan, O., and Griffin, D. E., 1985, Efficiency of *in situ* hybridization as a function of probe size and fixation technique, <u>J. Virol. Meth.</u>, 11:119.

Nojiri, H., Sato, M., and Urano, A., 1985, <u>In situ</u> hybridization of the vasopressin mRNA in the rat hypothalamus by use of a synthetic oligonucleotide probe, <u>Neurosci. Lett.</u>, 58:101.

Nojiri, H., Sato, M., and Urano, A., 1986, Increase in the vasopressin mRNA levels in the magnocellular neurosecretory neurons of water-deprived rats: <u>in situ</u> hybridization study with the use of synthetic oligonucleotide probe, <u>Zool. Sci.</u>, 3:345.

Nojiri, H., Ishida, I., Miyashita, E., Sato, M., Urano, A., and Deguchi, T., 1987, Cloning and sequence analysis of cDNAs for neurohypophysial hormones vasotocin and mesotocin for the hypothalamus of toad, <u>Bufo japonicus</u>, <u>Proc. Natl. Acad. Sci. USA</u>, 84:3043.

Ruppert, S., Scherer, G., and Schutz, G., 1984, Recent gene conversion involving bovine vasopressin and oxytocin precursor genes suggested by nucleotide sequence, <u>Nature</u>, 308:554.

Sherman, T. G., McKelvy, J. F., and Watson, S.J., 1986, Vasopressin mRNA regulation in individual hypothalamic nuclei: a northern and <u>in situ</u> hybridization analysis, <u>J. Neurosci.</u>, 6:1685.

Van Tol, H. H. M., Voorhuis, D. T. A. M., and Burbach, J. P. H., 1987, Oxytocin gene expression in discrete hypothalamic magnocellular cell groups is stimulated by prolonged salt loading, <u>Endocrinology</u>, 120:71.

Wallace, R. B., Shaffer, J., Murphy, R. F., Bonner, J., Hirose, T., and Itakura, K. (1979) Hybridization of synthetic oligodeoxyribonucleotides to phi-x174 DNA: the effect of single base pair mismatch, <u>Nucleic Acids Res.</u>, 6:3543.

THE RELEASE OF ENKEPHALIN-CONTAINING PEPTIDES FROM THE ADRENAL GLAND IN CONSCIOUS CALVES

Jean Rossier, E. Barrès, A. Cupo[1] and A.V. Edwards[2]

Laboratoire de Physiologie Nerveuse,
CNRS, 91190 Gif-su-Yvette, France,
Centre d'Immunologie de Marseille-Luminy, France[1] and
The Physiological Laboratory, University of
Cambridge, Cambridge CB2 3EG, England[2]

INTRODUCTION

A number of recent studies have established the fact that enkephalins are released, together with catecholamines, from the adrenal medulla, both in vitro and in vivo, in response to various stimuli (Kilpatrick et al., 1980; Stine et al., 1980; Livett et al., 1981; Hexum et al., 1980; Chaminade et al., 1984; Edwards et al., 1986). It further appears that these peptides are present in adrenergic, but not noradrenergic, chromaffin cells in the adrenal medulla (Livett et al., 1982; Pelto-Huikko et al., 1982; Roisin et al., 1983), but the functional significance of the phenomenon remains obscure.

Chaminade et al., (1984) showed that, in the isolated perfused adrenal gland of the cat, the release of free [Met]enkephalin-like immunoreactivity was invariably associated with that of much larger precursor molecules. The extent to which these latter were released depended upon the intensity of the stimulus and it appeared that the smaller molecules, presumably obtained by the more complete processing of precursors, were secreted preferentially during more moderate forms of stimulation. In the present study the "adrenal clamp" technique (Edwards et al., 1974) has been employed to examine the release of these peptides under the most physiological conditions that have yet been achieved i.e. in response to splanchnic nerve stimulation in conscious animals. The results show that large molecular weight precursor forms of enkephalin are indeed released under these conditions and in amounts that far exceed those of free [Met]enkephalin. However, the amount of free peptide in adrenal effluent plasma during nerve stimulation was invariably much higher than that in extracts of the gland, indicating that release is associated with a final processing step or steps. Alternatively, these results could be taken to indicate preferential release of granules containing more highly processed enkephalin-like material. The finding that the ratio of adrenaline to total enkephalin-like material that was released was almost three times higher in adrenal effluent plasma during stimulation of the splanchnic nerve at 15Hz than at 4Hz is consistent with the hypothesis that the population of releasable granules is heterogeneous.

METHODS

Animals: Pedigree Jersey calves were obtained from local farms shortly after birth and used at ages ranging between 25 and 52 days (26.4-31.2 kg body weight). Thereafter, they were kept in individual pens in the laboratory animal house and maintained on a diet of either cow's milk or artificial milk (Easy-Mix Volac; Volac Ltd.) at a rate of 2-4 l/day. Food was withheld for at least 14 h before surgery and each experiment.

Experimental Procedures: Anaesthetic, surgical, post-mortem and experimental procedures were closely similar to those described in detail previously (Edwards et al., 1974; Edwards et al., 1980). Briefly, preparatory surgery was carried out under halothane anaesthesia and involved removal of the right kidney followed by implantation of a specially designed clamp to permit collection of the whole of the effluent blood from the right adrenal gland, periodically, when required.

In each experiment, a standard 10-20 V square-wave stimulus (pulse width 1 msec) was delivered at a continuous frequency for either 4 or 15Hz for 10min and was invariably below behavioural threshold.

Analytical procedures: Arterial blood samples were collected into heparinized tubes for haematocrit estimations. Adrenal venous effluent blood was collected into heparinized tubes containing EDTA and aprotinin (Trasylol, Bayer; 1000 K.I.U./ml blood) for catecholamine and enkephalin estimations. They were centrifuged as soon as possible at $+4^{o}C$; the plasma was then stored at $-20^{o}C$. Adrenaline and noradrenaline were measured by a modification of Euler and Floding's (1955) trihydroxyindole method as described previously (Bloom et al., 1975).

Radioimmunoassays: [Met]enkephalin radioimmunoassays were performed with antiserum AC 2158 on unextracted plasma essentially as described by Patey et al., (1985) except that the separation of bound and free iodinated [Met]enkephalin was performed by double immunoprecipitation.

Synenkephalin radioimmunoassay were performed with antiserum AS 5 VII essentially as described by Liston et al., (1983).

Digestion of [Met]enkephalin containing peptides: Plasma was diluted 1:10 with 50mM Tris HCl containing 2mM $CaCl_2$ (pH 8.6) and boiled for 20min. After centrifugation at 5000g for 10min, $250\mu l$ of the supernatant was incubated overnight at $37^{o}C$ with $200\mu l$ of a solution containing $20\mu g/ml$ TPCK trypsin (Worthington) in the same buffer. The samples were boiled again and $50\mu l$ of a solution of carboxypeptidase B (Boerhinger) at a concentration of $1\mu g/ml$ in the same buffer was added. The mixture was then incubated at $37^{o}C$ for 1h, boiled and centrifuged. Standards and unknowns were added in the same buffer and aliquots of 10, 50 and $100\mu l$ were then assayed.

Adrenal extracts: Whole adrenal glands were dissected out and frozen directly after each experiment. After thawing, they were homogenized with a polytron in 10ml 1.0 M HAc, adjusted to pH 1.9 with HCl. The homogenates were boiled for 30min and then centrifuged at 5000g for 15min. The supernatants were then centrifuged again for 2h at 22000g and the resultant clear solutions were lyophilized in aliquots.

G-100 Sephadex: A column (65 x 1.6cm) of fine grade G-100 Sephadex (Pharmacia) was equilibrated at $4^{o}C$ with 1M HAc. Flow rate was regulated with a pump at 6ml/hr. 3ml fractions were collected and the aliquots were then evaporated with a Speedvac. Each sample was resuspended in 1ml H_2O and re-evaporated. This procedure was repeated twice in order to ensure the elimination of any trace of HAc. Loading the samples onto the column was always performed using aliquots comprising 2ml plasma, made up to 4ml with 2.0M HAc. Whenever extracts of adrenals were analysed, $20\mu l$ of the original extract was added to 2ml of arterial plasma, that had been collected before the start of the experiment, and made up to 4ml with 2.0M HAc. These precautions ensured that chromatography was carried out under as nearly identical conditions as possible, with regard to such variables as viscosity, protein concentration and the like.

Statistics: Statistical analyses were made according to the methods of Snedecor & Cochran (1967).

RESULTS

Stimulation of the peripheral end of the right splanchnic nerve elicited the release of adrenaline (Fig. 1) and noradrenaline (not shown) from the right adrenal gland. The amount of adrenaline released during stimulation at 4Hz continuously for 10min was 20 times greater than the basal output and the amount released during stimulation at 15Hz was eight times greater than that. When 10μl of unextracted adrenal effluent plasma was assayed for [Met]enkephalin by the highly sensitive radioimmunoassay described in "Methods", the basal levels were below the sensitivity of the assay (1 fmol/sample). During stimulation at 4Hz the [Met]enkephalin-like material was readily detected in the adrenal effluent plasma, and the output from the gland amounted to 0.10 ± 0.01pmol.min^{-1}.kg^{-1} (n = 16). During stimulation at 15Hz continuously the mean average right adrenal output of [Met]enkephalin rose to 1.00 ± 0.13pmol.min^{-1}.kg^{-1}.

These results indicate that most of the enkephalin-like material that was released from the adrenal during splanchnic nerve stimulation was in the form of large molecules whose [Met]enkephalin content could only be assayed after digestion. Synenkephalin, the N-terminal portion of proenkephalin, was also measured by radioimmunoassay in these samples and found to be released from the adrenal gland during stimulation of the peripheral end of the splanchnic nerve (Fig. 2). The molar ratio of [Met]enkephalin-like immunoreactivity after digestion to synenkephalin was 5.8 ± 0.8 (n=16), in samples obtained during splanchnic nerve stimulation at 4Hz and 3.4 ± 0.4 during stimulation at 15Hz. The difference between these two ratios was highly significant (p<0.001) and they are directly comparable with the theoretical molar ratio in the intact precursor of 6.0.

The nature of the enkephalin-like material was analyzed by gel filtration on a Sephadex G-100 column equilibrated with 1.0M HAc. In Fig. 3, a chromatogram of a

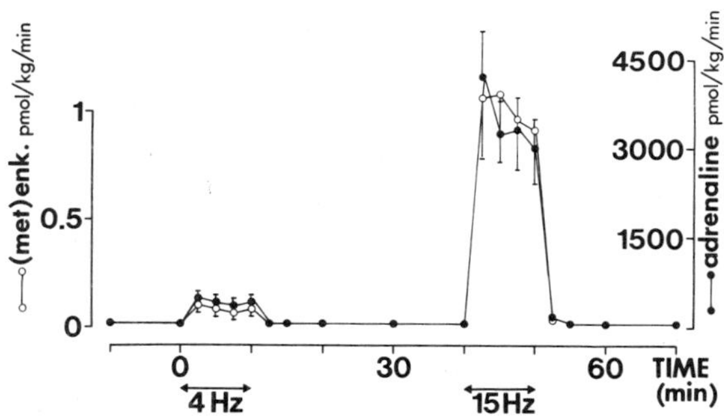

Fig. 1.   Changes in the mean output of adrenaline (o   o; right axis) and of [Met]enkephaline-like immunoreactivity (o   o; left axis) from the right adrenal gland in conscious, unrestrained, 3-7 week old calves (n=7) in response to stimulation of the peripheral end of the right splanchnic nerve at 4Hz (0-10min) and 15Hz (40-50min). Each sample was also assayed for [Met]enkephalin-like immunoreactivity after digestion with trypsin and carboxypeptidase B (Fig. 2), which generates [Met]enkephalin from larger enkephalin-containing peptides present in the sample. After digestion, [Met]enkephalin-like activity in the samples of adrenal effluent plasma collected during stimulation at 4Hz was 14.7 ± 2.2 times higher than prior to digestion (n=16). Similarly, in the samples collected during stimulation at 15Hz, enzymic digestion increased [Met]enkephalin activity 5.6 ± 0.8 fold (n=16). Each of these differences was found to be highly significant (p<0.001).

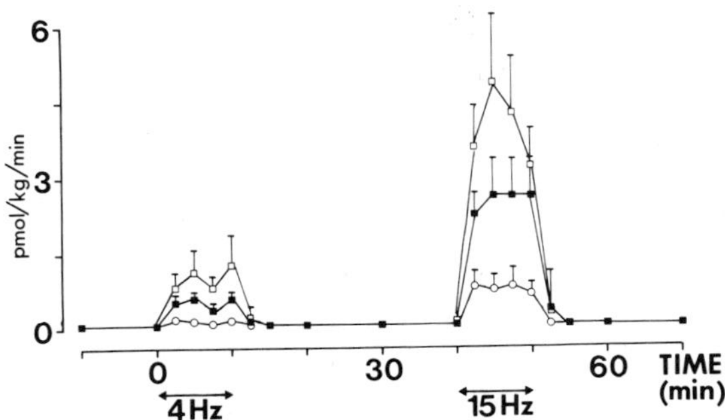

Fig. 2.   Changes in the mean output of [Met]enkephalin-like immunoreactivity (o   o),
[Met]enkephalin-like immunoreactivity after digestion (   ), and synenkephalin-
like immunoreactivity x 2 (   ) from the right adrenal gland in conscious,
unrestrained, 3-7 week old calves (n=4), in response to stimulation of the
peripheral end of the right splanchnic nerve at 4Hz (0-10min), and 15Hz (40-
50min).

tube was assayed for [Met]enkephalin-like immunoreactivity before digestion. The column
was calibrated with $[^{125}I]$ bovine proenkephalin (1-77) (8.6 KDalton) and ^{125}I-
[Met]enkephalin, the respective elution peaks of which are indicated on the chromatograms
by arrows.

Fig. 3 shows that all the [Met]enkephalin-like immunoreactivity in the adrenal extract
eluted before [Met]enkephalin standard. In the case of the sample of adrenal effluent plasma
a mere 6% of the total [Met]enkephalin immunoreactivity eluted at the position of authentic,
free [Met]enkephalin.

A major difference between the profiles of the chromatograms of adrenal extract and
adrenal effluent plasma was observed in fractions 46 to 53, in which materials with M_r
values of about 4000 elute. In the case of the extract from the gland, very little material was
eluted in these fractions. In contrast, a large amount of material, corresponding to 30% of
the total activity, was eluted in these fractions in samples of adrenal effluent plasma.
Otherwise, the two chromatograms were very similar. When [Met]enkephalin-like
immunoreactivity was assayed after digestion (not shown) the two profiles were almost the
same and the chromatograms of synenkephalin-like immunoreactivity in samples of gland
extract and effluent plasma (not shown) were also closely similar.

No difference was found in the general distribution of the two chromatograms of
samples collected during splanchnic nerve stimulation at the two different frequencies that
were employed.

DISCUSSION

These results indicate that most of the [Met]enkephalin immunoassayable material in
the adrenal gland of the calf and its effluent plasma is not free [Met]enkephalin but exists in
the form of much larger molecules. None of the [Met]enkephalin-like activity in extracts of
the gland itself corresponds to authentic free [Met]enkephalin and in the effluent plasma,
only 6% of the total immunoreactivity seems to correspond to free [Met]enkephalin. The
fact that the amounts available were so small prevented any further characterization of this
material, as might have been undertaken by h.p.l.c. for instance.

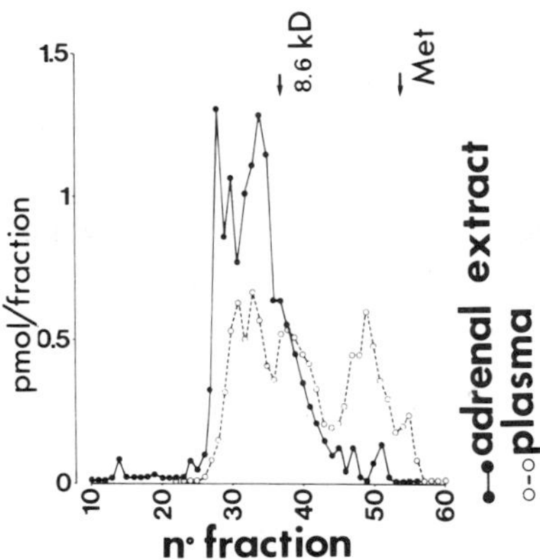

Fig. 3.   Comparison of the profile of [Met]enkephalin-like immunoreactivity before digestion of the column fractions on a G-100 Sephadex gel filtration column of an extract (o   o) of adrenal gland with that of an adrenal effluent plasma sample (o   o) obtained during stimulation of the peripheral end of the right splanchnic nerve at 15Hz. The adrenal extract and the plasma sample were prepared as described in "Methods". Arrows indicate the elution peak of ^{125}I-proenkephalin$_{(1-77)}$ (8.6 kD) and ^{125}I-monoiodo [Met]enkephalin (Met).

A further difference between the profiles of the chromatograms of the extract and the effluent plasma was observed. Thus, in the extract, virtually all the immunoreactivity corresponded to material with M_r values above 4000; in the effluent plasma, on the other hand, 30% of the total immunoreactivity was accounted for by molecules with M_r values of about 4000. This peak could correspond to peptides I,E,B,F, BAM 22 or BAM 18, all of which have been isolated from adult cow adrenal (Metters and Rossier, 1984). Moreover, Peptide E and BAM 18 have recently been shown to be released from cultured chromaffin cells in response to chemical stimulation (Boarder et al., 1987). The fact that there are relatively large amounts of enkephalin-like material with M_r values of about 4000 or less in adrenal effluent plasma during splanchnic nerve stimulation, and virtually none in the gland, suggests that hydrolysis of large enkephalin-containing peptides occurs during the release process, thereby generating free [Met]enkephalin and small enkephalin-containing peptides. It is also possible that splanchnic nerve stimulation releases a very small pool of granules which contain [Met]enkephalin and small enkephalin-containing peptides. This latter hypothesis was originally proposed by Chaminade et al. (1984) in a previous study on the perfused adrenal gland of the cat. In that preparation, it was observed that about 50% of the material released by splanchnic nerve stimulation represented final products of the maturation of proenkephalin i.e. [Met]enkephalin, [Leu]enkephalin, heptapeptide, octapeptide, and other molecules with M_r values below 4000. In contrast, most of the material in extracts of the cat's adrenal gland itself had M_r values above 4000. These previous findings, and those recorded in the present paper, could possibly be explained if the population of chromaffin granules was heterogeneous in relation to enkephalin content: a mature population in which proenkephalin had been processed, or partially processed, to molecules with M_r values of about 4000 and was preferentially released in response to splanchnic nerve stimulation, and another that was immature, containing exclusively peptides with high M_r values, and more resistant to release in response to splanchnic nerve stimulation.

Such a heterogeneity of the granules would also be consistent with the difference that we observed between the molar ratios of adrenaline to [Met]enkephalin, after digestion, in plasma samples collected during stimulation of the splanchnic nerve at the two different frequencies tested. The molar ratio during stimulation at 4Hz was 315 ± 36 (n=16) and at 15Hz was 877 ± 87 (n=16); a difference that was highly significant (p<0.001).

Taken together, these results suggest that increasing the frequency of splanchnic nerve stimulation from 4 to 15Hz not only increases the amount of enkephalin-like material that is released from the adrenal gland, but also modifies its composition. It may well be that granules containing different proportions of adrenaline and enkephalin are released in response to different frequencies of stimulation.

SUMMARY

Adrenal responses to stimulation of the peripheral end of the right splanchnic nerve at 4 and 15Hz continuously for 10min have been investigated in conscious calves by means of the "adrenal clamp" technique. The evoked release of catecholamines from the right adrenal gland was accompanied by the release of a substantial amount of enkephalin-like immunoreactive material i.e. [Met]enkephalin and synenkephalin immunoreactivities. Most of this enkephalin-like material was in the form of large enkephalin-containing peptides whose enkephalin content could only be assayed after digestion by sequential treatment with trypsin and carboxypeptidase B. Analysis of the enkephalin-like material in adrenal effluent plasma, by gel filtration revealed that 36% corresponded to peptides with M_r values of 4000 or less. This low M_r material was never detected in extracts of the gland itself.

ACKNOWLEDGEMENTS

This work was supported by grants from the British Heart Foundation, Fondation pour la Recherche Medicale (Paris), the C.E.E. (reseau neuropeptides, C.N.R.S.) and the Wellcome Trust. It is a particular pleasure to thank Mr. P.M.M. Bircham, Mr. M. Chaminade and Mrs. B.N. Daw for their skilled technical assistance.

REFERENCES

Bloom, S.R., Edwards, A.V., Hardy, R.N., Malinowska, K.W. and Silver, M.S., 1975, Endocrine responses to insulin hypoglycaemia in young calf, J. Physiol. (Lond.), 244:783.

Boarder, M.R., Evans, C., Adams, M., Erdelyi, E. and Barchas, J.D., 1987, Peptide E and its products, BAM 18 and Leu-enkephalin, in bovine adrenal medulla and cultured chromaffin cells: release in response to stimulation, J. Neurochem., in press.

Chaminade, M., Foutz, A.S., and Rossier, J., 1984, Co-release of enkephalins and precursors with catecholamines from the perfused cat adrenal gland in situ, J. Physiol. (Lond.), 353:157.

Edwards, A.V., Hardy, R.N. and Malinowska, K.W., 1974, The effects of infusions of synthetic adrenocorticotrophin in the conscious calf, J. Physiol. (Lond.), 239:477.

Edwards, A.V., Furness, P.N. and Helle, K.B., 1980, Adrenal medullary responses to stimulation of the splanchnic nerve in the conscious calf, J. Physiol. (Lond.), 308:15.

Edwards, A.V., Hansell, D., and Jones, C.T., 1986, Effects of synthetic adrenocorticotrophin on adrenal medullary responses to splanchnic nerve stimulation in conscious calves, J. Physiol. (Lond.), 379:1.

Euler, U.S. von and Floding, I., 1955, A fluorimetric micromethod for differential estimation of adrenaline and noradrenaline, Acta Physiol. Scand., 33, Suppl. 118:45.

Hexum, T.D., Hanbauer, I., Govoni, S., Yang, H.-Y.T., and Costa, E., 1980, Secretion of enkephalin-like peptides from canine adrenal gland following splanchnic nerve stimulation, Neuropeptides, 1:137.

Kilpatrick, D.L., Lewis, R.V., Stein, S. and Udenfriend, S., 1980, Release of enkephalins and enkephalin-containing polypeptides from perfused beef adrenal glands, Proc. Natl. Acad. Sci. USA., 77:7473.

Liston, D.R., Vanderhaeghen, J.J. and Rossier, J., 1983, Presence in brain of synenkephalin, a proenkephalin-immunoreactive protein which does not contain enkephalin, Nature, Lond., 302:62.

Livett, B.G., Dean, D.M., Whelan, L.G., Udenfriend, S. and Rossier, J., 1981, Co-release of enkephalin and catecholamines from cultured adrenal chromaffin cells, Nature, Lond., 289:317.

Livett, B.G., Day, R., Elde, R.P., and Howe, P.R.C., 1982, Co-storage of enkephalins and adrenaline in the bovine adrenal medulla, Neuroscience, 7:1323.

Metters, K.M. and Rossier, J., 1984, Biosynthesis of the opioid peptides, In: Les Collogues de l'INSERM, Spinal opioids and the relief of pain. Edited by Besson, J.M. and Lazorthes, Y., INSERM 127:31.

Patey, G., Cupo, A., Mazarguil, H., Morgat, J.L. and Rossier, J., 1985, Release of proenkephalin-derived opioid peptides from rat striatum in vitro and their rapid degradation, Neuroscience, 15:1035.

Pelto-Huikko, M., Salminen, T., and Hernoven, A., 1982, Enkephalin-like immunoreactivity is restricted to the adrenaline cells in the hamster adrenal medulla, Histochemistry, 73:493.

Roisin, M.P., Artola, A., Henry, J.P. and Rossier, J., 1983, Enkephalins are associated with adrenergic granules in bovine adrenal medulla, Neuroscience, 10:83.

Snedecor, G.W. and Cochran, W.G., 1967, Statistical Methods, 6th edn. Ames: Iowa State College Press.

Stine, S.M., Yang, H.-Y.T., and Costa, E., 1980, Release of enkephalin-like immunoreactive material from isolated bovine chromaffin cells, Neuropharmacology, 19:683.

MODERN MICROSCOPICAL IMAGING TECHNIQUES FOR THE STUDY OF THE NEUROENDOCRINE SYSTEM

Julia M. Polak

Department of Histochemistry
Royal Postgraduate Medical School
Hammersmith Hospital
Du Cane Road
London W12 OHS

The existence of a large, powerful neuroendocrine system has been recognised increasingly since the introduction of the neuroendocrine concept (Scharrer and Scharrer, 1940). Feyrter (1938) demonstrated the presence of specialised "clear" or "endocrine" cells diffusely dispersed throughout the body, often included in non-endocrine tissue (eg. the gastrointestinal and respiratory tracts) intermingled with other types of cells. The idea of a "diffuse endocrine system" was then proposed. Nowadays in view of a) the close anatomical relationships between nerves and endocrine cells, b) the existence of numerous active substances (eg. peptides) being produced and released from both endocrine cells and neurones and c) the various histological and histochemical features shared by both systems, Feyrter's diffuse endocrine concept has been expanded to include the neural controlling system. Thus, the name of neuroendocrine system has been adopted (Polak and Bloom, 1979). This "diffuse neuroendocrine system", together with the classical neuroendocrine glands, constitute the most powerful controlling system of the body. Its full recognition is largely due to a number of technical advances including the ability to extract, purify and synthesize active substances or to deduce their amino acid sequences by analysis of specific cDNA moieties, as well as the use of advanced analytical microscopical imaging methods which permit detailed study of the localisation, synthesis and receptor recognition of these active substances or regulatory peptides. This contribution will deal in some detail with a variety of imaging methods as used for light and electron microscopical studies of the diffuse neuroendocrine system.

HISTOLOGICAL/HISTOCHEMICAL STAINS FOR THE DIFFUSE NEUROENDOCRINE SYSTEM

These techniques were used in the early days for demonstration of the neuroendocrine system, including the use of vital dyes, methyl blue, masked metachromasia and silver impregnations. The latter can be broadly sub-divided into argentaffin and argyrophil methods (Grimelius and Wilander, 1985). Both techniques depend on the presence in the cell of biogenic amines which can either take up and reduce silver ions (argentaffin) or achieve reduction with a second step reducing agent e.g. formaldehyde. One of the most widely used silver impregnation methods is that proposed by Grimelius and Wilander, 1985 (Fig.1).

ELECTRON MICROSCOPY

The presence of electron-dense secretory granules (cytoplasmic organelles where mature hormones are stored) allows easy identification of endocrine cells at the electron microscopical level. In general, secretory granules are highly electron-dense, of variable size

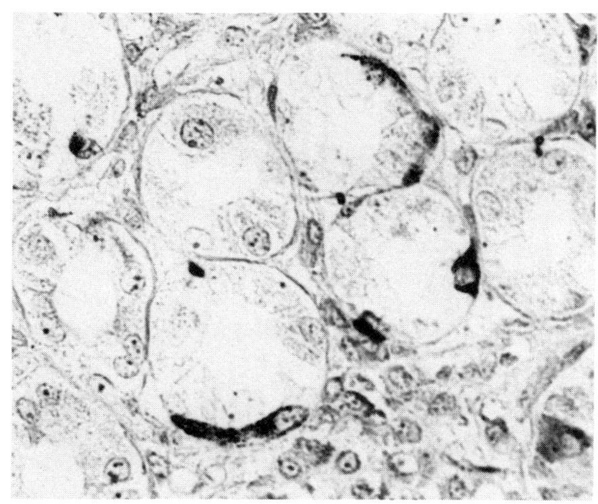

Fig. 1.   Enterochromaffin-like cells in human fundic mucosa stained by the silver
impregnation method of Sevier and Munger (x 250).

(100-500 nm) and frequently surrounded by a limiting membrane, specially visualised on
osmicated tissue (Fig. 2.)

IMMUNOCYTOCHEMISTRY

This is the technique most commonly used nowadays to investigate the
neuroendocrine system (Polak and Bloom, 1986). Numerous antibodies are available which
permit the demonstration of endocrine cells and sometimes associated nerves or the
innervation of its entirety (Fig. 3). These include antibodies to neuron-specific enolase (NSE)
an isomer of the glycolytic enzyme enolase (Polak and Marangos, 1985). NSE was originally
extracted from brain and localised to neurones, hence its name, but was later found in
endocrine cells also (Fig. 4). Chromogranin is a protein originally extracted from
catecholamine-containing granules of the adrenal medulla (O'Connor et al., 1983). Unlike
NSE, which is a cytosolic, non-granular protein, chromogranin antisera immunostain
secretory granules both at light (Fig. 5) and electron microscopical levels (Varndell et al.,
1985) (Fig. 6). A recently discovered peptide termed pancreastatin (Tatemoto et al., 1986)
named for its location in the pancreas and its ability to modulate insulin release was found in
the precursor molecule of chromogranin. Pancreastatin, however, is expressed within
endocrine tissue as a separate peptide but, like chromogranin, as a widespread distribution in
all classes of neuroendocrine tissue (Bishop et al., 1987). Synaptophysin is a glycosylated
polypeptide component of presynaptic vesicle membranes. It has been proposed as a general
neuroendocrine marker and has been cited as a particularly good indicator of endocrine
neoplasms, where its efficiency equalled that of NSE (Gould et al., 1987). 7B2 or APPG is a
large protein originally extracted from the anterior pituitary of the pig and later found to be
present in many other endocrine tissues, in particular the pituitary and pancreatic islets
(Suzuki et al., 1986). Antibodies to a sub-component of the cytoskeleton, the
neurofilaments, are excellent markers for central and peripheral neurones (Hacker et al.,
1988) and lately a protein termed protein gene product 9.5 (PGP) by its discovery (Doran et
al., 1983) was found to be present in large concentrations in peripheral nervous tissue
(Gulbenkian et al., 1987). Antibodies to PGP 9.5 are now used routinely and recognised to
be the best markers for all classes of central and, in particular, peripheral nerves (Fig. 7a
and b).

In recent years, peptide immunocytochemistry has been used to identify the various
cell types of the diffuse neuroendocrine system. A variety of antibodies are now available,
both polyclonal and monoclonal, capable of recognising various regions of the precursor and
bioactive forms of peptides (Hamid et al., 1986). The majority of these antibodies are now
available commercially and can be used on tissue fixed in a variety of ways, including
routine histological procedures. The antibodies demonstrate neurones (cell bodies and
terminals) (Fig. 8) and endocrine cells either separately or simultaneously (Fig. 9). The

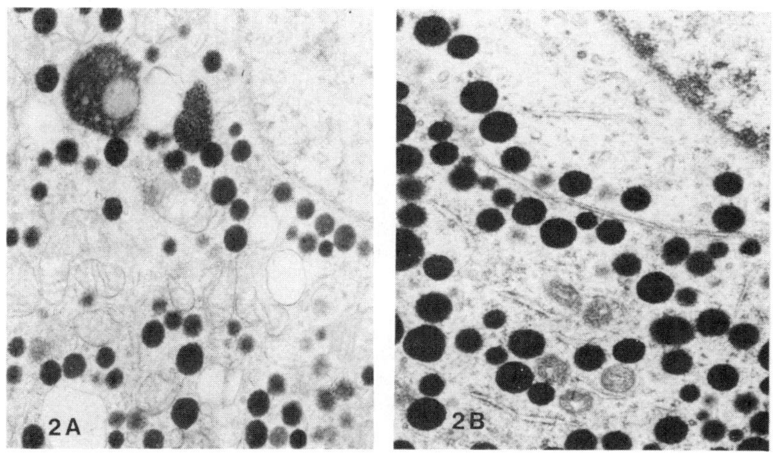

Fig. 2.   Electron micrographs of cells from anterior human pituitary A) ACTH cell (x 10,000) B) growth hormone (x 10,000).

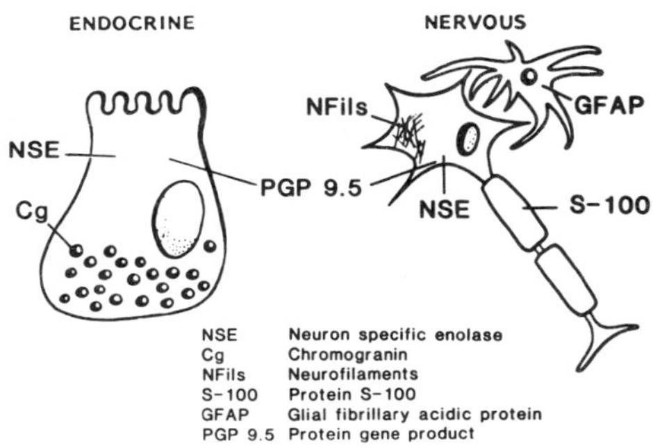

Fig. 3.   Diagram showing the subcellular localisation of five different neuroendocrine markers. Taken from Handbook of Physiol., The American Physiological Society, Section on the Gastrointestinal System Vol. 1, Neuroendocrinology of the Gut, Ed. Gabriel M. Makhlouf. In press 1987. Endocrine cells. J.M. Polak.

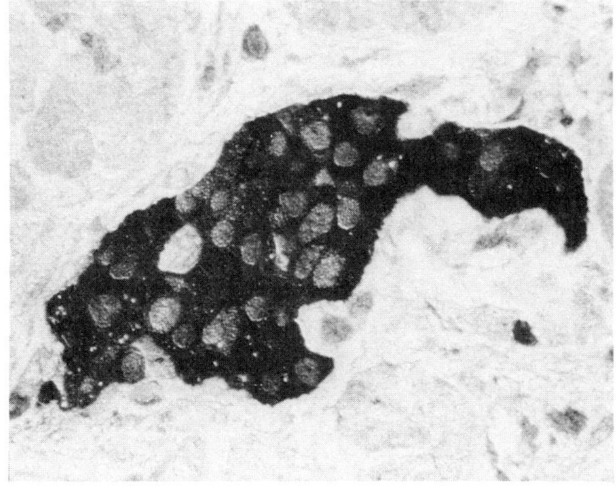

Fig. 4.   Neuron-specific enolase immunostained in human pancreatic endocrine cells using the peroxidase anti-peroxidase (PAP) method (x 250).

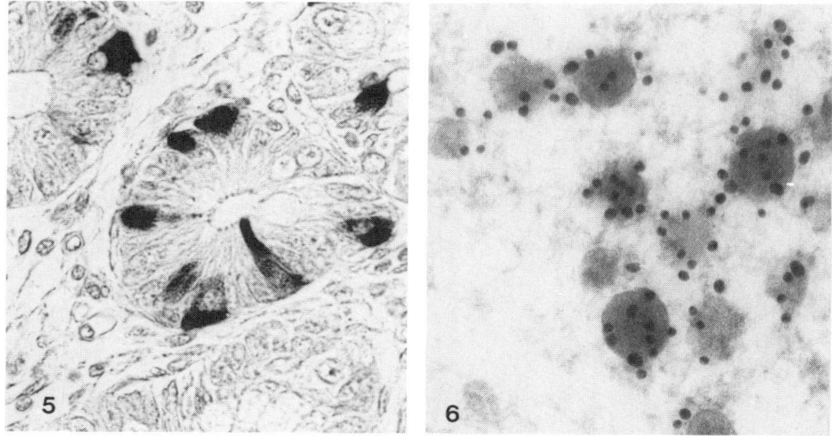

Fig. 5.   Human colonic mucosal epithelium with several chromogranin-immunoreactive cells visualised by the PAP technique (x 220).

Fig. 6.   Atypical granules in a human pancreatic A cell tumour immunostained for chromogranin using gold-labelled antibodies (x 12,000).

application of a number of neurobiological methods (Skirboll et al., 1984) in combination with peptide immunocytochemistry permits determination of the origin of specific nerve terminals, their pathways and anatomical relationship with each other and endocrine cells, their nature (sympathetic/sensory) (Fig. 10) as well as the connections between the peripheral and central nervous systems (Fig. 11 and 12). The application of immunocytochemistry at the electron microscopical level allows determination of the characteristics of the electron-dense secretory granules and the possible coexistence of neurochemicals in the same neurosecretory granules (Polak and Varndell, 1984; Gulbenkian et al., 1986).

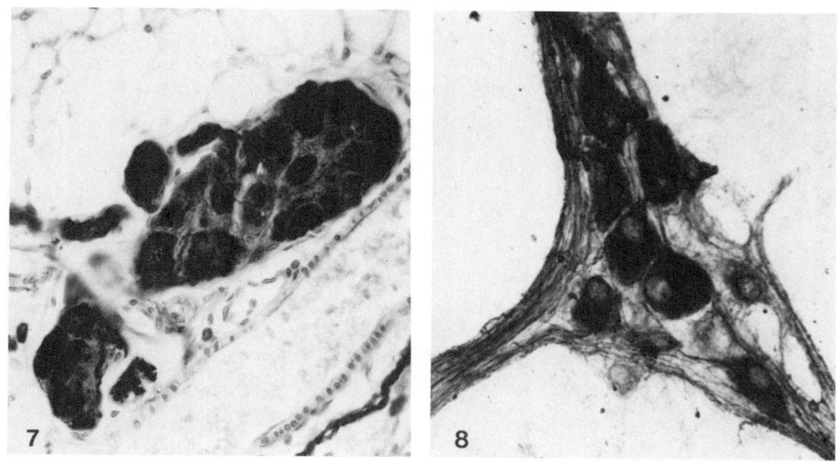

Fig. 7.   Small ganglion adjacent to blood vessel in the trachea of a rat. The ganglion cells and nerve fibres are strongly immunoreactive with antiserum to PGP 9.5 (x 360).

Fig. 8.   VIP-immunoreactive ganglion cells and nerve fibres in a whole mount preparation of human colonic submucosa (x 250).

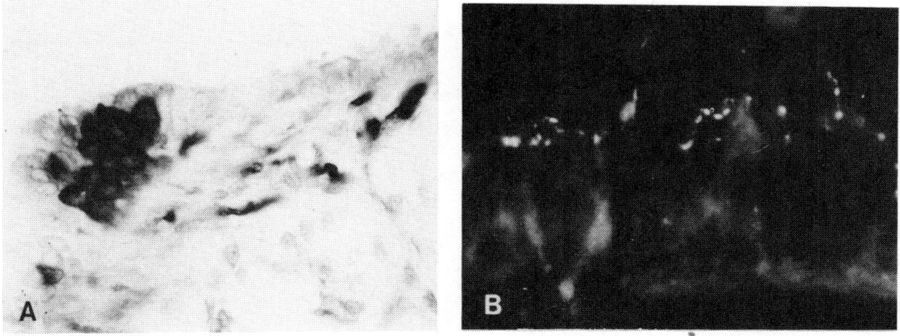

Fig. 9. A) Neuroepithelial body in bronchial epithelium of rat immunostained with antiserum to CGRP (x 140). B) Nerve fibres in the epithelium of rat trachea displaying immunoreactivity for CGRP (x 450).

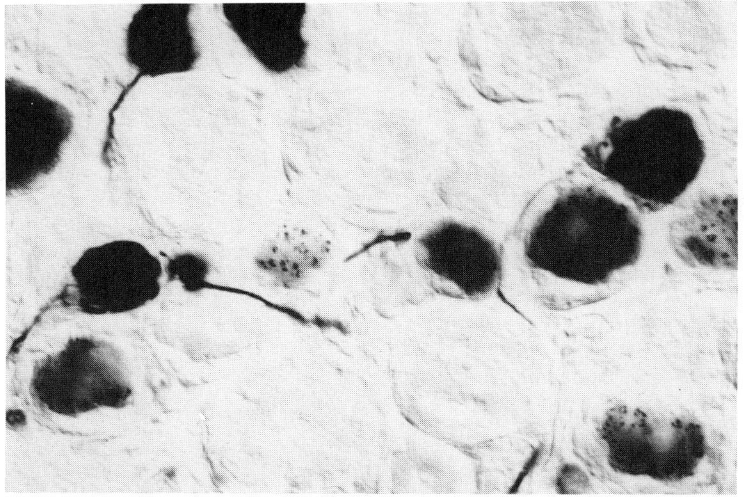

Fig. 10. CGRP immunostained in cells of rat dorsal ganglion (x 220).

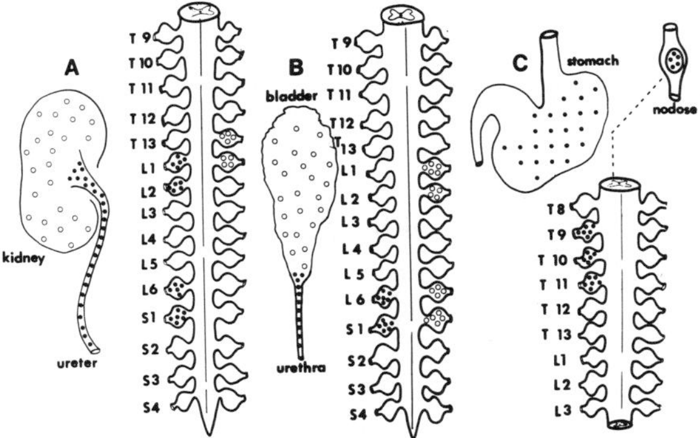

Fig. 11. Injection of peripheral terminal fields with True Blue results in retrograde transport of the dye via axons to the cells of origin where peptide transmitter can be visualised simultaneously. Thus major origins of CGRP-immunoreactive nerves were found to be from: L_1-L_2 and L_6-S_1 dorsal root ganglia in kidney (A); T_{13}-L_1 in ureter (A); L_1-L_2 and L_6-S_1 bladder (B); L_6-S_1 ureter (B); T_9-T_{11} and the nodose ganglia, stomach and pancreas (C).

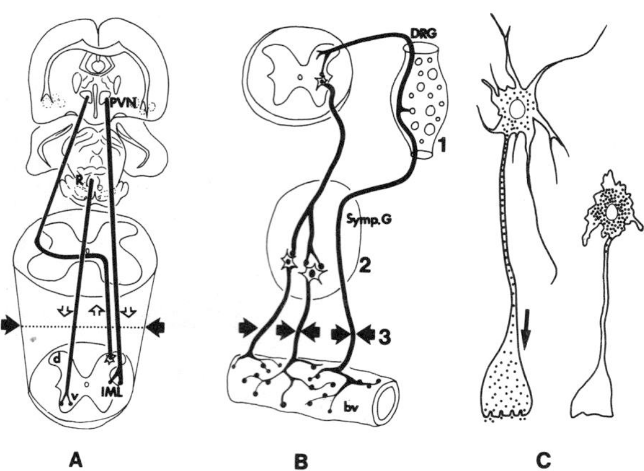

Fig. 12. Experimental manipulations for determination of the nature and origin of peptide-containing nerves. A) Surgical transection of spinal cord, to determine ascending and descending fibre projections. R raphe nuclei; PVN paraventricular nucleus; IML intermediolateral cell column. B) Combined surgical interruptions of neuronal pathways with use of neurotoxins for determination of the nature of peptide-immunoreactive nerve fibres. 1. Capsaicin selectively destroys small primary sensory neurones. 2. 6-Hydroxy-dopamine is a sympathetic nervous tissue toxin. 3. Denervations completely remove external sources of peptide. DRG., dorsal root ganglia; Symp.g., sympathetic ganglion; bv., blood vessel. C) Blockers of axonal transport eg., colchicine enhance peptide content within perikarya.

TOWARDS A "FUNCTIONAL" MORPHOLOGY

The ability to localise binding sites (receptors), the possibility to pinpoint intracellular events leading to peptide production (localisation of mRNA directing peptide synthesis) and the ability to monitor intra-cellular pH and hence follow biosynthetic peptide pathways all allow the transition from a static to a more "functional" morphology. In vitro autoradiography using radiolabelled peptides has recently been applied to tissue sections for the demonstration of peptide binding sites (Clark and Hall, 1986) (Fig. 13a and14). In addition, low numbers of binding sites can now be visualised at the electron microscopical level using either a specific receptor antibody as a divalent peptide which will attach to a receptor active site and to a specific antibody. Dimeric forms of bombesin have been used by Lackie et al., (1985) to localise receptors in growing neuroendocrine tumours (small cell carcinomas of the lung) known to contain receptor sites on the surface of the tumour cell demonstrated earlier by biochemical analysis of isolated membrane preparations (Fig. 13b and 15).

Hybridisation of complementary nucleotide sequences has been used for some time in molecular biology. A similar approach using labelled (radioactive and non-radioactive) nucleotide sequences has been proposed for the localisation of specific DNA or mRNA molecules (Hamid et al., 1987). This technique has been termed in situ hybridisation and permits demonstration of the intracellular machinery involved in the synthesis of a particular peptide or protein (Fig. 16 and 17a,b,c). A differential acidity is known to be present in separate intracellular compartments involved in peptide/protein synthesis (i.e. RER, Golgi area, secretory granules) (Fig. 18). The intracellular pH can be monitored by different means and now it is possible, using a specific antibody to DNP (dinitrophenol), to demonstrate low pH sites by the localisation of DAMP (3-2,4-Dinitroanilino-3' amino-N-methyl-diprophyl-amine) (Orci et al., 1987). This technique, in combination with the use of specific antibodies to preprohormones, permits demonstration of the various intracellular compartments involved in the separate stages of peptide synthesis.

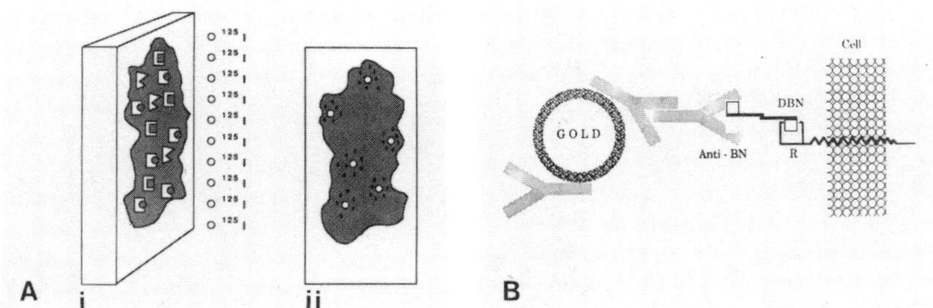

Fig. 13.   A) Diagrammatic representations of methods for binding site localisation. <u>In vitro</u> autoradiography. i) Radiolabelled ligand (^{125}I) is applied to unfixed section and attaches to binding site. ii) Binding site localisation is mapped on film or coverslip apposed to the section. B) Immunogold staining. A divalent ligand (DBN) attaches to bombesin receptors on the surface of a small cell carcinoma cell. The binding sites are visualised by immunostaining with the immunogold method. Taken from Lackie et al., <u>Histochemistry</u> 83: 57-59, (1984).

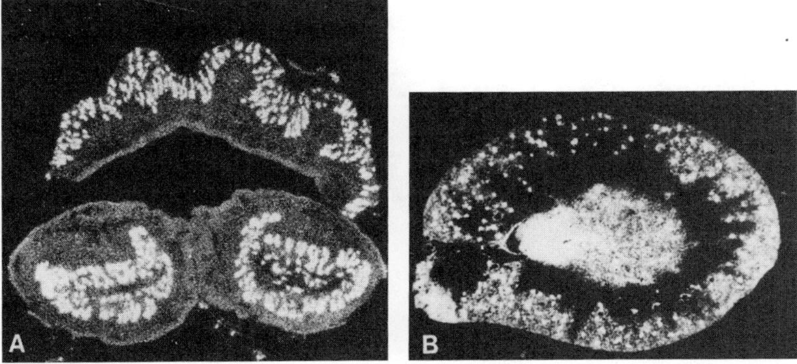

Fig. 14.   A) Autoradiograph of an unfixed cryostat section of guinea pig uterus incubated with ^{125}I-VIP. Dense labelling can be seen in the endometrium (Negative image x 800). Taken from C.O. Inyama et al., <u>Neurosc. Letts</u>. 1987 in press. B) Section of rat kidney incubated with ^{125}I-ANP.The autoradiograph shows dense labelling in glomeruli (Negative image x 150). Taken from "Some aspects of neuroendocrine pathology". J.M. Polak and S.R. Bloom. <u>J. Clin. Pathol</u>. 40, 1987 in press.

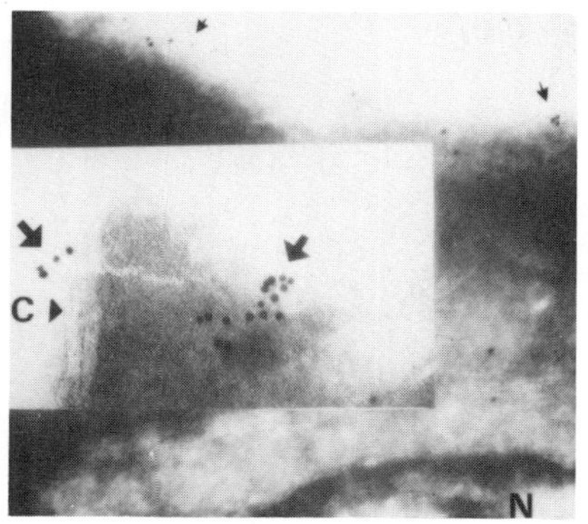

Fig. 15. Electron micrograph showing localisation of bombesin receptors on the surface of a small cell carcinoma cell, as described in Fig. 13 B) Taken from Lackie et al., _Histochemistry_,83, 57-69, (1984).

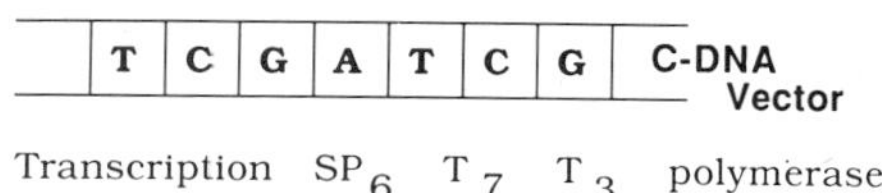

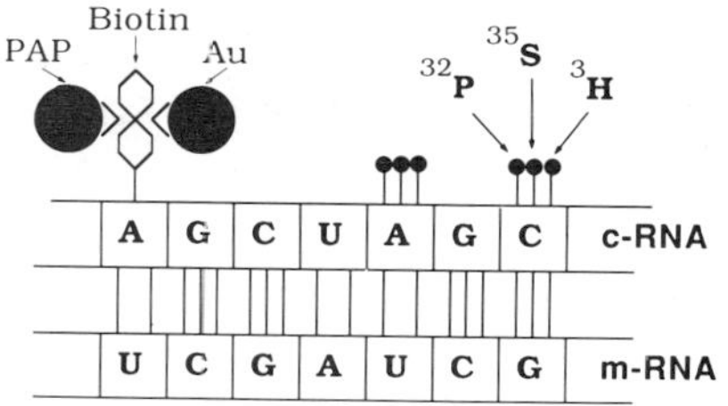

Fig. 16. Diagrammatic representation of the preparation and various labelling methods for c-RNA probes used _in situ_ hybridisation.

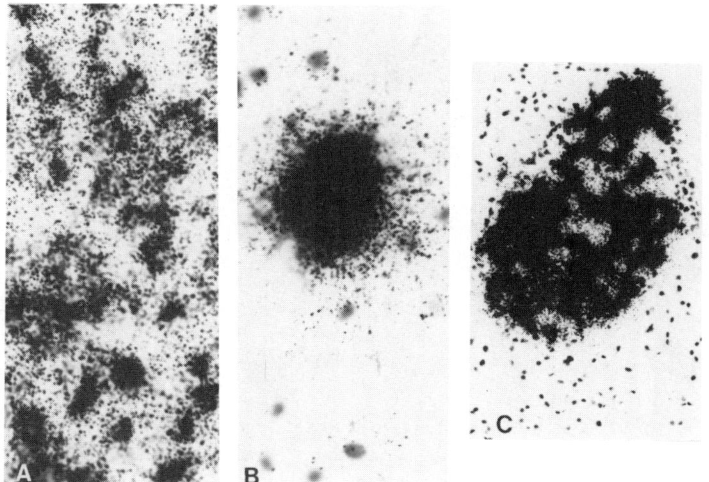

Fig. 17. A) _In situ_ hybridisation of prolactin m-RNA in normal rat pituitary. B) Localisation of NPY mRNA in human cerebral cortex by _in situ_ hybridisation. c) A rat pancreatic islet densely labelled with insulin c-RNA probe.

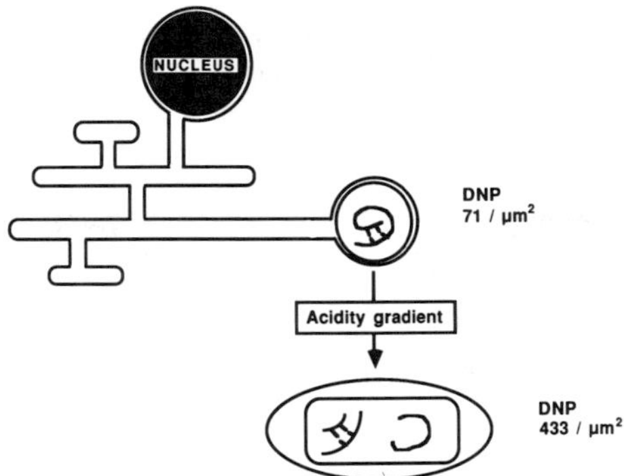

Fig. 18.   Diagram representing the maturation of β cell granules and the accompanying
rise in intragranular acidity. Intact proinsulin is seen in the immature granule on
the Golgi apparatus. The relatively high pH is shown by the low concentration
of DNP. As the granule matures, proinsulin is split and DNP concentration rises
indicating increased acidity. Taken from Orci et al., J. Cell Biol. 103: 2273,
(1987).

CONCLUSION

We have come a long way from the recognition of the neuroendocrine mode of
secretion to the full acceptance of the diffuse neuroendocrine system. These are stimulating
times for morphologists. Numerous techniques permit not only the visualisation of the
products of the various components but also monitoring of the stages from peptide formation
to release, as well as the localisation of the sites of peptide action (binding sites). Many
exciting years lie ahead in our quest to understand further the potential of the
neuroendocrine system in controlling bodily functions.

REFERENCES

Bishop, A.E., Bretherton-Watt, D., Ghatei, M.A. Bloom, S.R., Facer, P., Fahey, M.,
Valentino, K. and Polak, J.M., 1987, Pancreastatin its wide distribution in endocrine
cells and co-existence with chromogranin A, J. Pathol., 152:212A.

Clark, C.R. and Hall, M.D., 1986, Hormone receptor autoradiography: recent developments,
J.F.B.S. II, 195.

Doran, J.F., Jackson, P., Kynoch, P. and Thompson, R.J., 1983, Isolation of PGP 9.5, a new
human neuron-specific protein detected by high-resolution two-dimensional
electrophoresis, J. Neurochem. 40: 1542.

Feyrter, F., 1938, Uber diffuse endokrine epitheliale Organe, J.A. Barth, Ed., Leipzig,.

Gould, V.E., Weidenmann, B., Lee, I., Schwechheimer, K., Dockhorn-Dworniczak, B.,
Roserich, J.A., Moll R. and Franke, W.W., 1987, Synaptophysin expression in
neuroendocrine neoplasms as determined by immunocytochemistry, Am. J. Pathol.
126:243.

Grimelius, L. and Wilander, E., 1985, Silver impregnation and other non-
immunocytochemical staining methods, in: "Endocrine Tumours", J.M. Polak and
S.R. Bloom, eds., Churchill Livingstone, Edinburgh, London, Melbourne, New
York.

Gulbenkian, S., Merighi, A., Wharton, J., Varndell, I.M. and Polak, J.M., 1986,
Ultrastructural evidence for the coexistence of calcitonin gene-related peptide and
substance P in secretory vesicles of peripheral nerves in the guinea pig, J.
Neurocytol. 15:535.

Gulbenkian, S., Wharton, J. and Polak, J.M., 1987, The visualisation of cardiovascular
innervation in the guinea pig using an antiserum to protein gene produce 9.5 (PGP
9.5), J. Auton. Nerv. Syst. 18:235.

Hacker, G.W., Polak, J.M., Springall, D.R., Ballesta, J., Cadieux, A., Gu, J., Trojanowski, Q., Dahl, D. and Marangos, P.J., 1985, Antibodies to neurofilament protein and other brain proteins reveal the innervation of peripheral organs, Histochem., 82:581.

Hamid, Q.A., Bishop, A.E., Sikri, K.L., Varndell, I.M., Bloom, S.R. and Polak, J.M., 1986, Immunocytochemical characterization of 10 pancreatic tumours, associated with the glucagonoma syndrome, using antibodies to separate regions of the pro-glucagon molecule and other neuroendocrine markers, Histopathology 10:119.

Hamid, Q.A., Wharton, J., Terenghi, G., Hassal, C.J.S., Aimi, J., Taylor, K.M., Nakazato, H., Dixon, J.E., Burnstock G. and Polak, J.M., 1987, Localization of atrial natriuretic peptide mRNA and immunoreactivity in the rat heart and human atrial appendage, Proc. Nat. Acad. Sci., 84:6760.

King, W.J. and Greene, G.L., 1984, Monoclonal antibodies localize oestrogen receptor in the nuclei of target cells, Nature (Lond.) 307:745.

Lackie, P.M., Cuttitta, F., Minna, J.D., Polak, J.M. and Bloom, S.R., 1985, Localisation of receptors using a dimeric ligand and electron immunocytochemistry, Histochemistry 83:57.

O'Connor, D.T., Frigon, F.P. and Sokoloff, R.L., 1983, Human chromogranin A: purification and characterization from catecholamine storage vesicles of pheochromocytoma, Hypertension 6:2.

Orci, L., Ravazola, M., Amherdt, M., Madsen, O., Perrelet, A., Vassalli, J.D., Anderson, R.G.W., 1987, Conversion of proinsulin to insulin occurs co-ordinately with acidification of maturing secretory vesicles, J.Cell.Biol. 103:2273

Polak J.M. and Varndell, I.M., 1984, "Immunolabelling for Electron Microscopy," Elsevier Science Publishers, Amsterdam.

Polak J.M.and Marangos, P.J., 1984, Neuron-specific enolase, a marker for neuroendocrine cells, in: "Evolution and tumour pathology of the Neuroendocrine System," S. Falkmer, S. Hankanson R. and F. Sundler, Eds., Elsevier, Amsterdam.

Polak, J.M. and Bloom, S.R., 1979, The diffuse neuroendocrine system, J. Histochem. Cytochem 27:1398.

Polak, J.M. and Bloom, S.R., 1986, Immunocytochemistry of the diffuse neuroendocrine system, in: "Immunocytochemistry: Modern Methods and Applications, " J.M. Polak and S. Van Noorden, eds., John Wright & Sons Ltd., Bristol.

Scharrer, E., Scharrer, B., 1940,Secretory cells within the hypothalamus, in "The Hypothalamus," Vol. 20, Hafner, New York.

Skirboll, L., Hokfelt, T., Norell, G., Phillipson, O., Kuypers, M.G.J.M., Bentivoglio, M., Catsman-Berrevoets, C.E., Visser, T.J., Steinbusch, H., Verhofstad, A., Cuello, A.C., Goldstein, M. and Browstein, M., 1984, A method for specific transmitter identification of retrogradely labelled neurons: immunofluorescence combined with fluorescence tracing, Brain Res. Rev.8:99.

Suzuki, H., Ghatei, M.A., Williams, S.J., Uttenthal, L.O., Facer, P., Bishop, A.E., Polak, J.M. and Bloom, S.R., 1986, Production of pituitary protein 7B2 immunoreactivity by endocrine tumors and its possible diagnostic value, J. Clin. Endocrinol. & Metab. 63:758.

Tatemoto, K., Efendic, S., Mutt, V., Makk, G., Feistner, G.J.and Barchas, J.D., 1986, Pancreastatin, a novel pancreatic peptide that inhibits insulin secretion, Nature 324:476.

Varndell, I.M., Lloyd, R.V., Wilson, B.S. and Polak, J.M., 1985, Ultrastructural localisation of chromogranin: a potential marker for the electron microscopic recognition of endocrine cell secretory granules, Histochem. J. 17:981.

COMBINED USE OF LECTIN HISTOCHEMISTRY AND IMMUNOCYTOCHEMISTRY FOR THE STUDY OF NEUROSECRETION

E.M. Rodriguez, B. Peruzzo, L. Alfaro and H. Herrera

Instituto de Histologie y Patologia
Universidad Austral de Chile
Valdivia
Chile

INTRODUCTION

In 1971, on the occasion of a Symposium organized by Heller and Lederis, and also held in Bristol, I presented some histochemical evidence indicating that in the neural lobe of amphibians and reptilian only one of the two types of neurosecretory fibers contained carbohydrates. More substantial evidence was provided by Tasso (1974) who, working with rats at the ultrastructural level, demonstrated that glycoproteins were circumscribed to neurosecretory granules located in one of the two types of neurosecretory fibers. Later, Tasso et al., (1977), by studying the hypothalamo-neurohypophysial system of Brattleboro rats, showed that the neurosecretory granules containing glycoproteins were missing from homozygous rats and postulated that glycoproteins and arginine-vasopressin (AVP) coexisted within the same granule. Apparently, no further histochemical studies have been performed.

Parallel biochemical studies have led progressively to the conclusion that the AVP-Neurophysin precursor is glycosylated whereas the OXY-Neurophysin precursor is not. Holwerda (1972) isolated a glycopeptide of 39 amino acids from the pig neural lobe. A glycopeptide with the same or similar amino acid sequence was later isolated from bovine (Smyth and Massey 1979) and human (Seidah et al., 1981) neural lobe. Conclusive evidence that the AVP-Neurophysin precursor is glycosylated, and that the 39 amino acid glycopeptide is located at the carboxy terminus of the AVP-Neurophysin precursor, was published by Ivell et al., (1981) and Land et al., (1982).

An important achievement in the field of carbohydrate research has been the discovery of a series of lectins and the subsequent standardization of their use as probes for sugar identification. Lectins are sugar-binding proteins or glycoproteins of non-immune origin that are ubiquitous in plants, but that may also be found in animals (Goldstein and Hayes, 1978). Individual lectins bound to visualants have been used as histochemical markers to identify and localize specific sugar residues (Roth, 1978).

Although lectins have been widely used at the electron microscopic level for the detection of carbohydrate at the outer surface of the cell plasma membrane, their use to study intracellularly located carbohydrates has been restricted to only a few publications, due to technical difficulties. We have now standardized a number of procedures allowing lectins to be used to identify and localize sugar moieties along the different compartments of the secretory pathways.

The usefulness of lectins can be further increased by employing them in combination with a) specific glycosidases and, b) immunocytochemical techniques (Rodriguez et al., 1986, 1987).

Table 1. Occurrence of staining using different histochemical methods (RER = Rough Endoplasmic Reticulum, S.G. = Secretory Granules).

	Subcommissural Organ		Vasopressin Neurons	
	RER	SG	RER	SG
Concanavalin A (Con A) (man, glc)	+	-	+	+
Pisum sativum aggl. (PSA) (man. glc)	+	-	+	+
Wheat-germ aggl. (WGA) (GlcNAc, sialic acid)	-	+	-	+
Limax flavus aggl. (LFA) (sialic acid)	-	+	-	-
Limulus polyphemus aggl. (LPA) (sialic acid)	-	+	-	-
Lotus tetragonolobus aggl. (LTA) (fuc)	-	-	-	+
Ricinus communis aggl. II (RCA) (gal)	-	-	-	-
Arachis hypogaea, peanut, aggl. (PNA) (gal,Gal-NAc)	-	-	-	-
Dolichos Bifluorus aggl. (DBA) (Gal-NAc)	-	-	-	-

We have applied this combined methodological approach to take a second look at the hypothalamo-neurohypophysial system, with three aims in mind : i) to identify the sugar residues present in the neurosecretory granules of AVP-secreting neurons; ii) to obtain evidence concerning the probable configuration of the carbohydrate associated to the AVP-Neurophysin precursor; iii) to study the distribution of these sugar residues along the secretory pathway, including the neurosecretory granules of the cell body and those of the neural lobe. For discussion purposes, we shall include some of the results obtained in previous investigations on the subcommissural organ, using a similar methodological approach.

MATERIAL AND METHODS

Rats of the Holtzman strain were divided into the following groups: a) control; b) colchicine-treated (7μg into lateral ventricle); c) salt-loaded for 1 and 6 months (2% NaCl to drink); d) salt-loaded for 1 and 6 months and colchicine treatment (7μg colchicine 2d before killing). Untreated Brattleboro rats, both heterozygous and homozygous, were also used.

Lectin Histochemistry

Light microscopy. Fixation was in Bouin or in 0.25% glutaraldehyde and 2% acrolein followed by 1% OsO4. Embedding was in butyl-methyl methacrylate. 1 to 2μm thick sections were used for lectin binding after removal of methacrylate with xylene. The lectins used are listed in Table 1. For the visualization of the lectin binding sites, the following methods were used:

(a) Labelling of lectin with i) fluorescein isothiocyanate (FITC) or ii) peroxidase (Fig. 1).

(b) Immunoperoxidase staining using lectin antibodies (Fig. 2A). This involves incubating tissue sections consecutively with i) non-labelled lectin (i.e. Con A) at a high

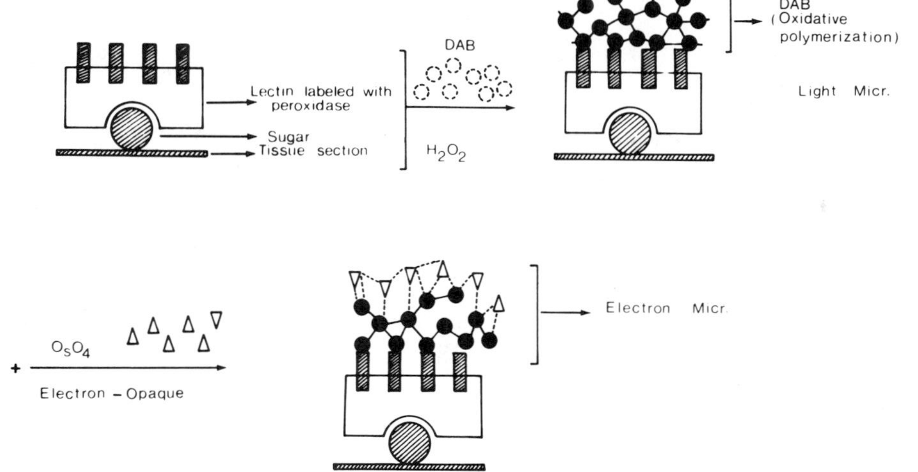

Fig. 1.   Method using a peroxidase-labelled lectin followed by DAB reaction (light microscopy) and OsO$_4$ treatment (electron microscopy).

dilution (0.3 to 1μg lectin/ml) for 45 min; ii) anti-lectin raised in rabbits (i.e. anti-Con A) (1:4000 dilution); iii) second antibody (1:50); iv) PAP complex followed by DAB reaction.

(c) Immunoperoxidase staining using a pre-prepared lectin-antilectin complex (Fig. 3).   The complex was prepared by incubating 60μg of WGA with 15 1 of anti-WGA serum in a final volume of 300μl of 0.01M PBS, pH 7.2, for 12h at 4°C.   Sections were incubated with the complex diluted 1:80 for 1h, and then processed for immunostaining as in (b).

The lectin binding sites where checked by coupling the lectin with the corresponding binding sugar (D-mannose for Con-A, glucosamine for WGA and sialic acid for LFA) before using these solutions to incubate the sections.

<u>Use of specific glycosidases</u>.   Separate sections were incubated with α-mannosidase (1.6 U/ml citric acid buffer, pH 5.6, for 18h at 37°C), neuroaminadase (1U/ml acetate buffer, pH 5.6, for 18h at 37°C), N-acetyl glucosaminidase H (1.5 U/ml sodium acetate buffer, pH 4.5, for 48h at 37°C).   Then sections were incubated with lectins according to the schedule shown in Table 2.

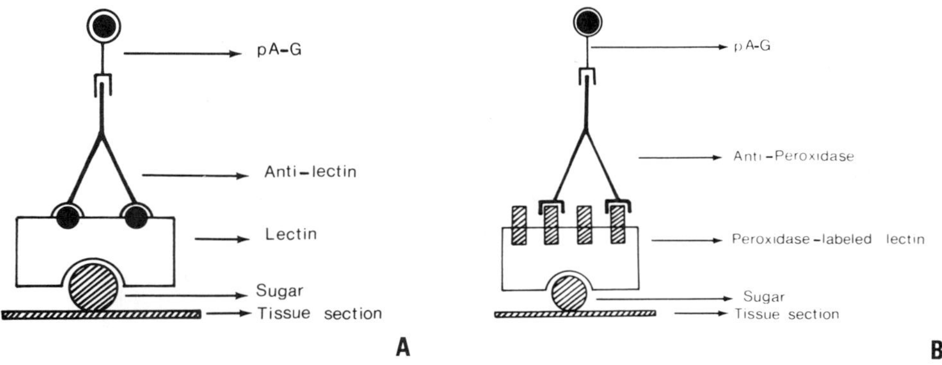

Fig. 2.   A Immunogold staining using lectin antibodies.
B. Immunogold staining using a peroxidase-labelled lectin and anti-peroxidase.

Electron microscopy. Three lectins were used: Con A, WGA and RCA. Pre- and post-embedding techniques were used for the demonstration of lectin binding sites.

Pre-embedding. Fixation was in Karnovsky fixative buffered to pH 7.4 with phosphate. Cryostat sections were incubated in the peroxidase-lectin solution (30μg/ml) in Iris buffer, pH 7.8, for 45min at 22°C. Then sections were processed for the demonstration of peroxidase (DAB), fixed in OsO_4 and embedded in Epon (for details see Rodriguez et al., 1986) (Fig. 1).

Post-embedding. Fixation was in 2% paraformaldehyde and 0.5% glutaraldehyde in 0.1M phosphate buffer, pH 7.4. Embedding was in Lowicryl. The following three staining procedures were designed:

(a) Immunogold technique using lectin antibody (Fig. 2A). Ultrathin sections were sequentially incubated with: 1) non-labelled lectin (1 g/ml) for 45min; ii) lectin antibody (1:1000); iii) protein A-gold complex (1:20).

(b) Immunogold technique using a pre-prepared lectin-antilectin complex (Fig. 3). Sections were incubated with the complex (1:80 dilution) for 45min and then with protein A-gold complex (1:20) for 1h.

(c) Immunogold technique using lectins labelled with peroxidase. Since lectins labelled with peroxidase are readily available or easy to prepare, whereas antibodies for each lectin are not, we designed this procedure which allows the immunocytochemical demonstration of the binding sites of lectins labelled with peroxidase. Sections were sequentially incubated with: lectin labelled with peroxidase, peroxidase antibody, protein A-gold complex (Fig. 2B).

Combined use of lectin histochemistry and immunostaining of peptides in tissue sections. The identification of the cell type, i.e. AVP- or OXY-secreting neuron, in which the presence of certain sugars is revealed by lectins, or the demonstration of the co-existence within subcellular structures (i.e. RER, secretory granules) of a given peptide and of sugar residues, may become useful tools. For this purpose we used three different protocols:

(a) Double sequential staining of the same 1μm methacrylate section (Rodriguez et al., 1986). Sections were first incubated with FITC-labelled lectin, studied under a fluorescence microscope and, after photographs had been taken, the lectin was eluted from the section. Then the section was processed for immunoperoxidase staining using the corresponding primary antibody.

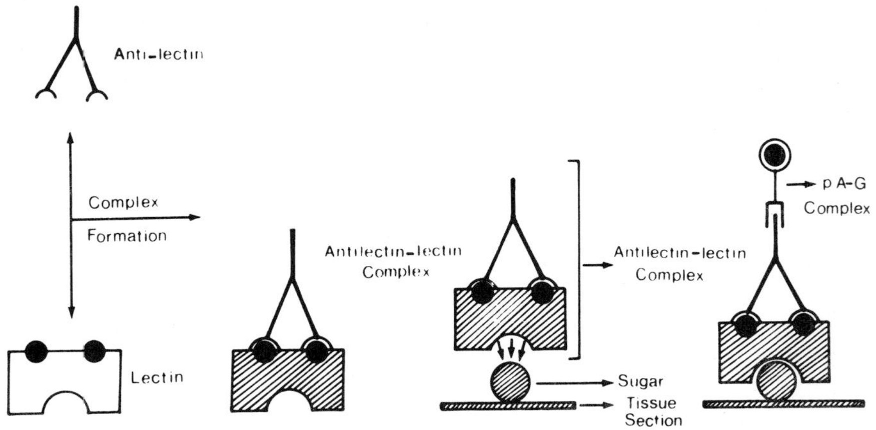

Fig. 3.   Immunogold staining using a pre-prepared antilectin-lectin complex.

(b) Staining of consecutive serial 1μm methacrylate sections: adjacent sections were separately processed for: i) immunoperoxidase staining for a given peptide; ii) lectin binding using one of the procedures mentioned above.

(c) Double immunogold staining of ultrathin Lowicryl sections: The sections were stained in sequence for: Avp or Oxy using gold particles 15 nm in diameter and then for WGA using 2-5 nm gold particles.

<u>Combined use of lectin binding and immunostaining of peptides (proteins) in nitrocellulose transfers.</u> This procedure was used for extracts of the bovine subcommissural organ. Nitrocellulose transfers of parallel runs were separately processed for: i) immunostaining using an antiserum reacting with the secretory compounds of the subcommissural organ; ii) lectin binding using the lectin-anti-lectin protocol. This procedure enabled us to identify the immunoreactive secretory products, to establish which one was glycosylated and to estimate roughly the sugar residues involved (Rodriguez et al., 1987).

RESULTS AND DISCUSSION

All magnocellular neurons of the supraoptic (SON) and paraventricular (PVN) nuclei were strongly Con A positive. In all of them Con A binding sites were found within the RER and certain cisternae of the Golgi apparatus. However, whereas the neurosecretory granules of AVP neurons were Con A positive, those of the OXY neurons did not bind Con A.

In the SON and PVN, WGA (glucosamine) and LTA (fucose) displayed affinity only for neurons which contained ir-AVP.

Although WGA has affinity for glucosamine and sialic acid, the following evidence clearly indicates that in the AVP neurons the WGA-binding sugar is exclusively glucosamine: the two lectins specifically binding to sialic acid (LFA and LPA) did not react with AVP neurons (Table 1); after neuroaminadase treatment, WGA continue to have affinity for AVP neurons (Table 2).

In contrast to the OXY-Np precursor, the AVP-Np precursor is glycosylated (Ivell et al., 1981). The oligosaccharide of the AVP-Np precursor has a high content of glucosamine (Smyth and Massey, 1979, Ivell et al., 1981). The following observations strongly indicate that the WGA-binding sugar in the SON and PVN is the glucosamine of the AVP-Np precursor: i) AVP neurons but not OXY neurons were labelled with WGA; ii) AVP and the WGA-binding sugar coexist within the same nsg (Fig. 6 insert) (OXY nsg did not bind

Table 2. Occurrence of sugar residues in rough endoplasmic reticulum (RER) and secretory granules (SG).

	Subcommissural organ		Vasopressin Neurones	
	RER	SG	RER	SG
Con A (man,glc)	+	-	+	+
α-Mannosidase --- Con A	-	-	-	+
α-Glucosidase --- Con A	+	-	ND	ND
β-N-Acetylglucosaminidase --Con A	-	-	+	+
WGA (GlcNAc, sialic acid)	-	+	-	+
Neuroaminidase --- WGA	-	+	-	+
LFA (sialic acid)	-	+	-	-
Neuroaminidase --- LFA	-	-	-	-
RCA (gal)	-	-	-	-
Neuroaminidase --- RCA	-	+	-	-

WGA); iii) the hypothalamo-neurohypophyseal system of homozygous Brattleboro rats lacked affinity for WGA.

<u>Sugar residues present in the AVP-Np precursor and in the AVP-glycopeptide</u>

Haddad et al., (1976) using tritium-labelled fucose and radioautography postulated that a glycoprotein synthesized in the SON and PVN was transported to the neural lobe. Ivell et al., (1981) demonstrated the incorporation of tritium-labelled mannose, glucosamine and fucose into the AVP-Np precursor. Similar findings were reported for the AVP-glycopeptide (Swann et al., 1983).

The use of a set of lectins with a whole range of affinities covering the seven monosaccharides occurring in glycoproteins, and the utilization of a few specific glycosidases also lead to the conclusion that mannose, glucosamine and fucose are present in the AVP granules (Tables 1 and 2), thus clearly indicating that these three sugar residues correspond to those of the AVP-Np precursor.

Another interesting conclusion of this study is that apart from mannose, glucosamine and fucose, no other sugar residues were found in the AVP neurosecretory granule (Tables 1 and 2). Specially relevant is the absence of galactose (lack of affinity for PNA and RCA and for RCA after neuroaminidase) and of sialic acid (no reaction with LFA and LTA). This makes the oligosaccharide of the AVP-Np precursor a rather exceptional type (Montreuil 1980, Staneloni and Leloir, 1982).

<u>Evidence suggesting probable configuration of the oligosaccharide of the AVP-Np precursor</u>

A good body of evidence indicates that the AVP-Np precursor is core-glycosylated i.e., a high mannose type core N-linked to asparagine: i) N-glycosidic linkage (Holwerda, 1972); ii) glucosamine linked to an asparagine residue (Holwerda, 1972; Smyth and Massey, 1979; Seidah et al., 1981); iii) typical N-glycosilation site Asn-X-Thr (Land et al., 1982); iv) inhibition of glycosylation by tunicamycin (Ivell et al., 1981); v) a high content of mannose (Ivell et al., 1981); iv) high content of mannose within RER, which is the site of core-glycosylation, and disappearance of Con A reaction within RER after α-mannosidase.

There is no evidence of the nature of the terminal sugar residues of the oligosaccharide of the AVP-Np precursor. Based on the present findings we shall discuss several possibilities.

Evidence that glucosamine is the terminal sugar residue: i) absence of galactose and sialic acid; ii) WGA has affinity for glucosamine added in the Golgi apparatus (terminal chain) and not for those added in the RER (core glucosamine); therefore the glucosamine detected in the nsg by WGA should correspond to residues conjugated to the mannose core at the Golgi membranes; iii) α-mannosidase, which is an exoglycosidase, abolishes Con A reaction in RER but not in the nsg, thus indicating that in the former location mannose residues are terminal but in the latter are not.

Some of the present results would support the possibility that mannose could also be a terminal residue, in addition to glucosamine. The strongest affinity of Con A is for terminal mannose residues (Goldstein and Hayes 1978), and the AVP granules do display affinity for Con A. At variance, the secretory granules of the subcommissural organ lack affinity for Con A and, in this case, the carbohydrate is of the complex type with galactose and sialic acid as the preterminal and terminal sugar residues (Table 1) (Rodriguez et al., 1986). On the other hand, other evidence from the present investigation would indicate that there are no terminal mannosyl residues. Thus, α-mannosidase does not abolish affinity of Con A for the AVP granules (Table 2). This suggests that there are no terminal mannose residues and that Con A is reacting with internal mannose residues. This latter possibility, although against what is generally accepted, is not unlikely. In fact, we have found that, when sialic acid is removed from complex-type oligosaccharides, Con A has affinity for internal mannose residues (Rodriguez et al., 1987). Since the carbohydrate of the AVP

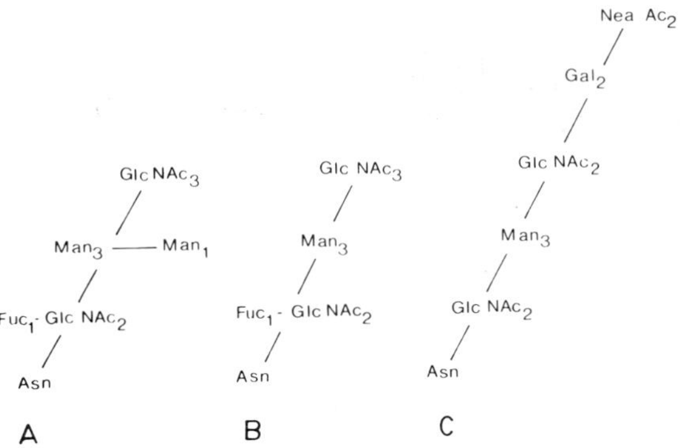

Fig. 4.   Probable configurations (A, B) of oligosaccharide of AVP-Neurophysin precursor. C configuration of oligosaccharide of secretory glycoproteins of the subcommissural organ (Rodriguez et al., 1987).

granules (AVP-Np precursor) lacks sialic acid, it seems likely that Con A is binding to internal mannose residues in the latter.

An alternative explanation would be that α-mannosidase does in fact remove terminal mannose residues from the nsg, but Con A continues to stain these nsg because it is binding to internal mannose residues (not removed by α-mannosidase) due to absence of sialic acid.

All these considerations lead us to suggest two possible configurations for the oligosaccharide of the AVP-Np precursor, as shown in Fig. 4.

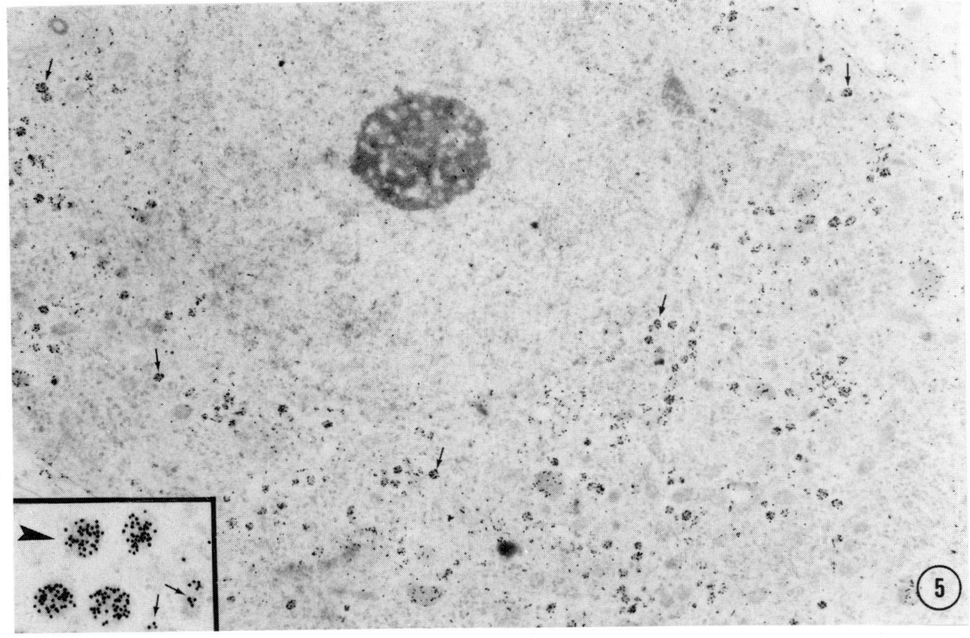

Fig. 5.   AVP neuron of the rat SON. Immunogold staining using WGA-anti-WGA. Secretory granules are homogeneously labelled. Insert: gold particles are found over the entire nsg (arrowhead) and in small vesicles (small arrows).

Configuration A considers glucosamine and mannose as terminal sugar residues and 5 glucosamine residues, as reported by Smyth and Massey 1979. A similar structure has been described for the oligosaccharide of ovalbumin (Staneloni and Leloir (1972). Configuration B considers glucosamine as the terminal sugar and, apparently, a similar configuration has not been described for any other N-linked carbohydrate (Montreuil 1980). Configuration B would correspond to the secretory glycoproteins of the subcommissural organ (Rodriguez et al., 1987).

Distribution of sugar residues within the AVP secretory granules

In the AVP granules located in the neural lobe, the Con A and WGA binding sugar were exclusively located at the periphery of the granule (Fig. 6). This is in agreement with the results reported by Tasso et al., (1977) using conventional histochemical methods for carbohydrates. However, in the perikaryon and in axons close to the cell body, the Con A and WGA binding sugars were homogeneously distributed in the entire AVP granule (Fig. 5). In colchicine treated rats, there was a large accumulation of neurosecretory granules in the cell body, especially in the region between the RER and the plasma membrane. The AVP granules that had accumulated in this latter region showed lectin binding sites only at their periphery, whereas the AVP granules close to the Golgi apparatus showed a homogeneous distribution of the lectin-binding sugars. It has been shown that processing of AVP and OXY precursors occur in neurosecretory granules which have been arrested in the cell body by colchicine treatment (Parish et al., 1981). All these findings lead as to suggest that, before processing of the AVP-Np precursor, the oligosaccharide (mannose, glucosamine) is homogeneously distributed in the neurosecretory granules, because the glycosylated precursor is also evenly distributed within the neurosecretory granules (Fig. 7). After processing of the AVP-Np precursor, the resulting products would take different locations within the neurosecretory granules, with AVP and Np occupying the core of the granule and the

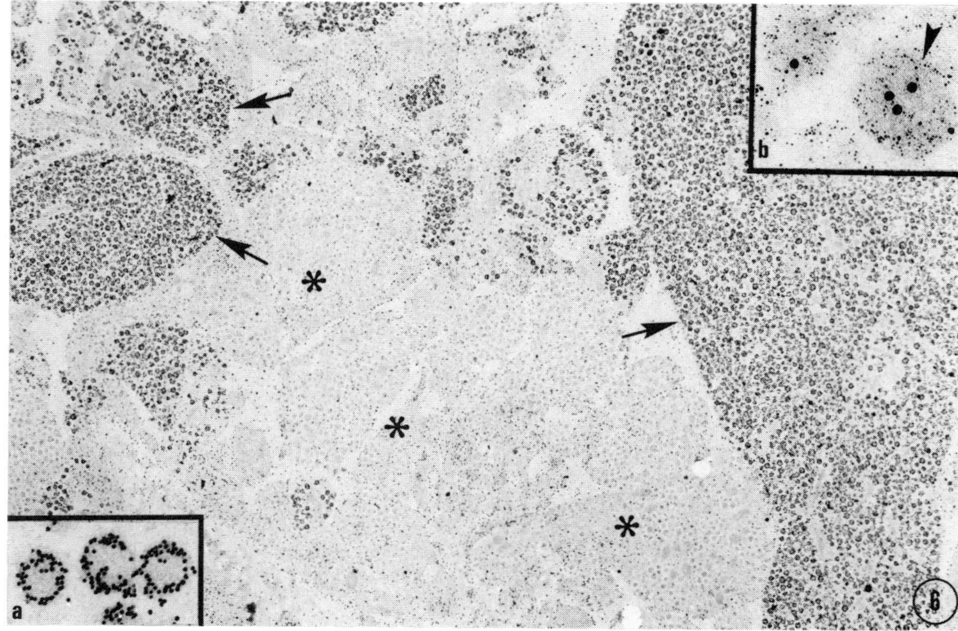

Fig. 6.   Rat neural lobe. WGA binding using same procedure as in Fig. 5. In one of ending (AVP) all granules are peripherally labelled (arrows). OXY endings (asterisks) are not labelled. Insert A: AVP granules showing the peripheral location of the WGA binding sites. Insert B: Double immunogold staining. WGA binding sites (small gold particles, arrowhead) are located at the periphery of nsg and AVP (large gold particles) are preferentially located in the core.

glycopeptide located at the periphery (membrane bound?). Since the 39 amino acid glycopeptide is apparently further cleaved into four sub-fragments, and only two of these bear the oligosaccharide side chain (Smyth and Massey 1979, Land et al., 1982), the possibility that these subfragments are the ones located peripherally should be borne in mind. The functional meaning of the "translocation" of the glycopeptide from the core to the periphery of the neurosecretory granules remains to be elucidated.

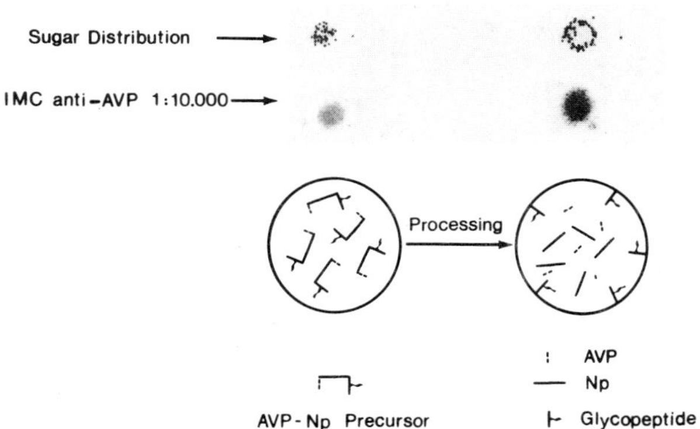

Fig. 7. Schematic representation of the AVP nsg. AVP-Neurophysin precursor (perikaryon) is homogeneously distributed within the granule, causing i) a homogeneous labeling with WGA and Con A, and ii) lack of affinity for a highly diluted anti-AVP. After processing (neural lobe): i) the glycopeptide molecules are "translocated" to the periphery (membrane bound?), thus explaining the peripheral distribution of WGA and Con A binding sugars; ii) AVP becomes reactive to highly diluted anti-AVP.

Whatever the explanation for the different distribution of sugar residues in neurosecretory granules of the cell body and those of the nerve terminal, it appears that these two patterns of distribution may be used as a morphological tool to identify neurosecretory granules containing processed or non-processed precursor molecules. In this respect, it is interesting to mention that all granules located within dendrites of the AVP neurons displayed the sugar binding sites at their periphery, thus suggesting that they contain processed material.

In a previous study (Rodriguez et al., 1984), we used a highly diluted AVP antibody to identify neurosecretory granules containing processed and non-processed material. Using this approach, we suggested that the neurosecretory granules located in the dendrites and axons of the neural lobe contain processed material, whereas those present in the perikaryon and the axon region close to the cell body mostly contain precursor forms. That the neurosecretory granules in the neural lobe contain processed material had been previously demonstrated by different approaches (see Pickering, 1981; Gainer, 1981). However, that dendrites of AVP neurons are loaded with granules containing processed material appears as a new finding whose functional significance needs to be investigated.

Supported by Grant S-85-39 from the Direccion de Investigaciones, Universidad Austral de Chile, and Grant I/60 935 from Stiftung Volkswagenwerk, F.R.G.

REFERENCES

Gainer, H., 1981, The biology of neurosecretory neurons, in: "Neurosecretion and Brain Peptides", J.B. Martin, S. Reichlin, and K.L. Bick, eds., Raven Press, New York.
Goldstein, I.J., and Hayes, C.E., 1978, The lectins: Carbohydrate binding proteins of plants and animals. Adv. Carbohydr. Chem. Biochem., 35:127.
Haddad, A., Pelletier, G., Marchi, F., and Brasileiro, L., 1976, Radioautographic study of glycoprotein biosynthesis migration and renewal in the hypothalamus-neurohypophysis system, Proc. 5th Colloq. Brazilian Soc. Elect. Microsc., p.62.

Holwerda, D.A., 1972, A glycopeptide from the posterior lobe of pig pituitaries, <u>Eur. J. Biochem.</u>, 28:340.

Ivell, R., Schmale, H., and Richter, D., 1981, Glycosylation of the arginine vasopressin/neurophysin II common precursor, <u>Biochem. and Biophys. Res. Commun.</u>, 102:1230.

Land, H., Schutz, G., Schmale, H., and Richter, D., 1982, Nucleotide sequence of cloned cDNA encoding bovine arginine vasopressin-neurophysin II precursor, <u>Nature</u>, 295:299.

Montreuil, J., 1980, Primary structure of glycoprotein glycans. Basis for the molecular biology of glycoproteins, <u>Adv. Carbohydr. Chem. Biochem.</u>, 37:157.

Parish, D.C., Rodriguez, E.M., Birbeck, S.D., and Pickering, B.T., 1981, Effects of small doses of colchicine on the components of the hypothalamo-neurohypophysial system of the rat, <u>Cell Tissue Res.</u>, 220:809.

Pickering, B.T., 1981, The neurosecretory neuron in the 1980s, <u>in</u>: "Neurosecretion. Molecules, Cells, Systems", D.S. Farner and K. Lederis, Eds., Plenum Press, New York, London.

Rodriguez, E.M., 1971, The comparative morphology of neural lobes of species with different neurohypophysial hormones. <u>in</u>: 19 Memoir of the Society for Endocrinology. Eds. H. Lederis, pp. 263-291, London: Cambridge University Press.

Rodriguez, E.M., Peruzzo, B. and Yulis, R., 1984, Immunocytochemical evidence of the processing of neurohypophysial peptides, <u>Proc. XV Meet. Lat. Amerc. Soc. Physiol. Sci.</u>, R105.

Rodriguez, E.M., Herrera, H., Peruzzo, B., Rodriguez, S., Hein, S., and Oksche, A., 1986, Light- and electron-microscopic immunocytochemistry and lectin histochemistry of the subcommissural organ: Evidence for processing of the secretory material, <u>Cell Tissue Res.</u>, 243:545.

Rodriguez, E.M., Hein, S., Rodriguez, S., Herera, H., Peruzzo, B., Nualart, F., and Okschu, A., 1987, Analysis of the secretory products of the subcommissural organ. <u>in</u>: Functional Morphology of Neuroendocrine Systems, B. Scharrer, H. Korf, H.-G. Hartwig (eds), Springer Verlag.

Roth, J., 1978, The lectin: Molecular probes in cell biology and membrane research, <u>Exp. Pathol.</u>, (Suppl.) 3:1.

Seidah, N.G., Benjannet, S., and Chretien, M., 1981, The complete sequence of the novel human pituitary glycopeptide. <u>Biochem. Biophys. Res. Commun.</u>, 100:901.

Smyth, D.G., and Massey, D.E., 1979, A new glycopeptide in pig ox and sheep pituitary, <u>Biochem. Biophys. Res. Commun.</u>, 87:1006.

Staneloni, R.J., and Leloir, L.F., 1982, The biosynthetic pathway of the asparagine-lined oligosaccharadies of glycoproteins. <u>CRC Critical Review in Biochemistry</u>, 12:289.

Swann, R.W., Gonzalez, C.B., Birkett, S.D., and Pickering B.T., 1983, Harbingers and hormones: inter-relationships of rat neurohypophysial hormone precursors <u>in vivo</u>. <u>in</u>: The Neurohypophysis: Structure, Function and Control, Progress in Brain Research, Vol. 60, B.A. Cross, G. Leng (eds.). Elsevier Science.

Tasso, F., 1974, Diversite d'aspect des fibres neurosecretrices revelees dans la posthypophyse du rat par latechnique de post-fixation combinant l'osmium et le ferrocyanure de potasium. <u>C. R. Soc. Biol. (Paris)</u>, 168:986.

MAPPING AND ANALYSIS OF RECEPTORS FOR NEUROHYPOPHYSEAL PEPTIDES PRESENT IN THE BRAIN

E. Tribollet[1], C. Barberis[2], S. Jard[2], J. Elands[2], M. Dubois-Dauphin[1], A. Marguerat[1] and J. J. Dreifuss[1]

[1]Department of Physiology
University Medical Center
1211 Geneva 4
Switzerland

[2]CNRS-INSERM Center of
Pharmacology Endocrinology
34033 Montpellier
France

INTRODUCTION

Vasopressin (AVP) and oxytocin (OT) meet most of the criteria which are usually applied to assign a role as a neurotransmitter or neuromodulator to a given compound. 1) In addition to their presence in the hypothalamo-hypophysial system, they are also found in neural pathways that project to various areas of the central nervous system (Sofroniew, 1981). 2) They can be released in a calcium dependent manner from several of these extra-hypothalamic areas (Buijs, 1983). 3) They affect the electrical activity of single neurones (Mühlethaler et al., 1985), as well as integrative processes such as temperature control, blood pressure regulation, avoidance behavior, etc. (Kovacs et al., 1987). The commercial availability of tritium-labelled AVP and OT has allowed the detection of high affinity binding sites for these peptides in the brain (Audigier and Barberis, 1985; Jard et al., 1987; Van Leeuwen, 1987). The experimental data presented in this chapter show the type of information which can be obtained by in vitro light microscopic autoradiography. Four points are considered. First, the distribution of high affinity binding sites is described for the rat and the guinea-pig brain; surprisingly, marked differences in their location apparently exist between these two species. Second, brain AVP and OT binding sites in the rat brain are compared to peripheral receptors, using synthetic structural analogues with enhanced selectivity for OT receptors and for the known subtypes of AVP receptors. Third, a case of long term receptor regulation is described, namely the effects of castration and of gonadal steroids on the density of [^{3}H]OT binding in the rat brain. Finally, preliminary results obtained with a newly synthetized and iodinated OT receptor ligand are presented and compared with those obtained with tritiated oxytocin.

VASOPRESSIN AND OXYTOCIN BINDING SITES IN THE RAT AND GUINEA-PIG BRAIN: TOPOGRAPHICAL DIFFERENCES.

In both species, AVP and OT binding sites were found widely distributed throughout the brain, but were not detected in the spinal cord.

[^{3}H]AVP. As a general rule, fewer areas were labelled in the guinea-pig than in the rat brain. A striking observation was the absence in the guinea-pig of detectable binding in regions which were among the most intensely labelled in the rat, i.e. the lateral septum, the nucleus of the solitary tract, the bed nucleus of the stria terminalis and the hypothalamic sigmoid nucleus (Fig. 1). In the guinea-pig, however, the entire cerebral cortex contained [^{3}H]AVP binding sites, located especially in layer V (Fig. 1 E), while only some very restricted cortical areas were found to be labelled in the rat.

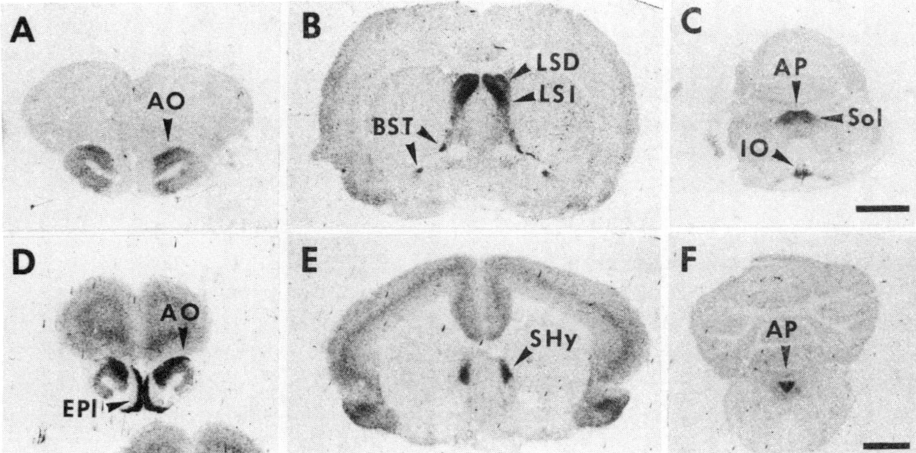

Fig. 1.   Comparative mapping of AVP binding sites in sections of rat (A-C) and guinea-pig brain (D-F). In this and in all following figures, autoradiography was performed as described elsewhere (Tribollet et al., 1987). Incubation was carried out in medium containing 1.5 nM [^{3}H]AVP. In both rat and guinea-pig, the anterior olfactory nucleus bound [^{3}H]AVP, but the external plexiform layer was labelled only in the guinea-pig (D). The lateral septum was intensely labelled in the rat (B) as was the septohypothalamic nucleus in the guinea-pig (E). In the rat brainstem, binding was detected in the area postrema, in the nucleus of the solitary tract and in the inferior olive (C), while it was restricted to the area postrema in the guinea-pig (F). Note also the specific cortical labelling in the guinea-pig (E) which was not seen in the rat (B). Abbreviations: AO, anterior olfactory nucleus; AP, area postrema; BST, bed nucleus of the stria terminalis; EPI, external plexiform layer of the olfactory bulb; IO, inferior olive; LSD, lateral septal nucleus, dorsal; LSI, lateral septal nucleus, intermediate; SHy, septohypothalamic nucleus; Sol, nucleus of the solitary tract. Calibration bars for A-C and D-F = 2.5 mm.

[3H]OT. When comparing rat and guinea-pig, binding sites were generally found in different areas, with the exception of the hypothalamic ventromedial nucleus, which was strongly labelled in both species (Fig. 2). In the rat brain, high densities of binding sites were also observed in the olfactory tubercle, in the bed nucleus of the stria terminalis, in the central amygdaloid nucleus, none of these areas were labelled in the guinea-pig. In contrast, intense labelling was found in the infralimbic and cingulate cortex, in the medial and cortical amygdaloid nuclei and in the pyriform cortex in the guinea-pig but not in the rat.

Such differences in receptor topography corroborate electrophysiological data which showed that homologous populations of central neurones from various species may contrast widely in their responsiveness to OT or AVP. For example, neurones of the dorsal motor nucleus of the vagus nerve (dmnX) of the rat, an area which contains OT binding sites (see Fig. 4), are excited by OT; in the guinea-pig, the dmnX is devoid of OT binding sites and vagal motoneurones are unresponsive to OT (Raggenbass et al., 1987). Similar results have been obtained in the hippocampus, which contains OT-sensitive neurones as well as OT binding sites in the rat, but not in the guinea-pig. Whereas a good correlation exists between autoradiographic and electrophysiological findings, indicating that the radioactive ligands bind to functional receptors, it is noteworthy that these findings do not accord with the results obtained by studying the distribution of endogeneous AVP and OT by immunocytochemistry. Our observations also suggest that caution should be exerted in extrapolating from one species to others.

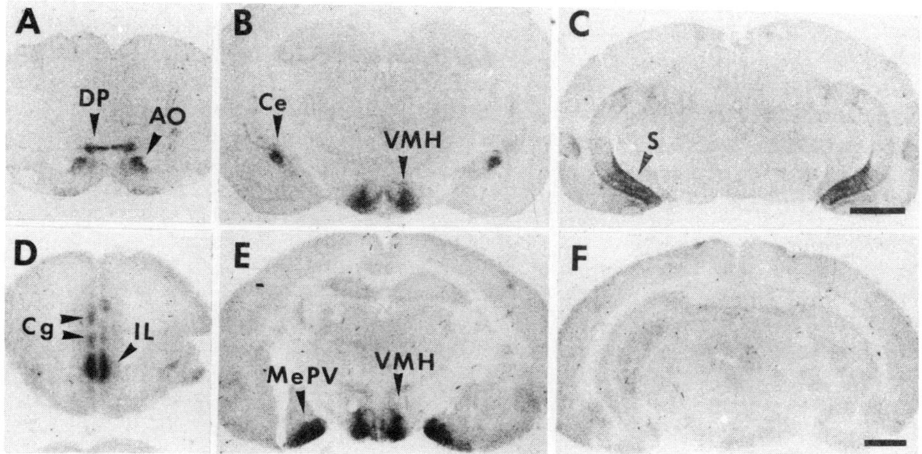

Fig. 2.   <u>Comparison of OT binding sites in brain sections from rat (A-C) and guinea-pig (D-F)</u>. Incubation was carried out in medium containing 3 nM [^{3}H]OT. Rostrally, labelling was present in the dorsal peduncular cortex in the rat (A), whereas in the guinea-pig other cortical fields contained binding sites (D). The ventromedial hypothalamic nucleus was densely labelled in both species, while binding to amygdaloid nuclei was species-dependent (B and E). The intense labelling present in the rat ventral subiculum (C) was not observed in the guinea-pig (F). Abbreviations: AO, anterior olfactory nucleus; Ce, central amygdaloid nucleus; Cg, cingulate cortex; DP, dorsal peduncular nucleus; IL, infralimbic cortex; MePV, medial amygdaloid nucleus, posteroventral; S, subiculum; VMH, ventromedial hypothalamic nucleus. Calibration bars for A-C and D-F = 2.5 mm.

OXYTOCIN AND VASOPRESSIN BINDING SITES IN THE RAT BRAIN: PHARMACOLOGICAL DISCRIMINATION

An overlap between the distribution of [^{3}H]AVP binding sites and of [^{3}H]OT binding sites was only rarely observed. Moreover when the same structure was labelled by both peptides, they were usually present in most cases in different parts of the structure. Thus, [^{3}H]AVP labelled the anterior part of the central amygdaloid nucleus, while [^{3}H]OT bound more posteriorly. Similarly, in the bed nucleus of the stria terminalis, [^{3}H]OT binding sites were located more caudally than [^{3}H]AVP binding sites.

Apparent overlap was found in the ventral subiculum and in the hypothalamic ventromedial nucleus, both of which were intensely labelled by [^{3}H]OT and lightly by [^{3}H]AVP. But this finding is probably explained by the attachment of [^{3}H]AVP to OT binding sites in these areas. This may occur since the OT receptor binds both OT and AVP with high affinity (Kds = 2 and 3.6 nM, respectively), while the AVP receptor binds only AVP with high affinity (Audigier and Barberis, 1985). Results obtained with a selective oxytocic agonist, OH[Thr^4,Gly^7]OT, substantiate this interpretation. This compound displaced both the strong [^{3}H]OT binding and the weak [^{3}H]AVP binding in the ventral subiculum, while it did not affect [^{3}H]AVP elsewhere in hippocampus (Tribollet et al., 1987, Fig. 4). These results are best interpreted by assuming that OT binding sites, but no AVP sites, are present in the ventral subiculum, while other parts of the hippocampus would contain AVP binding sites. We believe that in order to label selectively brain AVP binding sites in autoradiographic experiments, one must either use a low concentration of [^{3}H]AVP (a value close to the Kd of 1.7 nM of AVP for its binding site) or add a selective oxytocic agonist in the incubation medium. Otherwise, with [^{3}H]AVP concentrations of 5 nM or more, binding of this ligand to OT receptors is observed (De Kloet et al., 1985; Freund-Mercier et al., 1987; Van Leeuwen, 1987).

Competition experiments using synthetic structural analogues revealed that central AVP binding sites have a high affinity for V_1-selective peptides and a very low affinity for V_2-selective compounds. Thus, the V_1 antagonist desGly^{9}d(CH$_2$)$_5$AVP inhibited [^{3}H]AVP binding in the septum, but not in the kidney; in contrast, the V_2 agonist dDAVP displaced radioactivity in the kidney, but not in the septum (Tribollet et al., 1987, Fig. 7). The same ligand selectivity was found for all brain areas studied. These data confirm reports in which vasopressin binding to isolated membrane fractions was studied and of the stimulation of cyclic AMP production determined. These studies suggested that vasopressin receptors in the rat brain resemble V_1 receptors, but markedly differ from V_2 receptors, both in terms of ligand specificity and of the nature of the second messenger formed in target cells (Jard, 1983).

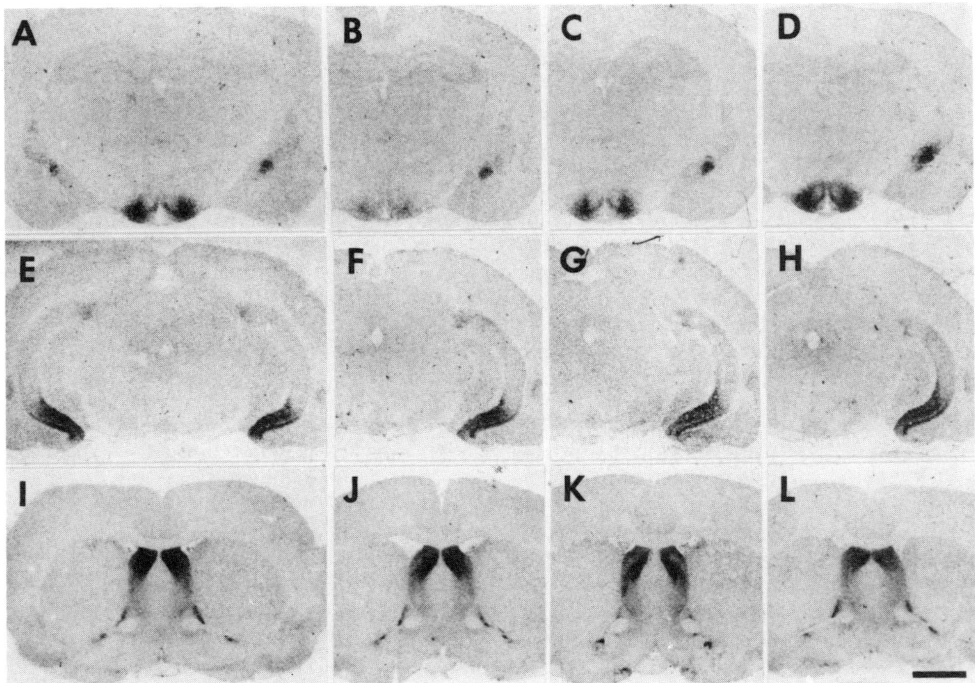

Fig. 3.   <u>Effect of castration and gonadal steroid hormones on OT and AVP binding in the rat brain</u>. Sections were obtained from 4 adult males: a normal rat (A, E and I), a castrated rat (B, F and J), a castrated rat treated one month with testosterone propionate (30μg/100g/day; C, G and K) and a castrated rat treated one month with oestradiol benzoate (10μg/100g/day; D, H and L). The upper two rows show [^{3}H]OT binding at the level of the ventromedial hypothalamus (A-D) and of the ventral subiculum (E-F), respectively. [^{3}H]AVP binding in septum is shown in the lower row (I-L). One month after castration, [^{3}H]OT binding sites had virtually disappeared in the ventromedial nucleus (B), unless the animals were treated with testosterone (C) or oestradiol (D). The density of binding was even higher in the oestradiol-treated castrate (D) than in the control rat (A). On the contrary, [^{3}H]OT binding sites present in the ventral subiculum were neither sensitive to castration (F) nor to hormonal treatment (G and H). Similarly, the density of [^{3}H]AVP binding sites in the septal area was similar in normal (I), castrated (J) and castrated-treated animals (K and L). Calibration bars = 2.5 mm.

EFFECT OF CASTRATION AND OF GONADAL STEROID HORMONES ON OXYTOCIN BINDING SITES IN THE RAT BRAIN

No obvious differences in the distribution and in the density of binding sites were detected between the brains of male and female rats.

In both sexes, castration caused a marked reduction of [^{3}H]OT binding in the olfactory tubercle (islands of Calleja and other cell groups), in the bed nucleus of the stria terminalis and in the hypothalamic ventromedial nucleus (Fig 3 B), and caused a partial reduction in the central amygdaloid nucleus. In contrast, labelling remained unchanged in the anterior olfactory nucleus and in the ventral subiculum (Fig. 3 F). The effects of castration on OT binding sites could be counteracted by treating the animals with either testosterone or oestradiol (Fig. 3 C and D). On the contrary, AVP binding sites were not influenced by either castration or gonadal steroid hormone levels (Fig. 3 I-L).

These results indicate that the synthesis of OT binding sites is dependent upon gonadal steroid hormones as has also been suggested by De Kloet et al., (1986). Those OT binding sites which disappear after castration may be part of neural circuits involved in reproductive functions. This hypothesis is supported by functional studies showing that centrally acting oxytocin facilitates sexual receptivity (Caldwell et al., 1986) and maternal behaviour (Fahrbach et al., 1985) in oestrogen-treated, female rats.

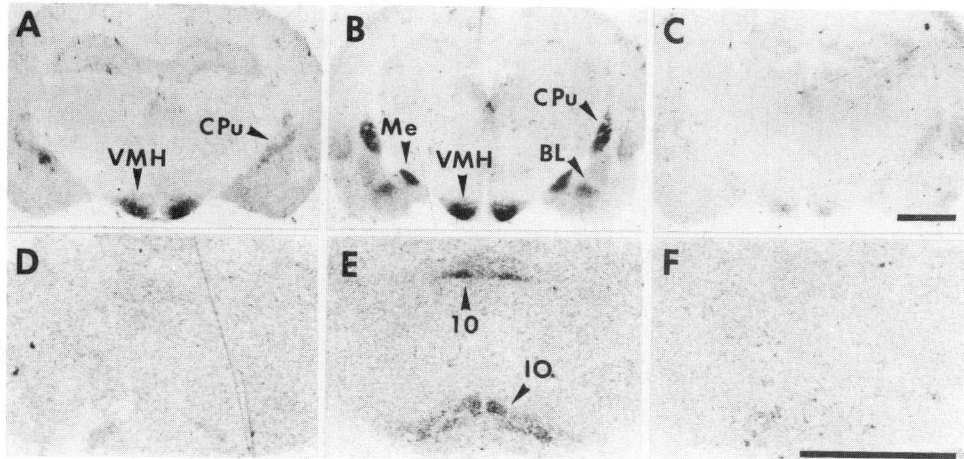

Fig. 4.   Comparison between [^{3}H]OT and a new iodinated OT receptor ligand. Autoradiographs are shown from sections at the level of hypothalamus (A-C) and of the brainstem (D-F). A and D were obtained, after 3 months of exposure, following incubation with 3 nM [^{3}H]OT. B and E were obtained after 3 days of exposure from sections incubated with 0.04 nM [^{125}I]-labelled d(CH$_2$)$_5$[Tyr(Me)^2,Thr^4,Tyr-NH$_2$^{9}]OVT. The ventromedial hypothalamic nucleus (VMH) was labelled with both ligands, while the caudoputamen (CPu) as well as the medial (Me) and basolateral (BL) amygdaloid nuclei were clearly evidenced only with the iodinated ligand. The latter labelled the dorsal motor nucleus of the vagus nerve (10) and the inferior olive (IO), while [^{3}H]OT did not (D). C and F show that the [^{125}I]-labelled ligand revealed specific OT binding sites, as label was displaced in the presence of 50 nM unlabelled OT. Calibration bars for A-C and D-F = 2.5 mm.

AN IODINATED LIGAND OF THE OXYTOCIN RECEPTOR : COMPARISON WITH [^{3}H]OT

Compared to [^{3}H]OT, a recently synthetized oxytocin analogue, [125 I]-labelled d(CH$_2$)$_5$[Tyr(Me)^2,Thr^4,Tyr-NH$_2$^{9}]OVT, has been shown to possess several advantages: its affinity for the uterine OT receptor is 440 times higher than its affinity for the V$_1$ vasopressin receptor and 330 times higher than its affinity for the V$_2$ vasopressin receptor (Elands et al., 1987). In the hippocampus, where both AVP and OT binding sites are present, the iodinated ligand binds only to the OT receptor, and with a dissociation constant close to that found in the uterus. Moreover, its affinity for synaptic membranes of hippocampus (Kd = 0.05 nM) is 40 fold higher than the affinity of [^{3}H]OT for the same preparation (Kd = 2.0 nM).

When this new ligand was used, 3 days only of exposure time gave a labelling which was as intense as that obtained with [^{3}H]OT after 3 or 4 months of exposure. In addition, as illustrated in Figure 4, due to the higher affinity, resulting in a more favourable ratio of specific/non specific binding of the iodinated ligand, additional structures could be detected which were either very faintly or not labelled by [^{3}H]OT.

CONCLUDING REMARKS

The demonstration that receptors specific for oxytocin and vasopressin are present in several areas of the brain strongly substantiates the concept that these peptides may act as central neurotransmitters. Other criteria in favour of this view include their calcium-dependent release and their electrophysiological action within the central nervous system (see Introduction). However, the overall functional significance of central neurohypophyseal peptides remains an open question. One way to come to tackle the problem is to study the long term changes in the distribution and density of AVP and OT receptors under various physiological or experimental conditions. The effect of castration illustrates the suitability of in vitro autoradiography for this kind of approach. Such studies will be much facilitated in the future by the availability of new radioactive ligands possessing a high specific activity, a marked affinity as well as an enhanced selectivity for the OT receptor (Elands et al., 1987) and for the V$_1$ vasopressin receptor.

ACKNOWLEDGMENT

This work was supported in part by the Swiss National Science Foundation (Grant 3.358.0.86).

REFERENCES

Audigier, S. and Barberis, C., 1985, Pharmacological characterization of two specific binding sites for neurohypophyseal hormones in hippocampal synaptic membranes of the rat, EMBO J., 4:665.

Buijs, R.M., 1983, Vasopressin and oxytocin. Their role in neurotransmission, Pharmacol. Ther., 22:127.

Caldwell, J.D., Prange, A.J., Jr., and Petersen C. A., 1986, Oxytocin facilitates the sexual receptivity of estrogen-treated female rats, Neuropeptides, 7:175.

De Kloet, E. R., Rotteveel, F., Voorhuis, Th. D. and Terlou, M., 1985, Topography of binding sites for neurohypophyseal hormones in rat brain, Eur. J. Pharmacol., 110:113.

De Kloet, E. R. , Voorhuis, Th. D., Boschma, Y. and Elands, J., 1986, Estradiol modulates density of putative 'oxytocin receptors' in discrete rat brain regions, Neuroendocrinology, 44:415.

Elands, J., Barberis, C., Jard, S., Tribollet, E., Dreifuss, J. J., Bankowski, K., Manning, M. and Sawyer, W. H., 1987, [^{125}I]-labelled d(CH$_2$)$_5$[Tyr(Me)^2,Thr^4,Tyr-NH$_2$^{9}]OVT: a selective OT receptor ligand, Eur. J. Pharmacol., (In press).

Fahrbach, S. E., Morell, J. L. and Pfaff, D. W., 1985, Possible role for endogenous oxytocin in estrogen-facilitated maternal behavior in rats. Neuroendocrinology, 40:526.

Freund-Mercier, M. J., Stoeckel, M. E., Palacios, J. M., Pazos, A., Reichart, J. M., Porte, A. and Richard, Ph., 1987, Pharmacological characteristics and anatomical distribution of [^{3}H]oxytocin-binding sites in the Wistar rat brain studied by autoradiography, Neuroscience, 20:599.

Jard, S., 1983, Vasopressin isoreceptors in mammals: relations to cyclic AMP-dependent and cyclic AMP-independent mechanisms, in: A. Kleinzeller (ed.), Membrane Receptors, Current Topics in Membranes and Transport, Vol. 18, Academic Press, New York, pp. 225-285.

Jard, S., Barberis, C., Audigier, S. and Tribollet, E., 1987, Neurohypophyseal hormone receptor systems in brain and periphery, in E. R. De Kloet, V. M. Wiegart and D. de Wied (eds.), Neuropeptides and Brain Function, Progress in Brain Research, Vol. 72, Elsevier, Amsterdam, pp. 173-187.

Kovacs, G. L., Szabo, Gy., Sastryal, Z. and Telegdy Gy., 1987, Neurohypophyseal hormones and behavior, in: E. R. De Kloet, V. M. Wiegart and D. de Wied (eds.), Neuropeptides and Brain Function, Progress in Brain Research, Vol. 72, Elsevier, Amsterdam, pp. 109-118.

Mühlethaler, M., Raggenbass, M. and Dreifuss, J. J., 1985, Oxytocin and vasopressin. in: M. A. Rogawski and J. L. Barker (eds.), Neurotransmitter Actions in the Vertebrate Nervous System, Raven Press, New York, pp. 431-450.

Raggenbass, M., Dubois-Dauphin, M., Charpak, S. and Dreifuss, J. J., 1987, Neurons in the dorsal motor nucleus of the vagus nerve are excited by oxytocin in the rat but not in the guinea-pig, Proc. Natl. Acad. Sci. USA., 84:3926.

Sofroniew, M. V., 1985, Vasopressin, oxytocin and their related neurophysins. in:A. Björklund and T. Hökfelt (eds.), GABA and Neuropeptides in the CNS, Handbook of Chemical Neuroanatomy, Vol. 4, Elsevier, Amsterdam, pp. 93-165.

Tribollet, E., Barberis, C., Jard, S., Dubois-Dauphin, M. and Dreifuss, J. J., 1987, Localization and pharmacological characterization of high affinity binding sites for vasopressin and oxytocin in the rat brain by light microscopic autoradiography, Brain Research, (In press).

Van Leeuwen, F. W., 1987, Vasopressin receptors in the brain and pituitary. in: D. M. Gash and G. J. Boer (eds.), Vasopressin: Principles and Properties, Plenum Press, New York, pp. 477-496.

SOMATOSTATIN AND NEUROPEPTIDE Y: COEXISTENCE IN THE HIPPOCAMPUS AND ALTERATIONS IN ALZHEIMER'S DISEASE

Victoria Chan-Palay

Neurology Clinic, University Hospital
Frauenklinikstrasse 26
8091 Zurich
Switzerland

INTRODUCTION

Alzheimer's type dementia (ATD) is characterized clinically by dementia, and histologically by the presence of numerous neuritic or senile plaques and neurofibrillary tangles in the neocortex and hippocampus (Tomlinson et al., 1970; Tomlinson and Corsellis, 1984). The degree of dementia is reported to be positively correlated with the number of cortical plaques (Blessed et al., 1968; Perry et al., 1978) and cortical and hippocampal neurofibrillary tangles (Wilcock and Esiri, 1982).

Of the numerous neurotransmitter systems studied in postmortem brains of patients with ATD, reduction of choline acetyl transferase (CAT) activity have been most consistently found (Bowen et al., 1976; Davies and Maloney, 1976; Perry et al., 1977; Rossor et al., 1982). This decrease of cortical and hippocampal cholinergic activity has been attributed to a loss of cholinergic innervation from the basal forebrain, the septal nuclei and the nucleus of Meynert (Whitehouse et al., 1982; Coyle et al., 1983; Rossor et al., 1982; Henke and Lang, 1983; McGeer et al., 1984). In addition to these well documented cholinergic abnormalities, noradrenergic and serotonin deficits (Hardy et al., 1985; Gottfries et al., 1983; Carlsson et al., 1980; Cross et al., 1983) have also been reported, indicating degeneration of projections from the locus coeruleus and the raphe nuclei. More recently, reductions of somatostatin concentrations have been demonstrated in the cerebral cortex (Hardy et al., 1985; Davies et al., 1980; Ferrier et al., 1983; Davies and Terry, 1981; Rossor et al., 1982) and somatostatin immunoreactive axons have been shown to participate in the formation of neuritic plaques (Chan-Palay, 1987; Armstrong et al., 1985; Morrison et al., 1985; Nakamura and Vincent, 1986) and neurofibrillary tangles (Roberts et al., 1985). However, reports of somatostatin reduction or loss in the hippocampus are less consistent; some studies report no change in ATD compared to normals (Ferrier et al., 1983) and others a dramatic reduction in the ATD condition (Davies et al., 1980; Chan-Palay, 1987).

Another peptide, Neuropeptide Y (NPY) has also been studied in relation to possible changes in ATD. NPY concentrations are unchanged or even increased in the substantia innominata (Allen, et al., 1984). Studies in the cerebral cortex of ATD patients indicated that the greatest severity of NPY loss parallels histopathological abnormalities shown by plaques and tangles. Of the areas studied, the temporal cortex was most severely involved (Chan-Palay et al., 1985a,b). NPY axons contribute to neuritic plaque formations (Dawbarn et al., 1984; Chan-Palay et al., 1985b). The hippocampal regions from the brains of ATD patients have been shown (Chan-Palay et al., 1986a,b) to have afflictions in NPY-immunoreactive (NPY-i) networks in the hilus, CA_1, the parasubiculum and entorhinal cortex. Parallel semi-quantitative estimates made of the numbers of neuritic plaques and neurofibrilllary tangles in the other hippocampi of the brains in every ATD case showed that the areas of heaviest pathological change by these indices are CA_1 and the entorhinal cortex, the subicular complex CA$_3$ and the area dentata being less affected. NPY-i axons were shown to participate in neuritic plaques. Thus areas with the most severe loss of NPY-i neurons and axons, CA$_1$ and the entorhinal cortex, are synonymous with those areas most severely affected by the other indices of ATD, thus confirming that NPY-i networks are involved in the ATD disease process. However, other NPY-i neurons survive, in some subfields better than others. The cumulative evidence suggests a population of hippocampal peptide neurons that are remarkably resistant in terminal neurological disease. The present study aims to clarify the changes encountered in somatostatin neurons in severe ATD and to compare them with the changes of NPY cells and fibers.

Studies of the possible coexistence of neuropeptides in the forebrain and cortex of animals and humans show that NPY and somatostatin (SOM) may be co-localized (Hendry et al., 1984; Köhler et al., 1986c; Chan-Palay et al., 1986c; Chan-Palay, 1987). We include in this paper studies using double label immunocytochemistry to demonstrate that SOM-i and NPY-i coexist in the hippocampal neurons, and the extent and regional differences inherent in this peptide-peptide coexistence.

The hippocampal region retains its reputation for the presence of an extensive and neurochemically varied neuronal population and extrinsic and intrinsic projections. There is the well documented catecholamine, serotonin, and cholinergic input from brain stem (Lewis et al., 1967; Fuxe, 1965; Köhler et al., 1981; Lidov, 1980; Swanson and Hartman, 1975) and septal nuclei (Alonso and Köhler, 1984; Amaral and Kurz, 1985; Armstrong et al., 1983; Oderfeld-Nowak et al., 1974; Shute and Lewis, 1967; Mellgren and Srebro, 1973). Neuropeptides are also well represented with (a) cholesystokinin octapeptide (CCK-8, Greenwood et al., 1982; Köhler and Chan-Palay, 1982; Stengaard-Pederson et al., 1983; Loren et al., 1979; Roberts, 1984), (b) corticotropin releasing factor (CRF, Swanson et al., 1983), (c) somatostatin (SOM, Finley et al., 1981; Köhler and Chan-Palay, 1982, 1983; Morrison et al., 1982; Roberts et al., 1984; Bakst et al., 1985; Vincent et al., 1985; Chan-Palay, 1987), (d) vasoactive intestinal polypeptide (VIP, Köhler, 1983; Köhler and Chan-Palay, 1983; Loren et al., 1979; Roberts et al., 1984, (e) Substance P (Davies and Köhler, 1986; Gall and Selawski, 1984), (f) dynorphin-A$_{1-17}$ (McGinty et al., 1983; Charkin et al., 1985), (g) alpha-melanocyte stimulating hormone (Köhler et al., 1984) and (h) NPY in extensive locations in the hippocampus of rats and monkeys (Köhler et al., 1986b) and for man (Chan-Palay et al., 1986a,b).

Antibodies against somatostatin (SOM) have made it possible to localize neurons containing this peptide possibly using immunohistochemical techniques. Numerous studies detail the presence of SOM neurons and their pathways in the brain (see review Nieuwenhuys, 1985). The present study focuses on the distribution of somatostatin in the neurons of the human hippocampus and temporal cortex by using immunocytochemical methods, several anti-somatostatin antibodies, and the immunogold silver method (Chan-Palay, 1987).

Double Labelling: SOM-i:NPY-i Coexistence

Cryostat sections, 10μm thick or vibratome sections not more than 20μm thick were prepared as pairs of sections. Strict attention was paid to keep track of the adjacent surfaces between the two sections. These exposures of the two adjacent surfaces were achieved by inverting one of the sections. For studying coexistence in the light microscope, cells lying in the plane of section could be cut into two parts, one part remaining within the exposed surface of the first section face, and the other part in the second exposed face. Other cells could remain unsectioned in one or the other face, of course. Each section was then treated separately but simultaneously, in individual wells with either a somatostatin or NPY antiserum in order to examine the coexistence of these two immunoreactivities in neurons. The following procedures were used. Pairs of sections were floated onto 1 % hydrogen peroxide in phosphate buffered saline for 2 h, in order to reduce or eliminate endogenous peroxidase reactions, and then reacted sequentially with the primary antibody (1:100 NPY or 1.1500 somatostatin) overnight, goat anti-rabbit antibodies, peroxidase and diaminobezidine (DAB) as previously described (Chan-Palay et al., 1985a,b, 1986a,b; Chan-Palay and Yasargil, 1986). Alternatively, the immunogold/silver reaction (Chan-Palay, 1987) was also used on paired sections with one antibody on each of the pairs. Or, one of a pair of sections was treated with the immunogold/silver method and the other with the peroxidase anti-peroxidase method. In another group of double label experiments, pairs of sections were simultaneously incubated with both antibodies on the same section in a mixture of the polyclonal NPY (1:1000) and the monoclonal somatostatin (1:1500) antibodies. Subsequently fluorescein (FITC) conjugated with goat anti-rabbit immunoglobulins and rhodamine conjugated anti-mouse immunoglobulins were applied. The preparations were mounted in a glycerine-buffer mixture (1.3) for examination in a bright field and fluorescence microscope (see Chan-Palay, 1987).

Mapping

In order to localize the neuronal elements accurately in the different hippocampal regions, maps were made of all the specimens, charting the lamination of the regions in relation to stained neurons. Identified SOM-i neurons were counted with attention to their specific location in the hippocampal layers, regions and the major fiber bundles. Because of the difficulty of maintaining consistent section thickness these counts were considered only semiquantitative. Black and white and colour photomicrographs were made on a Zeiss photomicroscope fitted with Nomarski optics.

Comparisons Between the Distribution of SOM-i and NPY-i Neurons in the Hippocampus and Temporal Cortex

SOM-i neurons are most abundant in the hilus of the area dentata, in strata oriens and pyramidale in the CA_1 subfields, and in the deeper layers of the entorhinal cortex (II through VI and white matter) and the perirhinal cortex. Small fusiform neurons occur in the white matter, fimbria, alveus and angular bundle, and white matter in the hippocampus, retrohippocampus and temporal cortex. NPY-i neurons are most abundant in the area dentata and the strata oriens and pyramidale of CA_1. The subicular complex, entorhinal cortex and the temporal cortex have numerous NPY-i neurons. Most prominent are the small fusiform neurons of the white matter in fimbria, alveus and the angular bundle (Chan-Palay et al., 1986a,b). The distributions of these two types of peptide immunoreactivities are remarkably similar, and the superimposition of the two maps compiled from the rostral hippocampus, one showing NPY-i neurons and the other showing SOM-i neurons support this impression. The major similarities of the two groups are those of like cell morphology and cell size and parallel locations of neuron distribution. The fact that neither cell population includes pyramidal neurons in our material is also relevant.

Several minor differences exist as well between these two neuronal immunoreactivities. The SOM-i neuron population by and large in every region of hippocampus and cortex is more numerous than the NPY-i population. This is especially evident in the hilar regions and in the entorhinal and temporal cortex. However, the reverse is true of the small, fusiform neurons in the white matter. Although common to both, these cells have a distinctly higher incidence in NPY-i material.

The parallel distribution of SOM-i and NPY-i cells raises the question of whether these neurons are separated neuron populations, or whether SOM-i and NPY-i coexist in some neurons. If SOM-i/NPY-i coexistence occurs, then to what extent and what are the differences in their regional distributions?

The subfields with the most severe loss in SOM-i and NPY-i neurons are the hilus of the area dentata, CA_1 and the entorhinal, perirhinal and temporal cortices although somatostatin loss is more severe than the NPY. The area with the highest densities of plaques and tangles are the hilus of the are dentata, CA_1 and the entorhinal and temporal cortices. The numbers of neuritic plaques and neurofibrillary tangles reported in these cases are in accordance with the guidelines recently established for the diagnosis of ATD (Katchaturian, 1985), and CA_1 and the entorhinal cortex are known to be sites of predilection for ATD morphological changes (Tomlinson and Corsellis, 1984; Ball, 1978; Hooper and Vogel, 1976; Wilcock and Esri, 1982; Hyman et al., 1984).

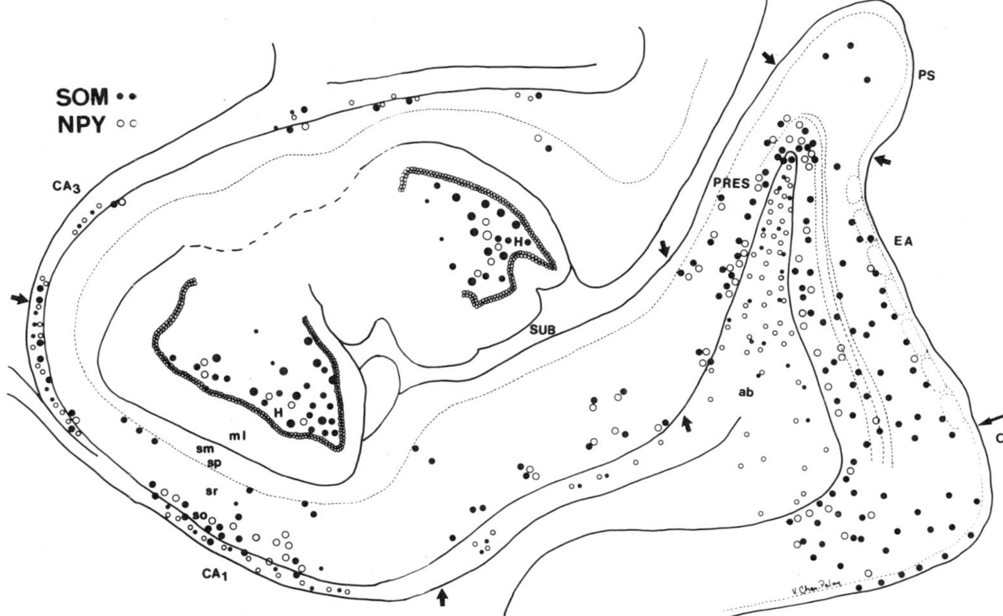

Fig. 1. Schematic representation of the rostral hippocampus with sites of SOM-i neurons (black dots in small, medium and large sizes) and NPY-i neurons (open circles) superimposed. Both peptides have comparable distributions and exhibit a significant proportion of SOM-i/NPY-i coexistence in the various hippocampal subfields. Both peptides are present in highest amounts in the area dentata, CA_1 and in the entorhinal and perirhinal cortices. <u>List of Abbreviations</u> AD - area dentata; C - perirhinal cortex; CA1, CA3 - subfields of Ammon's horn; EA - entorhinal area; H - hilus of area dentata; PRES - presubiculum; PS - parasubiculum; SUB - subiculum; ab - angular bundle; f - fimbria; ml - molecular layer of the area dentata; gl - granular layer of the area dentata; sm - stratum molecular; so - stratum oriens; sp - stratum pyramidale; sr - stratum radiatum.

Thus, the present study clearly indicates that in severe ATD, the predilection sites for highest incidence of neuritic plaques and neurofibrillary tangles are the sites of the severest loss of SOM-i neurons and networks. These regions are also the sites of the most severe loss and alterations in NPY-i neurons and networks (Chan-Palay et al., 1985a,b; 1986a,b,d), implicating both peptide neuron systems in the disease process.

Coexistence Between SOM-i and NPY-i in Single Neurons

The surgical biopsy material consistently provided the most reliable tissues for the double labelling experiments to demonstrate coexistence of SOM-i and NPY-i in neurons, compared to human postmortem material, regardless of postmortem delay or conditions of perfusion and fixation. The major factor was the clearly superior preservation and fixation consistently achievable. Vast numbers of axons are present and dendrites are fully stained. This experience corroborates similar observations from the demonstration of NPY-i in human cortical tissue in surgical and postmortem tissues (Chan-Palay and Yasargil, 1986). The parallel distributions of SOM-i and NPY-i neurons (Fig. 1) underscore the impression that these two peptides occur in similar neuronal populations in comparable subfields of the hippocampus and cortex. The results of the double labelling experiments demonstrate that about 30 % of the SOM-i neurons in the hilus of the area dentata are also NPY-i. However, since numerically, more SOM-i neurons are visualised by our methods than NPY-i neurons, the coexistence population constitutes approximately 50 % of all NPY-i neurons. NPY-i coexistence was found in several SOM-i neurons in CA_1, the dentate hilus and particularly

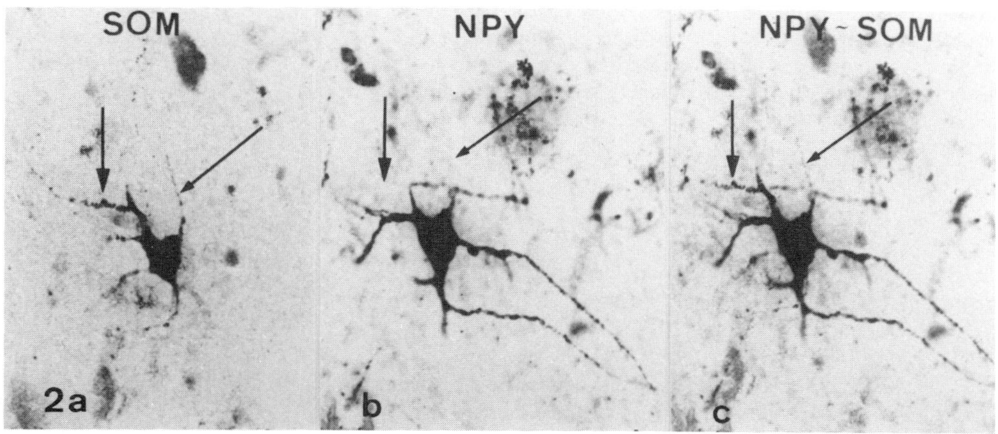

Fig. 2.   Coexistence of NPY-i and SOM-i in a single large multipolar neuron of the stratum oriens at the CA_1/subicular border. SOM-i by the immunogold/silver method and NPY-i is demonstrated by the peroxidase anti-peroxidase preparation method (b) on the adjacent section. The labelled neuron shown in these 2 figures comes from the adjacent opposing faces of 2 sections. The two faces are superimposed (c) x500. Note the fit of the dendritic branches indicated by 2 arrows with the blank spaces in the alternate profile (b), in the superimposed composite (c).

in the subicular complex. The highest coexistence coefficients were found in the deep layers of the entorhinal and perirhinal cortices where more than 50 % of the SOM-i neurons were NPY-i as well. Large neurons of the retrohippocampal areas and small neurons in the white matter were particularly noted as frequent examples of NPY-i and SOM-i coexistence.

In the ATD cases, both NPY-i and SOM-i in neurons and axons were seriously affected with a numerically higher survival of NPY-i neurons and axons than SOM-i neurons and axons in the cortical areas, and equally severe destruction of both peptide neurons in the area dentata and CA_1. Neurons with coexistence of NPY-i and SOM-i appear not to be better protected against ATD than the neurons with SOM-i alone, and remain or are destroyed in the same proportion as those with SOM-i alone. Because of the techniques that we use for the demonstrations of coexistence, we anticipate that our estimates are generally lower than might otherwise be found. Thus, the numerical estimates given here can be considered minimal.

The present study has also provided fundamental information on the phenomenon of peptide-peptide coexistence in human neurons. The extent of this distribution and the regional differences present in the various hippocampal subfields is also discussed. The importance of this observation is the fact that populations of SOM-i/NPY-i neurons exist in considerable numbers. Neurons with SOM-i and without NPY-i also exist as do their opposite counterparts. There is equally severe destruction of both peptide neuron groups in the area dentata and CA_1; although a higher incidence of NPY-i neurons and networks survive in the cortices compared to SOM-i neurons.

One would expect that this loss of a peptidergic neuron network would be important to the functioning of the hippocampal/limbic circuits perhaps as reflected in the severely decompensated state of these patients typically. Nevertheless a more significant observation may be that a considerable NPY-i intrinsic network of cells and fibers continues to exist into the terminal state at exitus. This unusual survivability points to the fact that these peptide neurons have participated in the maintenance of the neuronal networks required for whatever minimal function remains in the target area of the terminally diseased brain. As such they hold important links for our understanding of synaptic plasticity in neural reorganization and repair. This observation is true for NPY-i neuron networks not only in the hippocampus but in the cerebral cortex in ATD (Chan-Palay et al., 1985a,b), in the striatum in Huntington's Disease (Allen et al., 1984; Dawbarn et al., 1985), and in the region directly bordering on brain infarcts (Chan-Palay and Yasargil, 1986). The cumulative evidence suggests an NPY population throughout the hippocampus remarkably resistant to Alzheimer changes and a particularly susceptible somatostatin one.

ACKNOWLEDGEMENTS

This work was supported in part by United States AFOSR grant 86-0176.

REFERENCES

Allen, J.M., Fernier, I.N., Roberts, G.W., Cross, A.J., Adrian, T.E., Crow, T.J. and Bloom, S.R., 1984, Elevation of neuropeptide Y in substantia innominata in Alzheimer's type dementia, J. Neurol. Sci., 64:325.

Alonso, A. and Köhler, C., 1984, A study of the reciprocal connections between the sepotum and the entorhinal area using anterograde and retrograde axonal transport methods in the rat brain, J. Comp. Neurol., 225:327.

Amaral, D.G. and Kurz, J., 1985, An analysis of the origins of the cholinergtic and noncholinergic septal projections to the hippocampal formation of the rat, J. Comp. Neurol., 240:37.

Armstrong, D.M., Saper, C.B., Levey, A.I., Wainer, B.M. and Terry, R.D., 1983, Distribution of cholinergic neurons in rat brain demonstrated by the immunocytochemical localization of cholineacetyltransferase, J. Comp. Neurol., 216:53.

Armstrong, D.M., LeRoy, S., Shields, D. and Terry, R.D., 1985, Somatostatin like immunoreactivities within neuritic plaques, Brain Res., 338:71.

Bakst, I., Morrison, J.H. and Amaral, D.G., 1985, The distribution of somatostatin-like immunoreactivity in the monkey hippocampal formation, J. Comp. Neurol., 236:423.

Ball, M.J., 1978, Topographic distribution of neurofibrillary tangles and granovacuolar degeneration in hippocampal cortex of aging and demented patients, Acta Neuropath., 42:73.

Blessed, G., Tomlinson, B.E. and Roth, M., 1968, The association between quantitative measurements of dementia and senile changes in the cerebral grey matter of elderly subjects, Brit. J. Psychiat., 114:797.

Bowen, D.M., Smith, C.B., White, B. and Davison, A.N., 1976, Neurotransmitter related enzymes and indices of hypoxia and senile dementia and other abiotrophies, Brain, 99:459.

Braak, H., 1974, On the Structure of the Human Archicortex, Cell Tiss. Res., 152:349.

Carlsson, A., Adolfsson, R., Aquilonius, S.M., Gottfries, C., Oreland, L., Svennerholm, L. and Winblad, B., 1980, Biogenic amines in human brain in normal aging, senile dementia and chronic alcoholism, in: Goldstein, Ergot compounds and brain functions: Neuroendocrine and neuropsychiatric aspects, 295-304, Raven Press, New York.

Chan-Palay, V., 1987, Somatostatin immunoreactive neurons in the human hippocampus and cortex shown by immunogold/silver intensification on vibratome section: Coexistence with neuropeptide Y neurons, and effects in Alzheimer's type dementia, J. Comp. Neurol., in press.

Chan-Palay, V. and Yasargil, G., 1986, Immunocytochemistry of human brain tissue with a polyclonal antiserum against neuropeptide Y, Anatomy and Embryol., 174:27.

Chan-Palay, V., Allen, Y.S., Lang, W., Haesler, U. and Polak, J.M., 1985a, Cytology and distribution in normal human cerebral cortex of neurons immunoreactive with antisera against neuropeptide Y, J. Comp. Neurol., 238:382.

Chan-Palay, V., Lang, W., Allen, Y.S., Haesler, U. and Polak, J.M., 1985b, II. Corti neurons immunoreactive with antisera against neuropeptide Y are altered in Alzheimer's-type dementia, J. Comp. Neurol., 238:382.

Chan-Palay, V., Köhler, C., Haesler, U., Lang, W. and Yasargil, G., 1986a, Distribution of neurons and axons immunoreactive with antisera against neuropeptide Y in the normal human hippocampus, J. Comp. Neurol., 248:360.

Chan-Palay, V., Lang, W. Haesler, U., Köhler, C. and Yasargil, G., 1986b, Distribution of altered hippocampal neurons and axons immunoreactive with antisera against neuropeptide Y in Alzheimer's type dementia, J. Comp. Neurol., 248:376.

Charkin, C., Shoemaker, W.J., McCinty, J.F., Bayon, A. and Bloom, F.E., 1985, Characterization of the prodynorphin and proenkephalin neuropeptide systems in rat hippocampus, J. Neurosci., 5:808.

Coyle, J.T., Price, D.L. and Delong, M.R., 1983, Alzheimer's disease: A disorder of cortical cholinergic innervation, Science, 219:1184.

Cross, A.J., Crow, T.J., Johnson, J.A., Perry, E.K., Blessed, G. and Tomlinson, B.E., 1983, Monoamine metabolism in senile dementia of Alzheimer's type, J. Neurol. Sci., 60:383.

Davies, P. and Maloney, A.J., 1981, Selective loss of central cholinergic neurons in Alzheimer's disease, Lancet, ii:1043.

Davies, P. and Terry, R.D., 1981, Cortical somatostatin-like immunoreactivity in cases of Alzheimer's disease and senile dementia of Alzheimer type, Neurobiol. Aging, 2:9.

Davies, P., Katzmann, R., and Terry, R.D., 1980, Reduced somatostatin-like immunoreactivity in cases of Alzheimer's disease and Alzheimer senile dementia, Nature, 288:279.

Davies, S. and Köhler, C., 1985, The substance P innervation of the hippocampus in the rat, Anat. and Embryol., 173:45.

Dawbarn D., Hunt, S.P. and Emson, P.C., 1984, Neuropeptide Y: Regional distribution, chromatographic characterization and immunohistochemical demonstration in post mortem human brain, Brain Res. 296:168.

Ferrier, I.N., Cross, J.A., Johnson, J.A., Robers, G.W., Crow, T.J., Carsellis, J.A.N., Lee, Y.C., O'Shangoressey, A.T.E., McGregor, G.P., Bacarese-Hamilton, A.J. and Bloom, S.R., 1983, Neuropeptides in Alzheimer type dementia, J. Neurol. Sci., 62:152.

Finley, J.C.W., Maderdrut, J.L., Roger, L.J. and Petrusz, P., 1981, The immunocytochemical localization of somatostatin-containing neurons in the rat CNS, Neurosci., 6:2173.

Fuxe, K., 1965, Evidence for the existence of monoamine terminals in the central nervous system, Acta Physiol. Scand., 64:39.

Gall, C. and Selawski, L., 1984, Supramammillary efferents to guinea pig hippocampus contain substance P-like immunoreactivity, Neurosci. Lett., 51:171.

Gottfries, C.G., Adolfsson, R., Aquilonius, S.M., Carlsson, A., Ecernas, S.A., Nordberg, A., Oveland, L., Svennerholm, L., Wiberg, A. and Winblad, B., 1983, Biochemical changes in dementia disorders of Alzheimer type (AD/SDAT), Neurobiol. Aging, 4:261.

Greenwood, R.S., Godar, S., Reaves, T.A. and Hayward, J.N., 1982, Cholecystokinin in hippocampal pathways, J. Comp. Neurol., 203:335.

Hardy, J., Adolfsson, R., Alafusoff, I., Bucht,m G., Marcosson, J., Nyberg, P., Perdahl, E., Wester, P. and Winblad, P., 1985, Transmitter deficits in Alzheimer's disease, Neurochem. Int., 7:545.

Hendry, S.H.C., Jones, E.G. and Emson, P.C., 1984, Morphology, distribution and synaptic relations of somatostatin and neuropeptide Y-immunoreactive neurons in rat and monkey neocortex, J. of Neurosci., 4:2497.

Henke, H. and Lang, W., 1983, Cholinergic enzymes in neocortex, hippocampus and basal forebrain of non-neurological and senile dementia of Alzheimer type patients, Brain Res., 267:281.

Hooper, W.M. and Vogel, F.S., 1976, The limbic system in Alzheimer's disease, Amer. J. Path., 85:1.

Hyman, B.T., van Hoesen, G.W., Damasio, A.R. and Barnes, C.L., 1984, Alzheimer's disease: Cell specific pathology isolates the hippocampal formation, Science, 225:1168.

Katchaturian, Z.S., 1985, Diagnosis of Alzheimer's disease, Arch. Neurol., 42:1097.

Kemper, J., 1984, Neuroanatomical and neuropathological changes in nor aging and dementia, in: Albert, Clinical Neurology of Aging, 9-52, Oxford Univ. Press.

Köhler, C., 1983, A morphological analysis of vasoactive intestinal polypeptide (VIP)-like neurons in the area dentata of the rat brain, J. Comp. Neurol., 221:247.

Köhler, C., Chan-Palay, V., 1982, The distribution of cholecystokinin like immunoreactive neurons and nerve terminals in the retrohippocampal region in the rat and guinea pig, J. Comp. Neurol., 210:136.

Köhler, C. and Chan-Palay, V., 1982, Somatostatin-like immunoreactive neurons in the hippocampus: An immunocytochemical study in the rat, Anat. Embryol., 167:151.

Köhler, C. and Chan-Palay, V., 1983, Somatostatin and vasoactive intestinal polypeptide-like immunoreactive cells and terminals in the retrohippocampal region of the rat brain, Anat. Embryol., 167:151.

Köhler, C., Chan-Palay, V. and Steinbusch, H.W.M., 1982, The distribution and orientation of serotonin fibers in the entorhinal and other retrohippocampal areas: An immunohistochemical study with anti-serotonin antibodies in the rat's brain, Anat. Embryol., 161:237.

Köhler, C., Smialovska, M., Ericsson, L.G. and Chan-Palay, V., 1986b, Origin of neuropeptide Y in the rat retrohippocampal region, Neurosci. Lett., 65:287.

Köhler, C., Eriksson, L., Davies, S. and Chan-Palay, V., 1986c, NPY and somatostatin coexist in neurons of the rat hippocampus, in press.

Lewis, P.R., Shute, C.C.D. and Silver, A., 1967, Confirmation from choline acetyltransferase of a massive cholinergic innervation to the rat hippocampus, J. Physiol. Lond., 191:215.

Lidov, M.G.W., Grzanna, R. and Molliver, M.E., 1980, The serotonin innervation of the cerebral cortex. An immunohistochemical analysis, Neuroscience, 5:207.

Loren, I., Alumets, J., Hakanson, R. and Sundler, F., 1979, Immunoreactive pancreatic polypeptide (PP) occurs in the central and peripheral nervous system: Preliminary immunocytochemical observations, Cell Tiss. Res., 200:179.

McGeer, P.I., McGeer, E.G., Suzuki, J., Dolman, C.E. and Nagai, T., 1984, Aging, Alzheimer's disease and the cholinergic system of the basal forebrain, Neurology, 34:741.

McGinty, J.F., Hendriksen, S.J., Goldstein, A., Terenius, L. and Bloom, F.E., 1983, Dynorphin is contained within hippocampal mossy-fibers: Immunochemical alterations after kainic acid administration and colchicine induced neurotoxicity, Proc. Natl. Acad. Sci. USA, 80:589.

Mellgren, S.T. and Srebro, B., 1973, Changes in acetylcholinesterase and distribution of degenerating fibers in the hippocampal region after septal lesions in the rat, Brain Res., 52:19.

Morrison, J.H., Rogers, J., Scherr, S., Benoit, R. and Bloom, F.E., 1985, Somatostatin in neuritic plaques of Alzheimer's patients, Nature, 314:90.

Nakamura, S., Vincent, S.R., 1986, Somatostatin and neuropeptide Y-immunoreactive neurons in the neocortex in senile dementia of Alzheimer's type, Brain Res., 370:11.

Nieuwenhuys, R., 1985, Chemoarchitecture of the brain, 77 Springer Verlag, Stuttgart.

Oderfeld-Nowak, B., Narkiewicz, O., Bialowas, J., Dabrowska, J., Wieraszko, A. and Bradkowska, M., 1974, The influence of septal nuclei lesions on activity of acetylcholinesterase in the hippocampus of the rat, Acta Neurobiol. Exp., 34:787.

Perry, E.K., Perry, R.H., Blessed, G. and Tomlinson, B.E., 1977, Necropsy evidence of central cholinergic deficits in senile dementia, Lancet, i:189.

Perry, E.K., Tomlinson, B.E., Blessed, G., Bergman, K., Givson, P.H. and Perry, R.H., 1978, Correlation of cholinergic abnormalities with senile plaques and mental test scores in senile dementia, Brit. Med. J., II:1457.

Ramon y Cajal, 1911, Histologie du Systeme Nerveux de l'Homme et des Vértebrés, Paris, Norbert Maloine.

Roberts, G.W., Woodhams, P.L., Polak, J.M. and Crow, T.J., 1984, Distribution of neuropeptides in the limbic system of the rat. The hippocampus, Neurosci., 11:35.

Roberts, G.W., Crow, T.W. and Polak, J.M., 1985, Location of neuronal tangles in somatostatin neurons in Alzheimer's disease, Nature, 314:92.

Rosser, M.N., Svendsen, C., Hunt, S.P., Mountjoy, C.Q., Roth, M. and Iversen, L.L., 1982, The substantia innominata in Alzheimer's disease: A histochemical and biochemical study of cholinergic marker enzyme, Neurosci. Lett., 28:217.

Schaltenbrand, G. and Bailey, P., 1959, Introduction to stereotaxis with an atlas of the human brain. Vols. I-III, Georg Thieme Verlag, Stuttgart.

Shute, C.C.D. and Lewis, P.R., 1967, The ascending cholinergic reticular systems: Neocortical, olfactory, and subcortical projections, Brain, 90:497.

Stengaard-Pedersen, K., Fredens, K. and Larsson, L.I., 1983, Comparative localization of enkephalin and cholecystokinin immunoreactivities and heavy metals in the hippocampus, Brain Res., 27:381.

Swanson, L.W. and Hartman, B.K., 1975, The central adrenergic system. An immunofluorescence study of the localization of cell bodies and their efferent connections in the rat utilizing dopamine- -hydroxylase as a marker, J. Comp. Neurol., 163:487.

Swanson, L.W., Sawchenko, P.E., Rivier, J. and Vale, W.W., 1983, Organization of ovine corticotropin-releasing factor immunoreactive cells and fibers in the rat brain: An immunohistochemical study, Neuroendocrinology, 36:165.

Tomlinson, B.E. and Corsellis, J.A.N., 1984, Ageing and the dementias, in: Hume-Adams, Corsellis, Duchen, Greenfield's Neuropathology, 951-1025, Arnold, London.

Tomlinson, B.E., Blessed, G. and Roth, M., 1970, Observations on the brains of demented old people, J. Neurol. Sci., II:205.

Vincent, S.R., McIntosh, C.H.S., Buchan, A.M.J. and Brown, J.C., 1985, Central somatostatin revealed with monoclonal antibodies, J. Comp. Neurol., 238:169.

Whitehouse, P.J., Price, D.L., Struble, R.G., Clark, A.W. and DeLong, M.R., 1982, Alzheimer's disease and senile dementia: Loss of neurons in the central forebrain, Science, 215:1237.

Wilcock, G.K. and Esiri, M.M., 1982, Plaques, tangles and dementia: A quantitative study, J. Neurol. Sci., 56:343.

BIOACTIVE PEPTIDES AT THE NEUROMUSCULAR JUNCTION OF INSECTS

Michael O'Shea
Chair of Cell Biology
University of London
Royal Holloway and New Bedford College
Egham, Surrey
England

INTRODUCTION

In a symposium on neurosecretion it may seem inappropriate to discuss the physiology and pharmacology of the skeletal neuromuscular junction. It is becoming clear, however, that in both vertebrates and invertebrates active substances other than the primary neuromuscular transmitters can exert profound effects at the neuromuscular junction. Such substances may be secreted by the motoneurons themselves thereby acting as cotransmitters. They may also be secreted by specialized neuromodulatory neurons which innervate the muscle to regulate the actions of the classical motoneurons. Finally they may be humoral substances secreted at a distance and carried to the target muscle in the circulation. Thus the neuromuscular junction appears in some organisms, especially invertebrates, to be a site for the integration of various secreted chemical signals. Since it is also relatively accessible for physiological and biochemical studies, it can provide a model for understanding how multiple secreted products contribute to the control of physiological processes. In this brief essay I will describe how secreted peptides are involved in neuromuscular transmission in insects. There seem to be at least two known types of action, first as peptidergic cotransmitters and second as humoral peptides which may be released rather close to the target muscle or at a greater distance. Since more is known about peptidergic cotransmission, most of this paper will be devoted to this type of action.

TYPES OF NEURONS INVOLVED IN CONTROL OF INSECT MUSCLE

The classification of the types of neurons involved in the control of skeletal muscle in insects has in recent years become problematic. This is due in part to the recognition of a role for neuropeptides in neuromuscular physiology and by the realization that not all of the secretory cells which influence insect muscle are "true motoneurons". In fact a satisfactory definition of a "motoneuron" is, at least in insects, difficult to achieve and there is no consensus of opinion as to how this term should be used.

In contrast to the vertebrate neuromuscular system, we can identify in insects and some other invertebrates two basic types of motoneuron, excitatory and inhibitory. The excitatory motoneurons are thought to be glutaminergic and the inhibitory motoneurons are probably gaba-ergic. The excitatory motoneurons may be further classified into fast and slow according to the type of muscle fiber they innervate. These neurons have their somata on the ventral surface of central ganglia and send axon processes to form close junctions with specific muscle targets. The transmitters (glutamate and gaba) cause classical increased conductance transient junctional potentials, depolarizing for L-glu and usually hyperpolarizing for gaba. So, for the excitatory and inhibitory motoneurons there is little argument that the use of the term motoneuron is not appropriate (even though the concept of an inhibitory moto-neuron may be alien to the vertebrate physiologist).

Until the mid-1970's it was widely thought that the motoneuron pool in insects consisted only of excitatory and inhibitory motoneurons. In 1974, however, Hoyle et al., (1974) showed that skeletal muscle may also receive innervation from a type of neuron morphologically distinct from the excitatory and inhibitory motoneurons. The newly discovered class of putative "motoneuron" have their cell bodies on the dorsal surface of the central ganglia, forming a loose medial cell-cluster. Another morphological feature distinguishing them from previously described motoneurons is their bifurcated axons, such that each cell projects symmetrically to left and right pairs of muscles. By contrast the excitatory and inhibitory motoneurons are bilaterally paired, each member of the pair forming a contralateral mirror image of the other. The dorsal medial motoneurons are in this sense morphologically unpaired. In effect each cell embodies its own contralateral homologue in the form of its bilateral summetry. For the above reasons, Hoyle et al., (1974) decided to call these neurons Dorsal Unpaired Medial or DUM neurons (Figure 1).

What could be the physiological role of the DUM neurons? Because their stimulation resulted in neither direct muscle contraction nor inhibition of contraction, it seemed that they could be neither excitatory nor inhibitory motoneurons and therefore secreting neither glutamate nor gaba. By studying a single identified DUM neuron in the locust <u>Schistocerca gregaria</u>, the so-called DUMETi neuron (called DUMETi because it innervates the extensor tibialis or ETi muscle). Peter Evans and I (Evans and O'Shea, 1977, 1978; O'Shea and Evans, 1979) assigned a modulatory function to DUM cells and identified octopamine as their likely secretory product. The DUM cells can be classified as aminergic modulatory motoneurons. Probably the primary function of DUM motoneurons is to modulate the physiological actions of slow excitatory motoneurons. For example, when DUMETi is fired in a brief burst, subsequent twitch contractures induced by the slow extensor motoneuron (SETi) are increased in amplitude and their rate of relaxation is increased (Evans and Myers, 1986). These effects are produced by octopamine secreted by the DUM motoneuron.

To summarize, it appeared until recently that insect muscle was controlled by three types of motoneurons (excitatory, inhibitory and modulatory) which secrete respectively glutamate, gaba and octopamine.

This relatively simple view must now be elaborated further to accommodate the recently discovered role of secreted peptides in insect neuromuscular physiology (Figure 2).

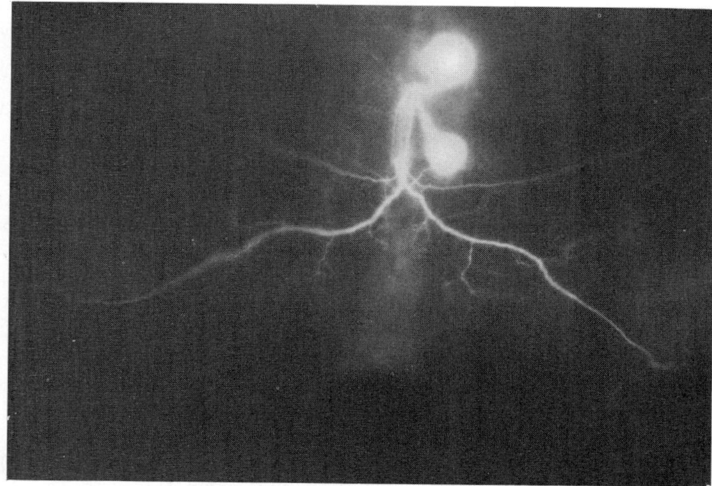

Fig. 1.   Two octopaminergic DUM neurons of the locust thoracic ganglion revealed by injection of lucifer yellow into each soma. Each gives rise to a bifurcated axon which innervate different bilateral pairs of skeletal muscles. The somata are about 60 μm diameter.

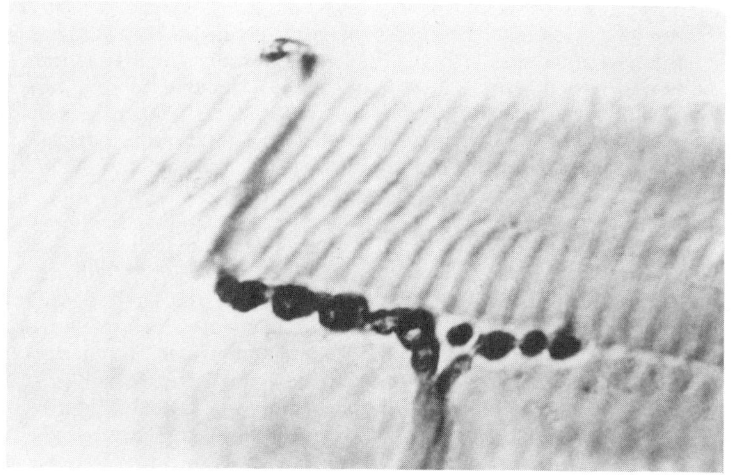

Fig. 2.   A peptidergic neuromuscular junction revealed by whole-mount immunocytochemistry using a proctolin antiserum. The nerve terminal shown here belongs to an identified skeletal motoneuron in the cockroach. This terminal employs two transmitters, glutamate and proctolin.

Neuropeptides in insects are now known to be delivered to skeletal muscle by some excitatory motoneurons and by neurosecretory cells which do not directly innervate muscle. We can define therefore at least two additional types of neurons which affect insect muscle. These are first glutaminergic excitatory motoneurons which are also peptidergic and second the neurons or neurosecretory cells which release peptides at sites too distant from the muscle to be defined as neuromuscular junctions. Little is known about this final class of non-motoneuronal muscle effector cells but there is growing evidence that it may be a large and diverse class exerting important influence on neuromuscular transmission and muscle performance.

The next part of this essay will be concerned with describing the roles of bioactive peptides in insect neuromuscular transmission.

PROCTOLIN: A NEUROMUSCULAR COTRANSMITTER

The only neuropeptide to be specifically associated with identified skeletal insect motoneurons is the pentapeptide proctolin (Arg-Tyr-Leu-Pro-Thr). It was discovered in the cockroach <u>Periplaneta americana</u> by B.E. Brown in 1967 (Brown, 1967). At that time it was referred to as "gut-factor" - a biologically active constituent of hindgut that could cause gut muscle contraction. In 1975 this factor was isolated in pure form from approximately 125,000 cockroaches (<u>Periplaneta americana</u>) and subjected to chemical analysis (Brown and Starratt, 1975; Starratt and Brown, 1975). Proctolin is widely distributed among the arthropods and there is some evidence for a proctolin-like peptide in non-arthropod invertebrates and in vertebrates (see O'Shea and Schaffer, 1985; O'Shea and Adams, 1986).

The history of establishing proctolin as a transmitter of skeletal motoneurons has its roots in the early studies of Brown (1967). He found the "gut-factor" in the peripheral skeletal motor nerves of the cockroach. When synthetic proctolin became available, its activity on insect skeletal muscle was discovered. For example, Piek and Mantel (1977) showed that at nanomolar concentrations proctolin induces and accelerates a myogenic rhythm of contraction and relaxation of the hindleg extensor tibialis muscle of the locust. This result and other examples of action on skeletal muscle could be interpreted in two ways: either the muscle is the target for circulating proctolin (hormonal action), or the muscle is the target of a proctolinergic motoneuron (transmitter action). Distinguishing between these possibilities for the extensor tibialis muscle had to await the development of immunocytochemical methods (see below). Physiological evidence for a transmitter role, however, was provided by the observations of May et al., (1979). These authors applied proctolin by microiontophoresis to the locust extensor tibialis muscle and recorded depolarizing iontophoretic potentials. Moreover, they demonstrated hot-spots of proctolin extra-sensitivity localized in the clefts between muscle fibres. Since it is in these clefts that nerve-muscle junctions were thought to be made, this observation suggested that proctolin may indeed be delivered to the muscle not through the haemolymph but by a peptidergic motoneuron.

The existence of proctolinergic motoneurons is now well established. They were discovered by a combination of immunocytochemistry, HPLC and electrophysiology. The following is a brief description of the work which led to the first identification of a dual transmitter peptidergic motoneuron. In an immunocytochemical survey of the cockroach CNS, Bishop and O'Shea (1982) described the presence of proctolin-immunoreactive axons in the peripheral nerves of thoracic ganglia. This study also revealed a bilaterally symmetrical pair of large immunoreactive neuronal cell bodies on the lateral posterior margins of the dorsal surface of the metathoracic ganglion. These cells were first named the giant dorsal bilateral or GDB neurons (Bishop and O'Shea, 1982). Subsequently it was realized that they were in the same place as the somata of the previously identified slow coxal depressor or Ds motoneurons (Pearson and Iles, 1971), which suggested that the GDB and the Ds might be identical. A combination of intracellular dye injection, immunocytochemistry and electrophysiology confirmed that the immunoreactive GDB neuron was indeed the Ds skeletal motoneuron of Pearson and Iles (O'Shea and Bishop, 1982). Moreover the presence of proctolin in this neuron was established by testing for proctolin bioactivity in extracts made from the somata of the Ds motoneuron. Reverse phase HPLC fractionation of the crude extract provided evidence that the immunoreactivity of the Ds neuron was due to the presence of proctolin (O'Shea and Bishop, 1982). Proctolin was shown subsequently to be released from the Ds motoneurons. This was established by electrically stimulating the neuron, collecting the superfusate from the coxal depressor muscle and applying it to a sensitive proctolin bioassay. Release was found to be calcium-dependent (Adams and O'Shea, 1983).

The final step in the identification of the proctolinergic nature of the Ds motoneuron was a demonstration of a correspondence between the effects of neuronal stimulation and the response to applied proctolin. When proctolin is applied at nanomolar concentrations the muscle develops a slow maintained dose-dependent contraction. This sustained catch-like proctolin-induced tension could be mimicked by stimulating the Ds motoneuron. Together, the evidence on presence, release and action of proctolin established Ds as the first identified peptidergic skeletal motoneuron. In this system proctolin functions as a cotransmitter. The experiments performed on the physiological action of the Ds neuron in fact provided the first example of a dual transmitter motoneuron (Adams and O'Shea, 1983). Stimulation of the individual Ds motoneuron produces a biphasic response consistent with the action of the two transmitters, L-glutamic acid and proctolin. The biphasic nature of the response was revealed by monitoring the membrane potential and force produced by a burst of Ds action potentials. Each action potential produces a transient depolarizing excitatory junctional potential (EJP) followed by a transient increase in force. A second type of force development, however, occurs which is characterized by a slow onset and persistence. It is initiated and maintained without a depolarization of the muscle. This type of contraction can be produced by application of nanomolar concentrations ofproctolin. Figure 3 summarizes some of these findings.

Another, more subtle consequence of stimulating the Ds neuron or of proctolin application is seen by close inspection of the fast transient contractions. By measuring the rate of relaxation of the muscle's fast response we see that proctolin not only causes a direct contraction, but that it also modulates the action of the transiently acting transmitter (L-glutamate). In the presence of proctolin the relaxation of the fast transient response is slowed. This effect can also be seen in response to Ds stimulation. Thus twitches later in a burst relax more slowly than early in a burst of Ds activity. With respect to this phenomenon, proctolin functions as a "homosynaptic modulator". That is, it modulates the post-synaptic action of a second transmitter with which it is co-released.

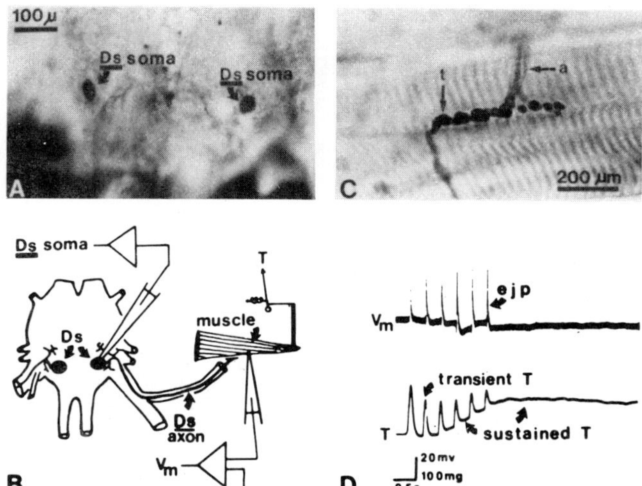

Fig. 3.   A dual transmitter skeletal neuromuscular system. A. Whole-mount immunocytochemical staining of the somata of a pair of proctolinergic skeletal motoneurons (Ds). The unstained cells on the midline of the ganglion are DUM neurons. B. Preparation used to study neuromuscular effects of activating the single identified Ds motoneuron. C. Proctolinergic motor nerve terminal revealed by proctolin immunocytochemistry. D. Biphasis effects of stimulating the dual transmitter motoneuron. Upper trace is muscle transmembrane voltage (V_m). The transient depolarizing potentials are glutaminergic ejps. The lower trace shows the muscle force (T) recorded simultaneously. Each ejp is associated with a transient tension. Released proctolin produces the delayed and sustained tension in the absence of a depolarization.

One other feature of the proctolin response in the cockroach coxal depressor muscle is noteworthy. This is that the proctolin-induced slow contraction can be transiently abolished by stimulation of an inhibitory motoneuron. The inhibitory motoneuron therefore can regulate the force of the catch-like tension in the coxal depressor muscle produced by the Ds neuron. We do not yet understand the mechanism of this transient loss of tension but it may be related to the hyperpolarizing effect of the inhibitory transmitter. But since the proctolin induced tension does not itself depend on a muscle depolarization this explanation seems paradoxical. It suggests that while not requiring depolarization the proctolin contraction is nevertheless voltage dependent, requiring the muscle resting potential to be within a certain range. The hyperpolarization produced by the inhibitory motoneuron may bring the resting potential more negative than the threshold for the proctolin effect. This hypothesis explains why the proctolin contracture returns at the termination of inhibitory motoneuron activity. Whatever the mechanism may be, the regulation of proctolin-induced tonus provides a new functional role for inhibitory motoneurons in insects.

NON-SYNAPTIC NEUROMUSCULAR ACTIONS OF PEPTIDES

The example given above provides a strong case for a direct role for peptides in neuromuscular transmission. It remains to be seen whether peptides other than proctolin are associated with true motoneurons and function as cotransmitters (see CONCLUSIONS). There is now however growing evidence that peptides can act non-synaptically at the insect neuromuscular junction, that is without being released from the skeletal motoneurons. A possible example of a peptide acting non- synaptically as a modulator of neuromuscular transmission is the molluscan cardioexcitatory peptide FMRFamide. Although this peptide was originally isolated from a molluscan nervous system there is evidence from its action in insects and from immunocytochemistry, that it or a similar peptide is present in insects. There are two types of modulatory effects of FMRFamide and relate peptides at the insect neuromuscular junction. Using the locust hindleg extensor muscle and stimulating the slow excitatory motoneuron Walther et al., (1984) showed that FMRFamide increases the size of the synapticpotential and also increases the efficiency-contraction coupling. The effect of FMRFamide therefore is to amplify the effect of the slow excitatory motoneuron by increasing the strength of the contractions induced by the motoneuron. Evans and Myers (1986) report similar potentiating actions of FMRFamide in the same neuromuscular preparation.

The physiological significance of the modulatory actions of FMRFamide are difficult to assess for two reasons. First, the sequence of the insect FMRFamide peptide has not yet been determined. The physiological evidence suggests strongly however that a sequence similar to that of the molluscan peptide must exist in insects. The second difficulty in under standing the significance of the findings is that the neurons which secrete the neuromuscular modulatory peptide have not been identified. Using immunocytochemistry, however, Walther and Schiebe (1987) and Evans and Myers (1986) have identified cells and processes which demonstrate FMRFamide-like immunoreactivity. The FMRFamide-like immunoreactivity is absent in the motoneurons innervating the extensor muscle. Thus it seems the FMRFamide-like peptide, in contrast to proctolin, is not contained in motoneurons and the release site of the insect FMRFamide peptide may therefore be somewhat remote from the site of physiological action. A possible source of the FMRFamide-like peptide is from FMRFamide immunoreactive (IR) neurons found in the CNS. In the thoracic ganglia for example there are FMRFamide-IR neurons which send processes into the medial neurohaemal nerve. Walther and Schiebe (1987) show that extracts of this nerve mimic in detail the action of an FMRFamide-related peptide. Also immunoreactive processes from the nerve may terminate close to skeletal muscles (Evans, 1987).

The FMRFamide peptides are not the only non-motoneuronal peptides to influence skeletal muscle performance. Another is adipokinetic hormone or AKH. This and similar related hormones regulate energy metabolism (Orchard, 1987) but are also known to have direct effects on skeletal muscle. The primary source of circulating AKH, and the primary site of AKH synthesis are the corpora cardiaca. These are large bilobed neurosecretory glands associated with the brain in locust and other insects. AKH-immunoreactivity can also however be found in the CNS and as for FMRFamide-IR, processes are found in the medial neurosecretory nerve. It is possible that muscles and motor nerve terminals can be affected by peptides released from terminals of neurosecretory cells of the CNS. Some of these neurosecretory cells may send processes into the connective tissue sheath of motor nerves and thereby obtain close access to the muscle targets. The presence of neurosecretory profiles in the sheath of motor nerves, close to the muscle target, has been recognised for some time. Perhaps these profiles belong to a class of neurons which in addition to the motoneurons can regulate muscle performance by the secretion of peptides.

CONCLUSIONS

Much remains to be understood about how secreted bioactive peptides affect the neuromuscular junction of insects. In this brief essay I have indicated that peptides may be secreted by excitatory motoneurons and also by neurosecretory cells which may closely approach the muscles in the nerve sheath. Many questions remain about both types of cell types. We can ask for example how many peptides might be involved in controlling muscle performance in insects ? The situation is already too complex to fully understand and I am compelled to ask why so many different modulators and transmitters appear to be needed to control insect muscle. This would seem to make insect neuromuscular mechanisms for more complex than their vertebrate counterparts. But is this likely to be the case ? Is it not possible that we have been oversimplifying vertebrate neuromuscular physiology and that here too there may be peptide cotransmitters and a variety of modulatory states determined by their action and balance of the hormonal environment?

REFERENCES

Adams M.E., and O'Shea, M., 1983, Peptide cotransmitter at a neuromuscular junction, Science, 221:286.

Bishop, C.A., and O'Shea, M., 1982, Neuropeptide proctolin (H-Arg-Try-Leu-Pro-Thr-OH): Immunocytochemical mapping of neurons in the central nervous system of the cockroach, J. Comp. Neurol., 207:223.

Brown, B.E., 1967, Neuromuscular transmitter substance in insect visceral muscle, Science, 155:595.

Brown, B.E., and Starratt, A.N., 1975, Isolation of proctolin, a myotropic peptide from Periplaneta americana, J. Insect Physiol., 21:1879.

Evans, P.D.,1987, Modulation of neuromuscular transmission in the locust by FMRFamide-like peptides, Symposium "Modulation of synaptic transmission and plasticity in nervous systems", Il Ciocco, Abst.

Evans, P.D., and O'Shea, M., 1977, An octopaminergic neurone modulates neuromuscular transmission in the locust. Nature, Lond., 270:257.

Evans, P.D., and O'Shea, M., 1978, The identification of an octopaminergic neurone and the modulation of a myogenic rhythm in the locust, J. exp. Biol., 73:235.

Evans, P.D., and Myers, C.M., 1986, Peptidergic and aminergic modulation of insect skeletal muscle, J. exp. Biol., 124:143.

Hoyle, G., Dagan, D., Moberly, B., and Colquhoun, W., 1974, Dorsal unpaired median insect neurons make neurosecretory endings on skeletal muscle. J. exp. Zool., 187:159.

May, T.E., Brown, B.E., and Clements, A.N., 1979, Experimental studies upon a bundle of tonic fibres in the locust extensor tibialis muscle, J. Insect Physiol., 25:169.

Orchard, I., 1987, Adipokinetic hormones - an update, J. Insect Physiol., 33:451.

O'Shea, M., and Evans, P.D., 1979, Potentiation of neuromuscular transmission by an octopaminergic neurone in the locust, J. exp. Biol., 79:169.

O'Shea, M., and Bishop, C.A., 1982, Neuropeptide proctolin associated with an identified skeletal motoneuron, J. Neurosci., 2:1242.

O'Shea, M., and Schaffer, M., 1985, Neuropeptide function: The invertebrate contribution, A. Rev. Neurosci., 8:171.

O'Shea, M., and Adams, M., 1986, Proctolin: From "gut factor" to model neuropeptide, in: "Advances in Insect Physiology", Vol. 19, P.D. Evans, and V.B. Wigglesworth, eds., Academic Press, London.

Piek, T., and Mantel, P., 1977, Myogenic contractions in locust muscle induced by proctolin and by wasp Philanthus triangulum venom, J. Insect Physiol., 23:321.

Pearson, K.G., and Iles, J.F., 1971, Innervation of coxal depressor muscle in cockroach, "Periplaneta americana", J. exp. Biol., 54:215.

Starratt, A.N., and Brown, B.E., 1975, Structure of the penta-peptide proctolin, a proposed neurotransmitter in insects, Life Sci., 17:1253.

Walther, C., Schiebe, M., and Voigt, K.H., 1984, Synaptic and non-synaptic effects of molluscan cardioexcitatory neuropeptides on locust skeletal muscle, Neurosc. Lett., 45:99.

Walther, C., and Schiebe, M., 1987, FMRFamide-like peptides modulate neuromuscular transmission at locust skeletal muscle, Society for Experimental Biology, Basel, Abst.

THE ROLE OF HEAD ACTIVATOR IN CELL GROWTH AND CONTROL PROCESSES

Graeme Bilbe and H. Chica Schaller

Zentrum für Molekulare Biologie Heidelberg
Im Neuenheimer Feld 282
D-6900 Heidelberg
F.R.G.

INTRODUCTION

Head activator (HA) is a neuropeptide isolated and sequenced from Hydra, a freshwater coelenterate. It is an undecapeptide of molecular weight 1124 and has the sequence p-Glu-Pro-Pro-Gly-Gly-Ser-Lys-Val-Ile-Leu-Phe (Schaller and Bodenmuller, 1981). In Hydra where its active concentration is 10^{-13}M, HA acts as a head-specific growth and differentiation factor (Schaller, 1973).

Head activator's action as growth factor is most easily demonstrated by stopping feeding which is the normal stimulus for growth in Hydra. Under normal feeding conditions interstitial cells, gland cells and epithelial cells of the animal's gastric region multiply constantly and, depending on the stimulus, can develop into head, foot or bud-specific cells. If feeding is interrupted, cell division is slowed down. Following addition of HA to such starved animals, more cells begin to divide as determined by an increase in the mitotic index, which begins 1 hour after treatment, and after a short lag phase an increase in the $[^3H]$thymidine labelling index is seen (Schaller, 1970a). This demonstrates that HA causes cell division and, after mitosis, cells immediately start a new round of DNA synthesis. All dividing cell types in Hydra, such as epithelio-muscular cells, gland cells and all interstitial cell types respond to HA in this way (Fig. 1).

Another role for HA is in the differentiation of interstitial stem cells. In Hydra, differentiation pathways are short. For example differentiation from interstitial stem cell to nerve cell requires one cell cycle, whereas differentiation to nematocytes takes 2-4 cell cycles. Using $[^3H]$thymidine pulse labelling, applied before or after treatment with HA, and assaying for labelled nerve cells, it was possible to demonstrate that HA is required in the early S-phase to cause determination of stem cells to nerve cells (Schaller, 1976b).

Thus one single peptide can interact with different target cells in Hydra and, depending on the cell's competence or state in the cell cycle, can have different effects on the same cell, thus influencing cell division and/or cell differentiation.

Head activator as a growth factor in mammalian systems

In parallel with its isolation from Hydra HA was isolated and fully sequenced from rat intestine (Bodenmuller and Schaller, 1981). Partial sequencing of bovine and human brain isolates combined with HPLC and immunological assays (radioimmunoassay and ELISA) confirmed that these isolates were also identical to the Hydra peptide (Bodenmuller and Zachmann, 1983; Schaller et al., 1984). Minor amounts of HA were also found in other neural and endocrine tissues and higher amounts in tumours of neuroendocrine origin. Table 1 indicates major production sites in the rat. Similar results were obtained from human tissues.

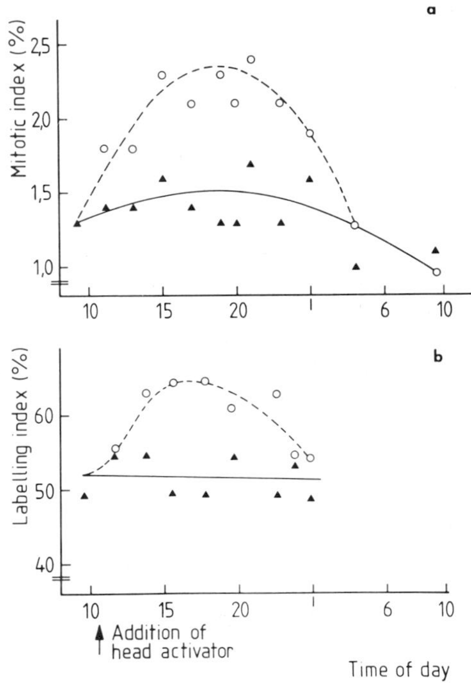

Fig. 1.   Increase in a) mitotic b) labelling index of interstitial cells after treatment with head activator. (▲) Untreated control animals, (○) animals treated with head activator (10^{-12}M).

Recently, a study of human brain tumours demonstrated that HA was over-produced up to 100-fold in astrocytomas and glioblastomas. Coincident with elevated tissue levels an increased level of HA was found in the blood which dropped to normal values following tumour removal, supporting the hypothesis that HA is involved in growth regulation.

Using the double hybrid neuroblastoma-glioma cell line NH15CA2, which overproduces HA, the growth-promoting property of HA was shown. When these cells are grown at low cell density in defined medium without HA-containing fetal calf serum and are then grown in HA-supplemented medium, cell division increases. At high cell density the effect of endogenous HA can be blocked by antibodies against HA. These effects can be explained by considering an autocrine action involving undifferentiated, or precursor, cells which respond to HA produced by a more differentiated cell. Similarly, the pituitary cell line AtT20 responds to HA by increased growth but, in contrast to NH15CA2 cells, produces only small amounts of HA.

We also found that HA appears early in development. For example, the rat embryo contains HA in both developing brain and intestine (Bodenmuller et al., 1980; Schaller et al., 1977). A teratoma cell line, P 19, derived from early mouse embryos can be stimulated by

Table 1 Head activator levels in non-tumour and tumour tissue from the rat.

Source of Tissue	Concentration of Head Activator (fmoles/mg protein)
Non-tumour tissue	
Cerebral cortex	10
Cerebellum	15
Hypothalamus	1000
Intestine	500
Retina	1000
Leg Muscle	13
Pancreas	10
Other tissue	<10
Tumour tissue	
Neuroblastoma-glioma (rat/mouse)	5000
Leg muscle tumour	5
Pancreas tumour	25

treatment with retinoic acid to differentiate into nerve cells. As confirmed by immunocytochemistry, only those cells with nerve-like processes produce HA (Fig. 2). This may suggest that HA has not only preserved its structure, but also its function from _Hydra_ to mammalian systems, i.e. it is produced by early nerve-like cells and then acts as a growth factor on its respective precursor cells.

Head activator as a neurohormone

In mammals, HA is found not only in growing tissue, as in embryos and tumours, but, additionally, in differentiated tissue such as the hypothalamus of an 86 year old man and in blood samples of all ages. Therefore, we assume that HA does not act only as a growth factor but, like other mammalian neuropeptides, has additional functions.

Methanol extraction and SepPak purification, followed by chromatographic and immunological analysis of samples from rat and human blood has revealed significant levels of HA (20-100 fmol/ml) (Bodenmuller and Roberge, 1985). In an attempt to determine its function in blood, tritium labelled synthetic HA was incubated with rat and human plasma and then submitted to HPLC separation. Interestingly, HA was rapidly degraded with a half life of 7min _in vitro_ and only seconds _in vivo_. The fragments found after HPLC separation indicated that the degradation was due to a dipeptidyl-carboxypeptidase, which was later shown to be angiotensin-converting enzyme (Roberge et al., 1984). In contrast, endogenous HA from plasma samples was found to be undegraded. This led to the discovery that endogenous HA in plasma is absorbed to a high molecular weight carrier, a situation similar to that found in _Hydra_ (Schaller et al., 1986). HA can be released from this carrier protein by methanol extraction, indicating that it is not covalently bound but only attached to the carrier. However, as endogenous HA in plasma is not degraded by angiotensin-converting

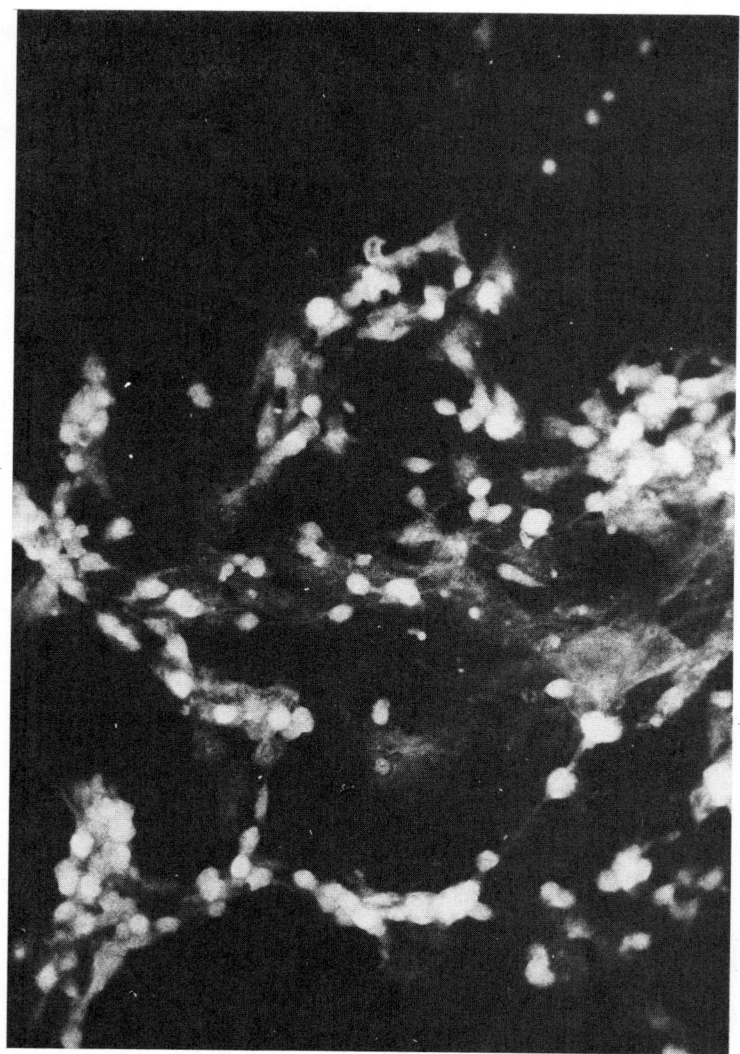

Fig. 2.   Head activator production in retinoic acid induced P19 teratoma cells 8 days after induction. Cells were fixed with 1% carbodiimide and 80% methanol, incubated with a 1:1000 dilution of antibody 12/3 and visualised with FITC labelled second antibody (Schaller et al., 1984).

enzyme, it would suggest that the carboxy end of HA is hidden in or protected by the carrier.

The occurrence of HA in blood, intestine and hypothalamus suggests a role as a neuromodulator. For example, the gastrointestinal peptides, secretin and cholecystokinin which also originate in the intestine, are known to stimulate pancreatic secretion. It was postulated therefore that HA may have a similar effect. Indeed, it was possible to stimulate amylase release from isolated rat pancreatic lobules after incubation with synthetic HA _in vitro_; a response which was time and dose-dependent, with an optimal concentration at 10^{-10}M (Feuerle et al., 1983). Another finding that, within 15-30min of a meal, HA levels in blood increase 30-200 fold also lends strong support to the hypothesis that HA has a modulatory function in the intestine.

Head activator as a neurotransmitter

For the possible action of a substance as a neurotransmitter at synapses, its fast degradation is necessary. Conventional transmitters are inactivated by enzymes or reuptake systems which are concentrated in the synaptic cleft to ensure a similar fast degradation. For HA no such enzyme is required because the molecule is able to inactivate itself by dimerisation.

The evidence for this was obtained from several observations. Synthetic head activator, or HA extensively purified from natural sources, was mostly inactive and more so at higher concentrations and the purer the molecule. Moreover, HA was found to form a stable dimer down to concentrations as low as 10^{-13}M. Only in the presence of salt (1M $(NH_4)_2SO_4$), or acid and low salt (0.1% TFA in 0.1M NaCl), was HA in a monomeric state. Assays were only possible when HA was diluted into a medium containing many other molecules (1% BSA and 0.1M NaCl), ensuring biological activity and/or interaction with the receptor (Bodenmuller et al., 1986). Since many peptides tend to form dimers, this may be a mechanism of general relevance for peptide inactivation at synapses.

By analogy with other peptides, the presence of high amounts of HA in hypothalamus indicates a HA function in the pituitary-hypothalamic axis. We have been able to demonstrate, in a collaborative experiment with Kordon's group in Paris, that HA blocks β-endorphin release from the pituitary intermediate lobe (unpublished results). By analogy with other peptides HA's action must be mediated via a specific receptor protein and much effort is now being concentrated on identifying and characterising this molecule. Preliminary work, using a ^{125}I-labelled HA-BSA conjugate, has demonstrated specific labelling of NH15CA2 cell membranes. In parallel experiments, an anti-idiotype antibody (raised against a monoclonal antibody to HA) appears to identify the same sites in these cell membranes, an effect which is abolished by co-incubation with HA and HA conjugate (unpublished results).

CONCLUSION

Hydra needs HA for cellular growth and differentiation processes. Similarly, a growth promoting effect of HA was found on mammalian cells. It may therefore be speculated that, as well as conserving its amino acid sequence during evolution, HA may also have preserved some of its functions. The importance of head activator in mammalian control processes is now under intense investigation.

We know that HA attached to a carrier is protected from degradation, which implies that delivery of intact head activator to its target cells is important to elicit an effect and to protect it from inactivation by dimerisation. We postulate that dimerisation of head activator (and likewise other neurotransmitter peptides) is a mechanism of general relevance for peptide function. At the present time we do not understand why this peptide-carrier-receptor system has been conserved during evolution but the fact that this is true for every system studied so far up to mammals makes a thorough analysis important.

ACKNOWLEDGEMENTS

We wish to thank the Deutsche Forschungsgemeinschaft (SFB 317) and the Fonds der Deutschen Chemischen Industrie for support and J. Rami and I. Nonnenmacher for typing the manuscript.

REFERENCES

Bodenmüller, H. and Roberge, M., 1985, The head activator: discovery, characterisation, immunoassays and biological properties in mammals, Biochim. Biophys. Acta., 825: 267.

Bodenmüller, H. and Schaller, H.C., 1981, Conserved amino acid sequence of a neuropeptide, the head activator, from coelenterates to humans, Nature, 293: 579.

Bodenmüller, H., Schaller, H.C. and Darai, G., 1980, Human hypothalamus and intestine contain a Hydra neuropeptide, Neurosci. Lett., 16: 71.

Bodenmüller, H., Schilling, E., Zachmann, B. and Schaller, HC., 1986, The neuropeptide head activator loses its activity by dimerisation, EMBO J., 5: 1825.

Bodenmüller, H. and Zachmann, B., 1983, A radioimmunoassay for the head activator from hydra, FEBS Lett., 159: 237.

Feuerle, G.E., Bodenmüller, H. and Bacca, I., 1983, The neuropeptide head activator stimulates amylase release from rat pancreas in vitro, Neurosci. Lett., 38: 287.

Roberge, M., Escher, E., Schaller, H.C. and Bodenmüller, H., 1984, The Hydra head activator in human blood circulation, FEBS Lett., 173: 307.

Schaller, H.C., 1973, Isolation and characterisation of a low molecular weight substance activating head and bud formation in Hydra, J. Embryol. Exp. Morphol., 29: 27.

Schaller, H.C., 1976a, Action of the head activator as a growth hormone in Hydra, Cell Differ., 5: 1.

Schaller, H.C., 1976b, Action of the head activation on the determination of interstitial cells in Hydra, Cell Differ., 5: 13.

Schaller, H.C. and Bodenmüller, H., 1981, Isolation and amino acid sequence of a morphogentic peptide from Hydra, Proc. Natl. Acad. Sci. USA., 78: 7000.

Schaller, H.C., Bodenmüller, H., Zachmann, B. and Schilling, E., 1984, Enzyme-linked immunosorbent assay for the neuropeptide head-activator, Eur. J. Biochem., 138: 365.

Schaller, H.C., Flick, K. and Darai, G., 1977, A neurohormone from Hydra is present in brain and intestine of rat embryos, J. Neurochem., 29: 393.

Schaller, H.C., Roberge, M., Zachmann, B., Hoffmeister, S., Schilling, E. and Bodenmüller, H., 1986, The head activator is released from regenerating Hydra bound to a carrier molecule, EMBO J., 5: 1821.

RELEASE OF NEUROPEPTIDES FROM MAGNOCELLULAR NEURONES: DOES ANATOMICAL COMPARTMENTATION HAVE A FUNCTIONAL SIGNIFICANCE ?

John Morris, David Pow and Fraser Shaw*

Department of Human Anatomy
University of Oxford
Oxford, U.K.

*Present address
Anatomy Department
Cambridge University
Cambridge, U.K.

INTRODUCTION

The mammalian magnocellular neurosecretory system which projects to the capillaries of the neural lobe has, for many years, served as a useful and archetypal model in the study of mechanisms of neuropeptide secretion. The magnocellular neurones produce either vasopressin (VP) or oxytocin (OT), and other peptides including dynorphin, from precursors synthesized in the cell bodies and packaged into neurosecretory granules (NSG). Many of the typically varicose axons transport the maturing neurosecretory product in NSG through the median eminence to the neural lobe where it is either stored, released by exocytosis into the systemic circulation, or degraded.

The ultrastructure of the neurones has been investigated extensively and morphological similarity between the neurosecretory nerve terminals and classical synaptic boutons has fuelled the assumption that release of the neuropeptides occurs from this limited compartment. However, this apparently simple situation is deceptive and we must now ask fundamental questions about the site of neuropeptide release. It has been shown recently that the neuropeptides VP and OT are released within the hypothalamus (Moos et al., 1984; Freund-Mercier and Richard, 1984; Mason et al., 1985) and median eminence (Holmes et al., 1986; Verbalis et al., 1986), where OT at least has important functions (Freund-Mercier and Richard, 1984; Frawley et al., 1985; Theodosis et al., 1986). Is hormone released from classic synapses (Theodosis, 1985) which are sparse, or is it released from dendrites which are packed with NSG and often give rise to the varicose axons at some distance from the cell? Dendritic release of amine and peptide is well documented in the substantia nigra (Greenfield, 1985) and could be the source of VP, OT and NP in the CSF (Robinson, 1983) since dendrites of the many PVN neurones run close to the ependyma of the third ventricle and those of many SON cells run in the subpial region of the base of the brain.

The precise anatomy of the neurones in the median eminence is poorly understood and release studies are complicated by release of vasopressin from parvocellular neurones (Verbalis, 1986). The magnocellular axons have few typical nerve endings in the median eminence, so is the hormone released from axons and their dilatations en passant? Recently, (Buma and Nieuwenhuys, 1987) have shown that exocytosis of both OT and VP from varicose fibres can be visualised in the median eminence, but nothing more is known of the origin and nature of release in other hypothalamic sites. Our studies concern release of hormone in the neural lobe, but the results have clear implications also for release within the hypothalamus.

THE ANATOMY OF NEUROSECRETORY AXONS WITHIN THE NEURAL LOBE

Within the neural lobe the axons have undilated regions, and dilated regions of two types: nerve 'endings' which resemble synapses by their content of microvesicles similar in appearance to synaptic vesicles and by the presence of 'presynaptic' densities, and which abut the perivascular basement membrane; and nerve 'swellings' often much larger than the endings which lack significant numbers of microvesicles and usually lie distant from the perivascular basement membrane, separated from it by the pituicytes (Morris, 1976). However, although these morphologically specialised compartments of the axons can be recognised in ultrathin sections, the precise inter-relationship between undilated axons, endings and swellings is poorly understood. The interrelationships are, however, important for the interpretation of the movement of NSG through the system. The classical studies of Howard Sachs showed that the newly-formed vasopressin was preferentially released (Sachs and Haller, 1968). Acute stimulation (56mM K$^+$) causes detectable loss of granules from only the nerve endings (Morris and Nordmann, 1980) and autoradio-graphic transport studies show a quantitatively preferential movement of newly-formed (labelled) granules into the nerve endings (Heap et al, 1975) which seems to result from the availability of space within this compartment as granules are released (Chapman et al, 1982).

With Dr David Chapman, the anatomical relationship between nerve endings and swellings has been determined by analysis of serial sections. These (Figs. 1,2) reveal a lobular architecture in which undilated axons (A) lie in the core of the lobules with the pituicytes (P). The axons are dilated by small and large swellings (S); the nerve endings (E) lie peripherally within the lobules and arise primarily as small excrescences from the surface of the large swellings. This topographic organisation is confirmed by scanning microscopy of cut surfaces of neural lobe tissue (F D Shaw & R E J Dyball, unpublished data) (Fig.3) which reveals cauliflower-like excrescences (E)(the endings) on the surface of large rounded structures (the swellings); presumptive undilated axons (A) can also be seen. Where this arrangement obtains, NSG must traverse the swellings to reach the endings.

The relationship between neurosecretory axons, extracellular space and pituicytes changes subtly with increasing secretion. In the homozygous Brattleboro rat the nerve endings are not larger, but their contact with the perivascular basement membrane is significantly increased (Morris, 1982). In lactating and saline-drinking animals the proportion of the perivascular basement membrane that is contacted by neurosecretory processes is increased and in saline-drinking but not lactating animals the length of contact of individual nerve terminals with the basement membrane is increased (Tweedle and Hatton, 1987). It is not known whether increased secretion also results in any change in the proportion of the processes that have the form of nerve endings and nerve swellings.

VISUALISATION OF EXOCYTOSIS

The stabilisation of exocytosed granule cores by tannic acid in a variety of techniques pioneered by Buma and co-workers (Buma et al., 1984; Buma and Roubos, 1986) has permitted visualisation of the process in a way that was not possible before. Initially it was used largely in invertebrate systems (Roubos and van der Wal-Divendal, 1980) but, by application of the technique to the magnocellular neurosecretory system, Buma's group (Buma and Nieuwenhuys, 1987; van Putten et al., 1987) have shown that exocytosis is not restricted to nerve endings, but also occurs at the surface of undilated axons and axonal swellings in both the neural lobe and median eminence.

Our own studies which combine a tannic acid perfusion technique with immunogold detection of OT- and VP-associated neurophysin by the monoclonal antibodies PS38 and PS41 confirm, amplify and quantify these findings. Figs. 4, 5 and 6 illustrate exocytosis from nerve endings, axonal swellings and undilated axons respectively. Exocytosis occurs from nerve endings both at the surface facing the basement membrane and more centrally

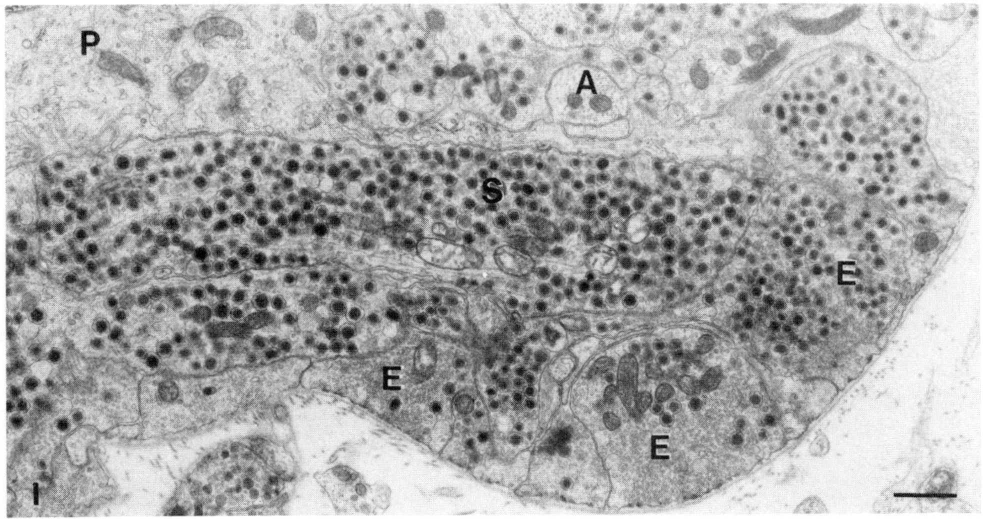

Fig. 1.   Transmission electron micrograph of part of a lobule in the posterior pituitary. Bar 1μm.

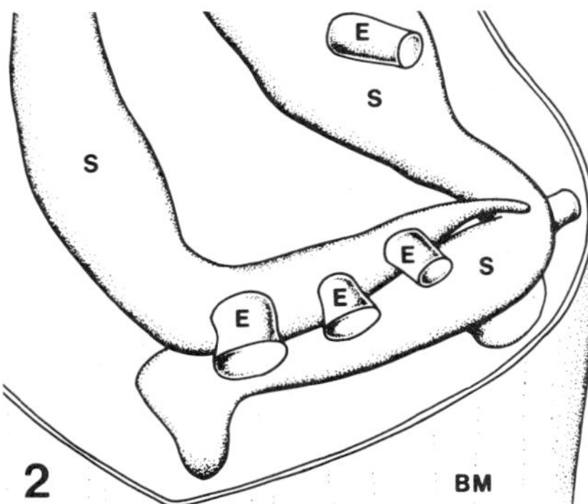

Fig. 2.   Reconstruction of part of the lobule shown in Fig. 1 showing nerve endings arising from the surface of nerve swellings.

Fig. 3.   Scanning micrograph of cut surface of posterior pituitary. Bar 1μm.

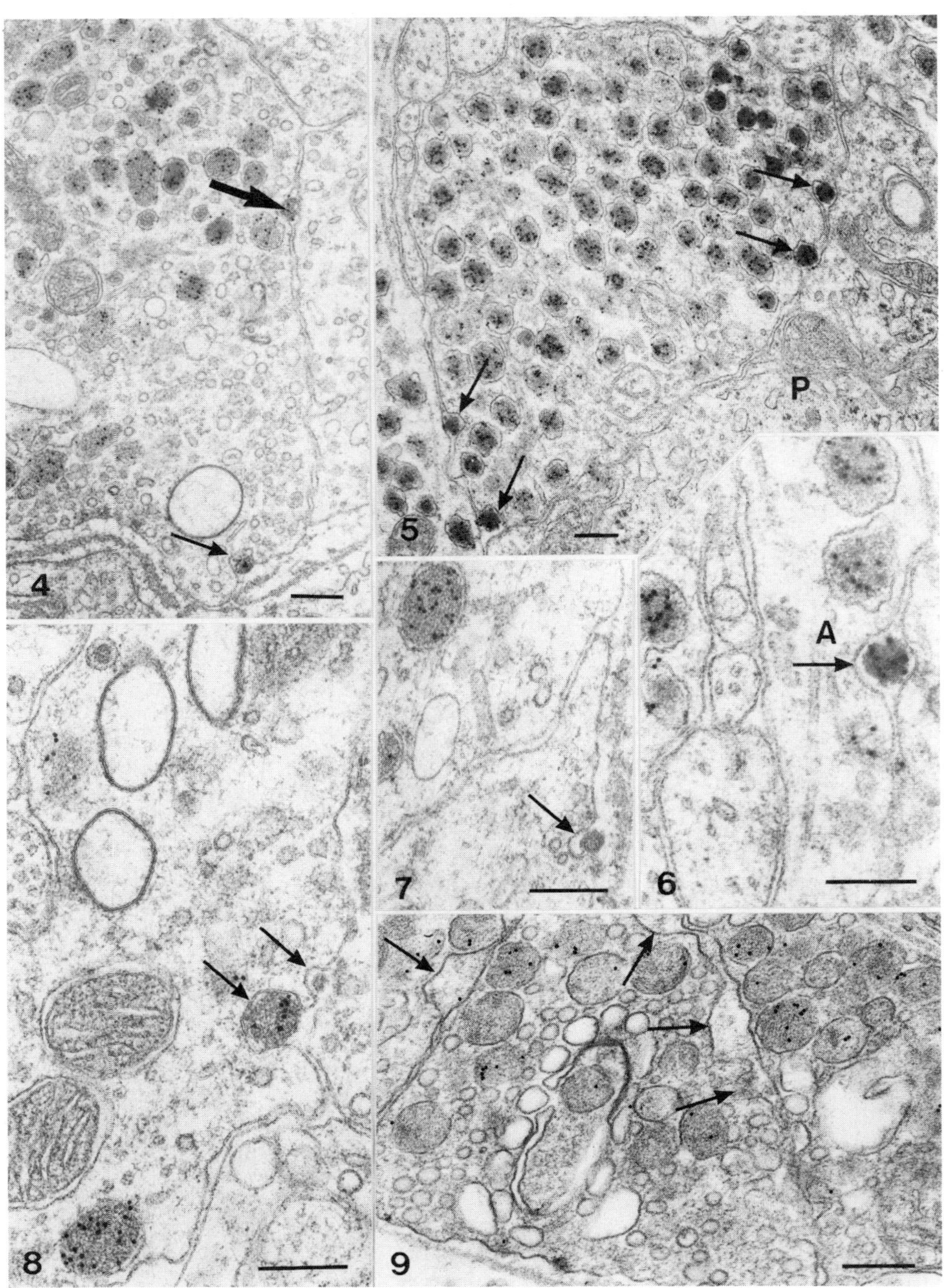

Figs. 4-9.   Exocytoses (arrowed) in the posterior pituitary visualised by perfusion of the animal with tannic acid. NSG are immunogold labelled for oxytocin-neurophysin except in Fig. 8. Bars 200nm. Fig. 4. One exocytosis (↓) occurs at the surface abutting the basal lamina, another (↘) between two endings. Fig. 5. Exocytoses around a nerve swelling; pituicyte (P). Fig. 6. Exocytosis from an undilated axon (A). Fig. 7. Exocytosis of a small dense granule from the ending of an abnormal neurone in a homozygous Brattleboro rat. Fig. 8. Side-by-side exocytoses of a 160nm VP-neurophysin labelled NSG and of a small 100nm granule. Fig. 9. Exocytoses of oxytocin granules from a nerve ending of a naloxone-treated, morphine-dependent rat.

within the lobules (Fig. 4); it occurs at any point of the periphery of a swelling, often abutting a pituicyte (Fig. 5), and within bundles of undilated axons.

Tannic acid also reveals exocytosis of the small dynorphin-containing (Whitnall et al., 1985) granules in the abnormal neurones of homozygous Brattleboro rats (Fig. 7) and the 100nm dense granules that are found in small numbers with the 160nm NSG in magnocellular neurones of normal rats (Chapman and Morris, 1985) (Fig. 10) and oxytocin neurones of Brattleboro homozygotes.

In unstimulated tissue, exocytosis of VP- and OT-containing granules occurs with similar frequency. Stimulation of the tissue by vascular perfusion of anaesthetised rats with 56mM K^+ in the presence of 0.2% tannic acid increases the incidence of exocytosis of both gold-labelled OT and unmarked (VP) granules, but our initial quantitative data (Fig. 10) indicate that the release of VP granules predominates 3-fold. This is surprising because assay of hormone released from neural lobes by electrical stimulation (R E J Dyball, unpublished observations) or 56mM K^+ (F D Shaw, unpublished observations), and from isolated neurosecretosomes by 56mM K^+ (Cazalis et al., 1987) shows that the release of oxytocin predominates. At present we have no explanation for this discrepancy. It is unlikely that we have failed to label OT-containing cores; the immunocytochemistry produces clearcut results which concur with granule morphology (Castel et al., 1985; Castel et al., 1986) and application of the tannic acid method to naloxone-treated morphine-dependent rats (Bicknell et al., 1984) reveals the expected predominance (by 14:1) of exocytoses of OT NSG cores (Fig. 8). Exocytosis of VP granules also predominates in normal mice and, as expected in mice with hereditary nephrogenic diabetes insipidus (di/di). It is possible that the presence of tannic acid in the extracellular space during release of VP and OT has mordanted and thereby rendered inoperative other peptides/proteins which are either in the membranes, or are secreted, and which control the amount of vasopressin or oxytocin released.

The total number of exocytoses was clearly increased by perfusion of the animal with 56mM K^+ and this effect was abolished by pre-perfusion with, and addition of, 1mM EGTA (Fig. 12). In conditions of normal hormone release Wistar rats would be expected to release, during 5min, 0.17mU hormone equivalent to 2×10^{-4} of the total hormone in the gland (data derived from Morris, 1976). If the same rate of release obtains during the 5min perfusion with tannic acid of Long Evans rats and all exocytoses are captured then, since pituitary tissue of Long Evans rats contains ~9.5 NSG/μm^2 ultrathin section (Morris, 1982), we might expect to find ~19 exocytoses in every $10^4 \mu m^2$ grid space. In fact, in two animals we found (average) 16 and 17 exocytoses per grid space. The remarkable similarity between expected and observed suggests that release of neuropeptide continues 'normally' during

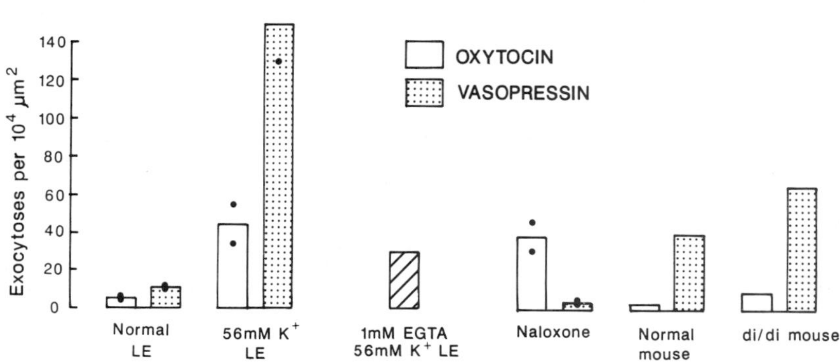

Fig. 10. Incidence of exocytosis in posterior pituitary tissue from normal Long Evans (LE) rats; LE rats stimulated by perfusion with 56mMK$^+$ for 5min without, and in the presence of 1mM EGTA (data for oxytocin and vasopressin combined); in morphine-habituated rats stimulated to release oxytocin by administration of naloxone; and in normal and nephrogenic diabetes insipidus (di/di) mice.

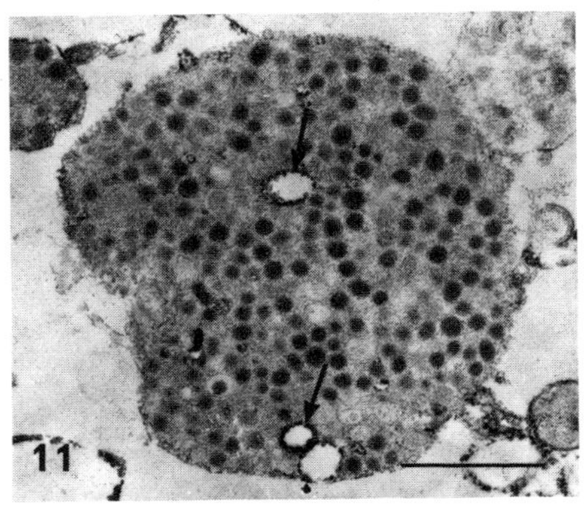

Fig. 11.   Isolated nerve swelling that has been stimulated by 56mM K$^+$ in the presence of horseradish peroxidase. Release of hormone has occurred and HRP has been endocytosed into vacuoles (arrows) in the nerve swelling. Bar = 1μm

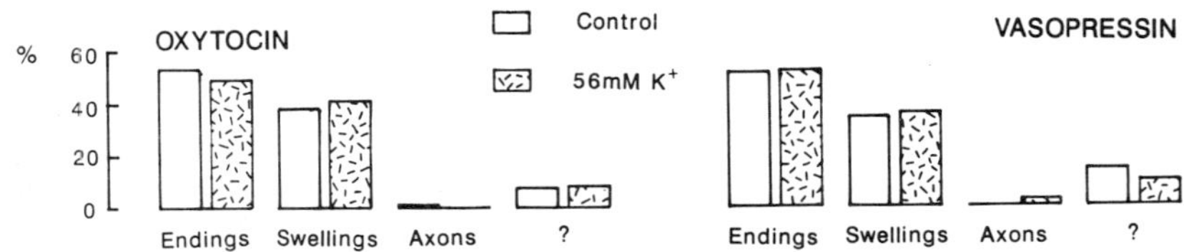

Fig. 12.   Proportion of exocytotic figures occurring from nerve endings, nerve swellings, undilated axons and unidentified sites (?) in unstimulated tissue and tissues stimulated with 56mM K$^+$.

perfusion with tannic acid. Furthermore, 56mM K$^+$ markedly increases the number of exocytoses, indicating that the tissue still responds to depolarisation, and an increased frequency of exocytosis is seen in naloxone-treated, morphine-dependent rats,and in di/di mice, both of which have endogenous stimuli which increase the rate of release of neuropeptide.

THE LOCATION OF EXOCYTOSES AND ITS IMPLICATIONS

Exocytoses therefore occur, not only from nerve endings where they were expected, but also from undilated axons and nerve swellings. Exocytosis from nerve swellings and undilated axons are not an artefact of stimulation with 56mM K$^+$ since they are found in unstimulated tissue and in tissue stimulated _in vivo_ by naloxone-treatment of morphine-dependent rats. Further evidence that exocytosis from nerve swellings is a physiological process comes from separation of neurosecretosomes into fractions that contain predominantly either nerve endings or nerve swellings (Fig. 11) as judged by their size (mostly 0.5-2.0μm and 2-10μm respectively) and organelle content. This separation was achieved by a two-step differential centrifugation of homogenised rat neurohypophyses. Preliminary results show that stimulation by 56mM K$^+$ provoked the release of vasopressin and oxytocin from both fractions, but release from the fraction enriched with swellings was less (when expressed as a proportion of total hormone, by ~5:1) than was release from the fraction enriched with nerve endings.

Preliminary quantitative analysis of the tannic acid-treated tissue (Fig. 12) shows that the proportions of granules exocytosed at these different sites is not altered by stimulation with 50mM K+ and is similar for both OT- and VP-containing granules. If exocytosed cores of uncertain origin (e.g. those lying between an ending and a swelling) are excluded, then ~55% of exocytoses occur from nerve endings and ~40% from nerve swellings. Exocytoses at the border of undilated axons is less common (~5%). Considered in terms of granule turnover, since the nerve endings contain ~30% of all NSG and the nerve swellings ~60% (Morris, 1976), this means that exocytosis of a granule in a nerve ending is about 3 times more likely than exocytosis of a granule in a nerve swelling, a finding consistent with the data from neurosecretosomes (above) and one that would explain the earlier finding that significant loss of NSG was only detected in the nerve endings. Thus, turnover of NSG by exocytosis is more common in the nerve endings, but occurs also to a significant extent in the nerve swellings. This is consistent with our model (Morris and Nordmann, 1980) of the distribution of NSG within the system based on autoradio-graphic tracing (Heap et al., 1975) and the refilling of the depleted neural lobe (Chapman et al., 1982) and with recent data on HPLC analysis of neurophysin processing (Newcomb and Nordmann, 1987) and, most importantly, with the observed preferential release of newly-formed hormone. Exocytosis of granules from undilated axons is clearly also quite common when considered as a proportion of the granules in the compartment, but would contribute only marginally to the total amount of hormone released in the neural lobe. However, hormone released from the axons and endings would have to diffuse through the intercellular clefts between the neurosecretory processes and glia and could there exert a paracrine or autocrine role more easily than hormone released from the nerve endings. The extent to which release from undilated or dilated axons could explain observed release of hormone in the hypothalamus has yet to be determined.

Exocytoses were not evenly distributed among the neurosecretory processes. Many exocytotic profiles were present at the surface of some profiles (e.g. Fig. 5), and not at the surface of others of similar size and type. This observation implies heterogeneity of release among neurosecretory processes of the same basic type.

Calculation of release rates in terms of the membrane available for exocytosis leads to an interesting result which has fundamental implications for the mechanism of exocytosis. If the neural lobe of Long Evans rats is similar to that in Wistars (the distribution of granules is analogous: compare data from (Morris, 1976 and Morris, 1982) then the gland contains 3.4×10^7 endings of average surface $7.5 \mu m^2$ and 7.1×10^6 swellings of average surface $25.5 \mu m^2$ - a total of $2.6 \times 10^8 \mu m^2$ membrane attributed to the endings and $1.8 \times 10^8 \mu m^2$ membrane attributed to the swellings (data from Nordmann, 1977). Thus there is 1.4 x more 'ending' than 'swelling' membrane, and 1.5 x more exocytosis from nerve endings. The similarity between these two figures could imply that exocytosis can occur equally at the membrane of either compartment; i.e. that nerve endings are not specialised for release in any way other than a high surface area/volume ratio and their position close to the perivascular space, which will minimise the time taken by hormone to diffuse into the vascular system. If this is correct it implies either that exocytosis does not require specialised 'docking sites' in the plasmalemma or that any exocytosis-specific sites are not restricted to or even concentrated in the nerve endings.

THE MICROVESICLES OF THE NERVE ENDINGS - A CONTINUING ENIGMA

The role of the microvesicles, which characterise the nerve endings but not swellings, is rendered even more enigmatic if we accept that exocytotic hormone release is equally likely from the surface of either the endings or the swellings. Although similar in appearance to cholinergic and other 'synaptic' vesicles, the microvesicles are not associated with any known neurotransmitter, although improvements in both immunocytochemical techniques and the spectrum of available antibodies almost certainly justifies a careful re-examination of this problem. There is no good evidence that microvesicles represent granule membrane retrieved after exocytosis (Morris and Nordmann, 1980), and studies on the ultrastructural localisation of calcium (Shaw and Morris, 1980) and of accumulation of calcium by subcellular fractions enriched in microvesicles (Nordmann and Chevalier, 1980) have indicated that the microvesicles are calcium-accumulating organelles.

Although the volumetric density of microvesicles remains unchanged during stimulation with potassium _in vitro_ (Morris and Nordmann, 1980), further analysis of stimulated endings has demonstrated that the microvesicles do show dynamic morphological changes in response to stimulation of hormone release. In unstimulated tissue, microvesicles are present in both clustered and dispersed forms; clustered microvesicles often occupy a central position in the endings and have a smaller average diameter than the dispersed microvesicles which tend to occupy the region adjacent to the basement membrane. Although it has been suggested that these appearances imply two separate organelle types (Theodosis et al., 1977), it is striking that in stimulated tissue most of the microvesicles appear dispersed and large (Figs. 4,8,9,13) whereas in tissue treated with the metabolic inhibitor dinitrophenol (Fig. 14), or perfused with 1mM EGTA (Fig. 15) the microvesicles are nearly all clustered and small. Quantitative comparison of microvesicles in control and potassium-stimulated neural lobes confirms that stimulation results in the majority of microvesicles being larger and dispersed (Fig. 16). We therefore suggest that the clustered and dispersed microvesicles represent the two extremes of a continuum of functional states of a single calcium-accumulating organelle. Swelling of microvesicles could result from

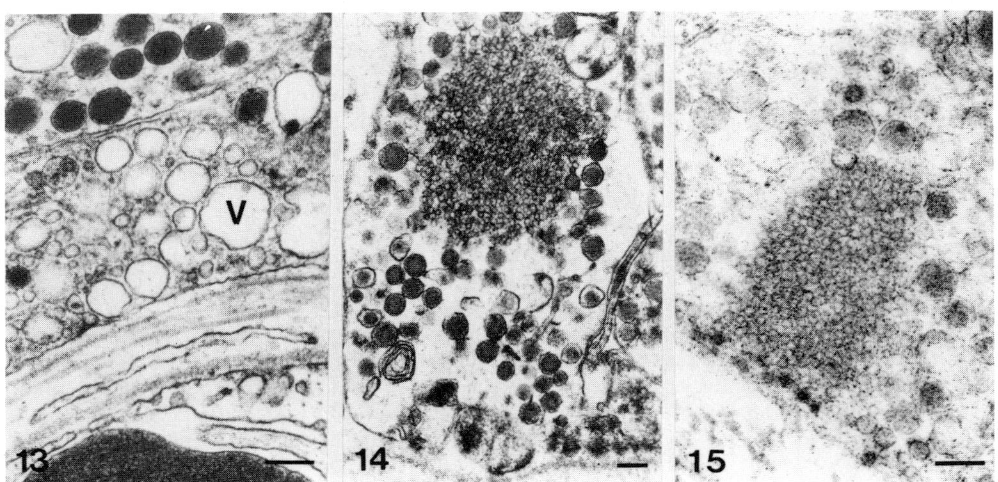

Fig. 13.   Enlarged, dispersed microvesicles in 56mM K$^+$-stimulated neural lobe. Note also the large vacuoles (V). Bar = 200nm.

Figs.14-15.   Small, clustered microvesicles in neural lobe tissue exposed to dinitrophenol (Fig. 14) or 1mM EGTA (Fig. 15). Bar = 200nm.

accumulation of calcium (plus water), and dispersion could result from electrostatic repulsion as the microvesicles acquire a net positive charge. The presence of a major population of calcium-accumulating organelles in the endings but not the swellings could result from the greater surface area/volume ratio of the endings which, if their plasma membranes were otherwise similar with respect to their calcium-gating properties, would result in a greater change in intracellular calcium during stimulation, and a need to control this in some way.

These observations raise many questions. If the plasma membrane of magnocellular neurones is relatively unspecialised in terms of the sites in the neural lobe at which

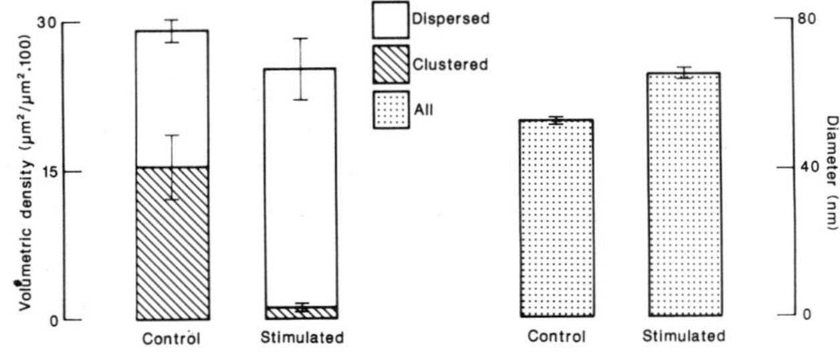

Fig. 16.   The extent of clustering and dispersal of neural lobe microvesicles and changes in their size as a result of stimulation with 56mM K$^+$ for 30min.

exocytosis can occur, then does the same apply to the entire surface of the cells?   The regenerative action potential could spread to invade the entire surface membrane. Since an increase in the local intracellular concentration of calcium appears to be the key intermediate trigger for exocytosis, it could be that the calcium-handling properties of the different parts of the membrane will be a major determinant of the location of hormone release.   The amount of hormone released from a given site would depend also on the availability of granules at that site.   Two immediate tasks are, therefore, to determine the way in which different parts of the neuronal cell membrane control calcium fluxes, and to determine how the availability of granules in different parts of the system is controlled.

ACKNOWLEDGEMENTS

We gratefully acknowledge financial support from AFRC grants AG43/94 and 43/140 and MRC grants G608/236 and G8411/610N.   The Royal Society provided the image quantitation equipment.

REFERENCES

Bicknell, R.J., Leng, G., Lincoln, D.W. and Russell, J.A., 1984, Activation of putative oxytocin neurones following naloxone administration to rats treated chronically with intracerebroventricular (i.c.v.) morphine, J. Physiol. 357:97P.

Buma, P, Roubos, E.W. and Buijs, R.M., 1984, Ultrastructural demonstration of exocytosis of neural, neuroendocrine and endocrine secretions with an in vitro tannic acid (TARI-) method, Histochemistry, 80:247.

Buma, P. and Nieuwenhuys, R., 1987, Ultrastructural demonstration of oxytocin and vasopressin release site in the neural lobe and median eminence of the rat by tannic acid and immunogold methods, Neuroscience Letters, 74:151.

Buma, P. and Roubos, E.W., 1986, Ultrastructural demonstration of non-synaptic release sites in the central nervous system of the snail Lymnaea stagnalis, the insect Periplaneta americana, and the rat, Neuroscience, 17:867.

Castel, M., Morris, J.F., Ben-Barak,Y., Timberg, R.,Sivan, N. and Gainer, H., 1985, Ultrastructural localization of immunoreactive neurophysins using monoclonal antibodies and Protein A-gold, J. Histochem. Cytochem., 333:1015.

Castel, M., Morris, J.F., Whitnall, M. and Sivan, N., 1986, Improved visualization of the immunoreactive hypothalamo-neurohypophysial system by use of immuno-gold techniques, Cell Tissue Res., 243:193.

Cazalis, M, Dayanithi, G. and Nordmann, J.J., 1987, Hormone release from isolated nerve endings of the rat neurohypophysis, J. Physiol. 390:55.

Chapman, D.B. and Morris, J.F., 1985, Granule populations in oxytocin and abnormal perikarya of the supraoptic nucleus of homozygous Brattleboro rats: effects of colchicine administration, Cell Tissue Res., 241:435.

Chapman, D.B., Morris, J.F. and Valtin, H., 1982, How do granules distribute between nerve endings and nerve swellings in the neural lobe? Evidence from Brattleboro rats, in: "Neuroendocrinology of Vasopressin, Corticoliberin and Opiomelanocortins," A.J. Baertschi and J.J. Dreifuss, eds., Academic Press Inc., London.

Frawley, L.S., Leong, D.E. and Neill, J.D., 1985, Oxytocin attenuates TRH-induced TSH release from rat pituitary cells, Neuroendocrinology, 40:201.

Freund-Mercier, M.J. and Richard, Ph., 1984, Electrophysiological evidence for facilitatory control of oxytocin neurones by oxytocin during suckling in the rat, J. Physiol., 352:477.

Greenfield, S.A., 1985, The significance of dendritic release of transmitter and protein in the substantia nigra, Neurochem. Internat., 7:887.

Heap, P.F., Jones, C.W., Morris, J.F. and Pickering, B.T., 1975, Movement of neurosecretory product through the anatomical compartments of the neural lobe of the pituitary gland: an electron microscopic autoradiographic study, Cell Tissue Res., 156:483.

Holmes, M.C., Antoni, F.A., Aguilera, G. and Catt, K., 1986, Magnocellular axons in passage through the median eminence release vasopressin, Nature, (Lond.), 319:326.

Mason, W.T., Hatton, G.I., Ho, Y.W., Chapman, C. and Robinson, I.C.A.F., 1985, Central release of oxytocin, vasopressin, and neurophysin by magnocellular neurone depolarization: evidence in slices of guinea pig and rat hypothalamus, Neuroendocrinology, 42:311.

Moos, F., Freund-Mercier, M.J. Guerne, Y., Stoeckel, M.E. and Richard, Ph., 1984, Release of oxytocin and vasopressin by magnocellular nuclei in vitro: specific facilitatory effect of oxytocin on its own release, J. Endocr., 102:63.

Morris, J.F. and Nordmann, J.J., 1980, Membrane recapture after hormone release from nerve endings in the neural lobe of the rat pituitary gland, Neuroscience, 5:639.

Morris, J.F., 1976, Distribution of neurosecretory granules among the anatomical compartments of the neurosecretory process of the pituitary gland, J. Endocr., 68:225.

Morris, J.F., 1976, Hormone storage in individual neurosecretory granules of the pituitary gland, J. Endocr., 68:209.

Morris, J.F., 1982, The Brattleboro magnocellular neurosecretory system: a model for the study of peptidergic neurons, Ann. N.Y. Acad. Sci., 394:54.

Newcomb, R.W. and Nordmann, J.J., 1987, Quantitative HPLC analysis of rat neurophysin processing, Neurochem. Int., In Press.

Nordmann, J.J. and Chevallier, J., 1980, The role of microvesicles in buffering [CA]i in the neurohypophysis, Nature (Lond.), 287:54.

Nordmann, J.J., 1977, Ultrastructural morphometry of the rat neurohypophysis, J. Anat., 123:213.

Robinson, I.C.A.F., 1983, Neurohypophysial peptides in cerebrospinal fluid, Prog. Brain Res., 60:129.

Roubos, E.W. and van der Wal-Divendal, R.M., 1980, Ultrastructural analysis of peptide-hormone release by exocytosis, Cell Tissue Res., 207:267.

Sachs, H. and Haller, E.W., 1968, Further studies on the capacity of the neurohypophysis to release vasopressin. Endocrinology, 83:251.

Shaw, F.D. and Morris, J.F., 1980, Calcium localization in the rat neurohypophysis, Nature, (Lond.), 287:56.

Theodosis, D.T., 1985, Oxytocin-immunoreactive terminals synapse on oxytocin neurones in the supraoptic nucleus, Nature, (Lond.), 313:682.

Theodosis, D.T., Dreifuss, J.J. and Orci, L., 1977, Two classes of microvesicles in the neurohypophysis, Brain Res., 123:159.

Theodosis, D.T., Montagnese, C., Rodriguez, F., Vincent, J.-D. and Poulain, D., 1986, Oxytocin induces morphological plasticity in the adult hypothalamo-neurohypophysial system, Nature, (Lond.), 322:738.

Tweedle, C.D. and Hatton, G.I., 1987, Morphological adaptability at neurosecretory axonal endings on the neurovascular contact zone of the rat neurohypophysis, Neuroscience, 20:241.

van Putten, L.J.A., Kiliaan, A.J. and Buma, P., 1987, Ultrastructural localization of exocytotic release sites in immunocytochemically characterized cell types. A combination of two methods, Histochemistry, 86:375

Verbalis, J.G., Baldwin, E.F., Ronnekleif, O.K. and Robinson, E.G., 1986, In vitro release of vasopressin and oxytocin from rat median eminence tissue, Neuroendocrinology, 42:481.

Whitnall, M.H., Gainer, H., Cox, B.M. and Molineaux, C.J., 1985, Dynorphin-A-(1-8) is contained in vasopressin vesicles in rat pituitary, Science, 222:1137.

BIOSYNTHESIS AND RELEASE OF MULTIPLE PEPTIDES BY THE CAUDODORSAL

CELLS OF LYMNAEA STAGNALIS

E.W. Roubos

Department of Biology, Vrije Universiteit
P.O. Box 7161, 1007 MC Amsterdam
The Netherlands

The peptidergic Caudodorsal Cells (CDC) of the freshwater snail Lymnaea stagnalis control egg laying and egg-laying behaviour by releasing various peptides that act upon different targets (e.g., Roubos, 1984; Geraerts et al., 1987). Egg laying lasts about 2 hours and involves ovulation of up to 200 oocytes from the ovotestis, packaging of these cells by various types of female accessory sex gland into an egg mass, and oviposition. Overt egg-laying behaviour consists of a number of stereotyped behavioural acts (Goldschmeding et al., 1983). It begins with cessation of locomotion and posture changes. After about one hour the animal starts crawling about and cleans the substrate by rasping with its buccal mass before depositing the egg capsule. Actual oviposition takes 10-20 minutes depending on the size of the egg mass (Dogterom et al., 1983). Finally, the animal crawls back along the egg mass, touching it with the lip, before moving off. In this paper a brief survey will be given of the mechanisms by which CDC activity is controlled, of the way the CDC synthesize and release their peptides, and of the effects of the CDC peptides upon their targets. This may help to clarify how the CDC initiate an egg-laying behaviour that is coordinated with other important physiological processes such as feeding, locomotion and copulation (Fig. 1). Particular attention will be paid to structure-function relationships of the CDC-system.

STRUCTURE OF THE CDC

The structure of the CDC has been studied in much detail (for a review see Roubos, 1984). The CDC occur in two clusters in the cerebral ganglia (left: c. 25 cells, right: c. 75 cells) (Joosse, 1964). Each cluster contains c. 7 ventral CDC that have an axon branch running through the cerebral commissure. The "crossing axons" make electrotonic contacts with the contralateral CDC, thereby enabling the cells of both clusters to function as one unit (de Vlieger et al., 1980; Roubos et al., 1985). The CDC are characterized by a well-developed rough endoplasmic reticulum (RER) and Golgi apparatus and by electron-dense secretory granules with a mean diameter of about 150 nm (Wendelaar Bonga, 1971). The granule contents, including the ovulation-inducing peptide Caudodorsal Cell Hormone (CDCH), are released into the haemolymph by exocytosis from neurohaemal axon terminals located in the periphery of the cerebral commissure (Roubos and van der Wal-Divendal, 1980; Roubos and van de Ven, 1987; Roubos et al., 1987a). Furthermore, secretion occurs into the intercellular space of the central nervous system, from nonsynaptic release sites in the cerebral commissure (Fig. 2) (Roubos et al., 1983; Buma and Roubos, 1986; Schmidt and Roubos, 1987a,b).

GENES, PRECURSORS AND PEPTIDES

The recombinant DNA technique has been the method of choice to demonstrate that CDCH and related peptides in the CDC are initially synthesized as parts of large precursor proteins. Among the available recombinant DNA methods, the one that makes use of differential (or +/-) screening has proved particularly suitable for the isolation of cDNA

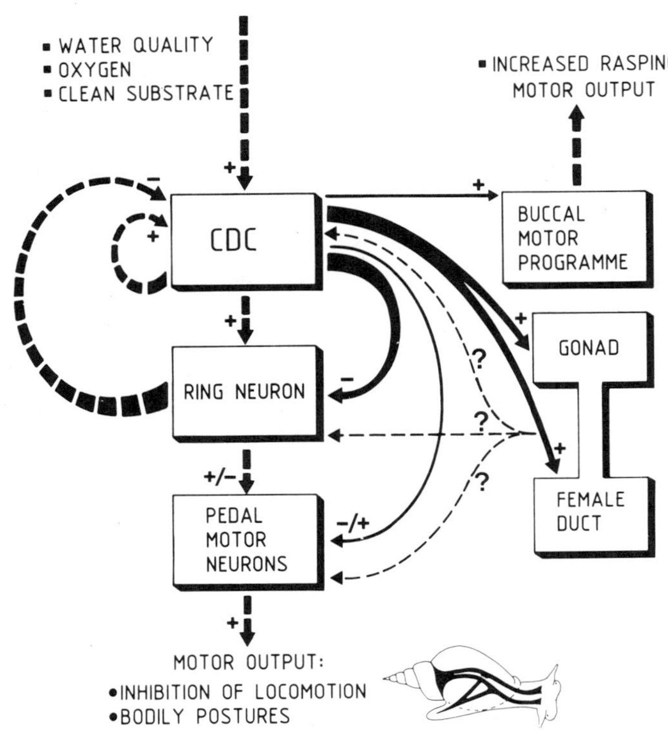

Fig. 1.   Scheme summarizing the organization of egg laying in <u>L. stagnalis</u>. External stimuli trigger, via an auto-excitatory peptide, the CDC-discharge, which induces ovulation and causes excitation of the Ring neuron. This neuron affects the firing of pedal motor neurons. In addition, the CDC control, possibly nonsynaptically, motor neurons in the pedal and buccal ganglia. Motor neuron activities can explain the elevated rate of rasping and the shell forward position, which are characteristic components of egg-laying behaviour. (----) neural pathway, (----) blood borne pathway (- - -), neural or blood borne pathway (modified after Geraerts et al., 1987).

clones encoding the precursors of CDCH (like) peptides. Furthermore, results have been obtained indicating that the CDCH gene is a member of a multigene family consisting of a small number of highly homologous, yet distinct, genes that are expressed in a tissue-specific fashion (Vreugdenhil et al., 1985; Geraerts et al., 1987; E. Vreugdenhil et al., unpublished results). Two members of the gene family are expressed in the CDC. One gene encodes the CDCH-precursor, the other encodes a CDCH-like precursor (Fig. 3).

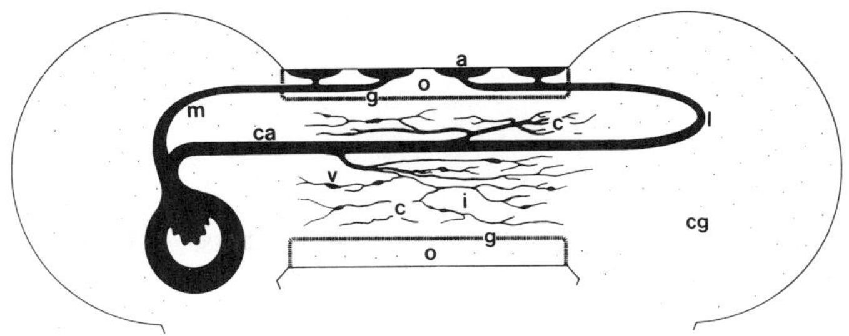

Fig. 2. Outer (o) and inner (i) compartments of the cerebral commissure (COM) of <u>L. stagnalis</u>. The compartments are separated by a glial sheath (g). Only one ventral CDC is shown. Its main axon (m) runs through the loop area (l) to the neurohaemal area, the outer compartment, where the axon terminals (a) end blindly. A branch of the main axon, the crossing axon (ca), runs through the inner compartment (i), passes through the contralateral loop area, and then runs to the outer compartment. In the inner compartment the crossing axon gives rise to the collaterals (c). cg, cerebral ganglion; v, varicosity.

A cDNA clone encoding the CDCH precursor has been completely sequenced. The precursor consists of 259 amino acids and contains 11 predicted peptides, one of which is CDCH. This 36 amino acids long peptide has an amidated carboxyterminal, a calculated molecular weight of 4529 Da and a pI of 9 (Ebberink et al., 1985). Another peptide has been identified as calfluxin (14 amino acids). This peptide stimulates the influx of calcium into the mitochondria of the albumen gland <u>in vitro</u> (Dictus et al., 1987). The influx may be related to the stimulation of glandular synthesis and/or release of perivitelline fluid, which surrounds the egg cells. Furthermore, 4 peptide regions occur with a marked repetitive character, viz. α Caudodorsal Cell Peptide (αCDCP) and the β_{1-3} Caudodorsal Cell Peptides (β_{1-3}CDCP's). Possibly, αCDCP is the auto-excitatory messenger (Moed et al., 1987) that induces high electrical CDC activity (CDC discharge; see below).

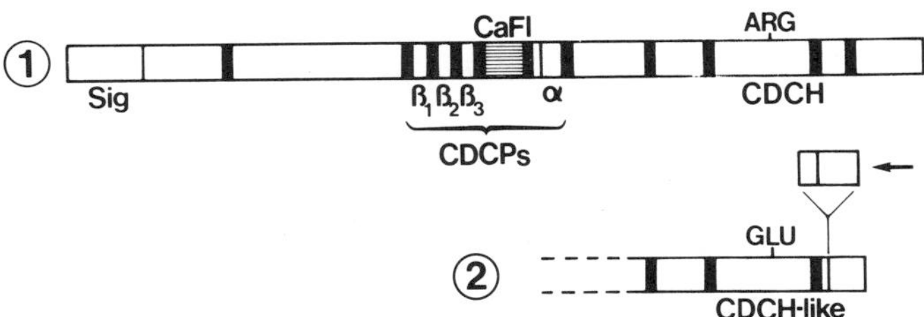

Fig. 3. Structure of the CDCH precursor (1) and the CDCH-like precursor (2) in the CDC. Potential cleavage sites are indicated by vertical bars, arrow points to deletion. CaFl calfluxin, Sig signal peptide (courtesy of E. Vreugdenhil).

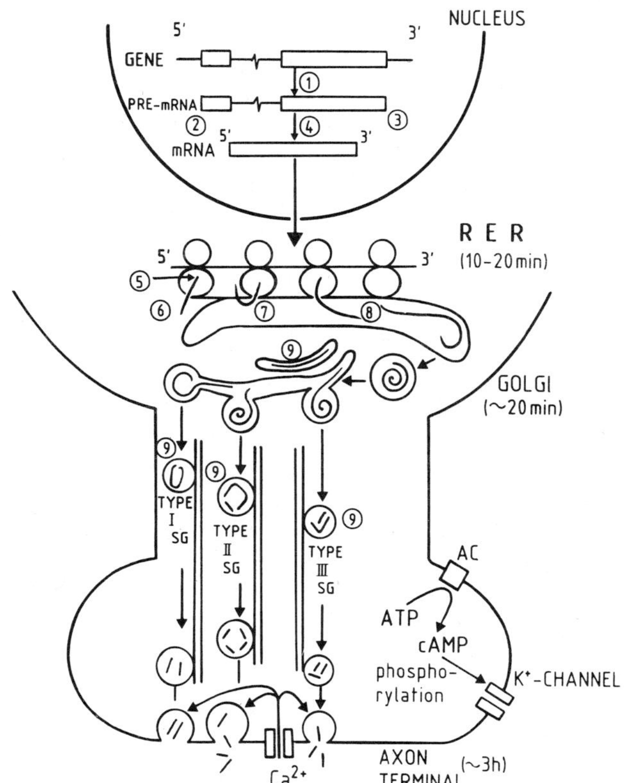

Fig. 4.   Partly hypothetical scheme of biosynthesis, transport and release of CDC-peptides controlling egg laying and associated behaviours in <u>L. stagnalis</u>. Approximate times required for newly synthesized peptides to reach various cell organelles have been indicated. In the gene and pre-mRNA the boxes represent exons and the lines represent the intron and the flanking regions of the gene. Enzymes, possibly involved in the processing of mRNA templates and of the CDCH- and CDCH-like precursors, are indicated by circled members. 1=RNA polymerase II, 2=enzymes involved in capping and methylation of the 5' terminus of mRNA, 3=polyadenylate polymerase, 4=RNA-splicing enzymes, 5=peptidyl transferase, 6=initiator methionyl aminopeptidase, 7=signalase, 8=enzymes that cleave the signal sequence from the precursor, 9=endo- and exopeptidases, and enzymes that modify the amino- and/or carboxyterminus of the peptides. Differential processing of secretory granule (SG) contents may led to exocytosis of different (sets of) peptides from morphologically different granule types (I, II, III). AC = adenylate cyclase. (For detailed information, see Geraerts et al., 1987 and Roubos et al., 1987b).

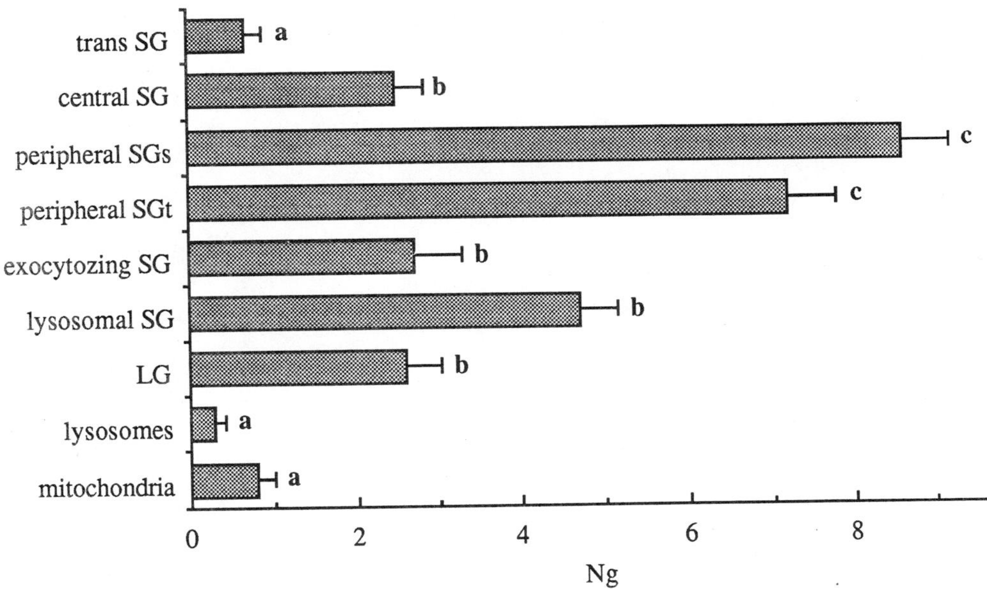

Fig. 5.   Intensity of immunolabelling with anti-CDCH of various organelles in CDC expressed in bar diagram as numbers of gold particles per organelle per sampling site (Ng) ± standard error of the mean. Statistically significant differences are denoted by different (bold) characters (P<0.05). Peripheral secretory granules were studied in the soma (SGs) and in the axon terminals (SGt) (after Roubos et al., 1987a).

The overall homology of the CDCH precursor with the egg-laying hormone (ELH) of <u>Aplysia californica</u> is very low (29%). However, a detailed comparison (cf. Chiu et al., 1979; Mahon et al., 1985) reveals a highly differential conservation of the encoded peptides. Significant homology exists in the CDCH/ELH region (48%), the αCDCP/α-/Bag cell peptide (BCP) (70%), the β_{1-3}CDCP/BCP-region (80%) and the calfluxin region (57%) (Vreugdenhil et al., 1985; Geraerts et al., 1987; E. Vreugdenhil et al., unpublished results; W.J.A.G. Dictus and R.H.M. Ebberink, unpublished results).

A cDNA clone encoding the CDCH-like precursor has been sequenced partially. It shows strong homology as well as clear differences with the CDCH-precursor (Fig. 3) (E. Vreugdenhil et al., unpublished results).

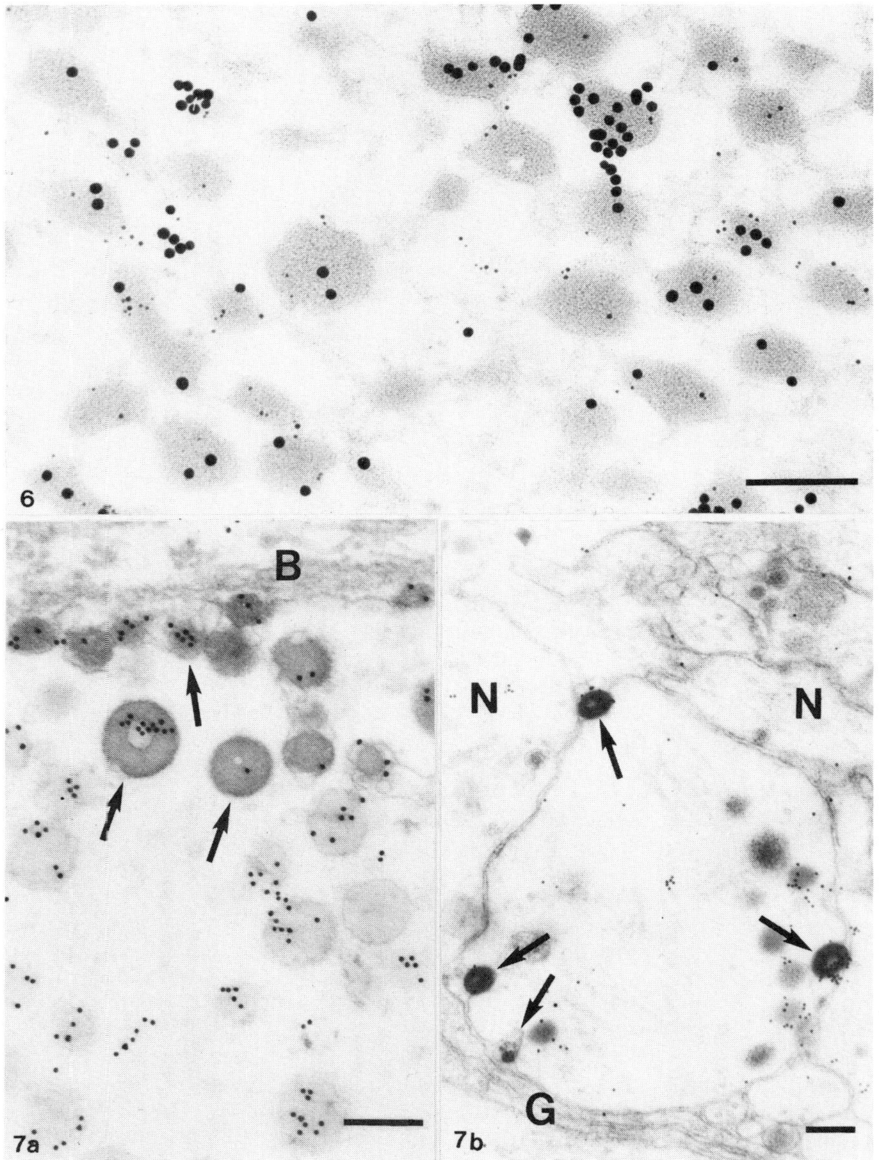

Fig. 6.   Immunoelectron microscopy (gold technique, double labelling) with anti-CDCH (small, 5nm gold particles) and anti-calfluxin (large, 15 nm gold particles). some secretory granules show colocalization of the respective peptides. Fixation with 0.5% glutaraldehyde. Bar: 0.2 μm.

Fig. 7.   Immunoelectron microscopy (gold technique) with anti-CDCH. Neurohaemal axon terminals (a) and collaterals (b) of CDC showing exocytosis of immunopositive contents of secretory granules (arrows). B basal lamina, G glial cell processes, N unidentified (non-CDC) neural element. Fixation with 0.5% glutaraldehyde and 1% OsO_4. Bar: 0.2 μm.

Pulse-label and pulse-chase studies have indicated that the CDC synthesize a 35 kDa protein that is cleaved into three intermediates (20, 10 and 7 kDa) and 5 end products of 6, 4.5, 3.5, 2 and 1.5 kDa molecular weight classes. The 4.5 kDa class contains CDCH-activity (Geraerts et al., 1985). Probably, the 35 kDa protein is the CDCH- (and/or CDCH-like) precursor, whereas the other end product classes may represent other CDC peptides.

CDCH is relatively rich of arginine (6 residues). Ultrastructural radioautographic (Roubos, 1985; Roubos et al., 1987b; E.W. Roubos and W.R.A. van Heumen, unpublished results) and biochemical pulse-chase studies with ^{3}H-arginine (Geraerts et al., 1987) show that CDCH is synthesized within 20 minutes by the RER and packed by the Golgi apparatus into secretory granules (Fig. 4). The cellular dynamics of CDCH have been studied in some detail using an antiserum raised to a synthetic portion of CDCH comprising the 20-36 amino acid sequence. With the ultrastructural secondary antibody-immunogold technique, specific immunoreactivity was found in all CDC. RER and Golgi apparatus showed very little reactivity as did secretory granules that were in the process of being budded off from the Golgi apparatus. However, secretory granules that were being discharged from the Golgi area showed more immunoreactivity, whereas apparently mature secretory granules, located distantly from the Golgi apparatus, were strongly reactive (Fig. 5). Apparently, CDCH is cleaved from the CDCH-precursor within secretory granules during granule migration from the Golgi apparatus; subsequently, the mature granules are transported towards the neurohaemal axon terminals where they release CDCH into the haemolymph (Fig. 4; Roubos et al., 1987b) (see also below).

Preliminary immunoelectron microscopy studies (poly- and monoclonal antibodies) indicate that secretory granules of the CDC contain not only CDCH but also CDCP-like peptide and calfluxin. Double-labelling experiments using differently sized gold probes show that at least some of the peptides are co-localized within the same granule (Fig. 6). This does not only hold for peptides derived from the same precursor; CDCH and the CDCH-like peptide are colocalized in at least ca 20% of all secretory granules (W.R.A. van Heumen and E.W. Roubos, unpublished results).

AXONAL TRANSPORT

On the basis of autoradiographic evidence it appears that secretory granules in the CDC travel at a speed of 15-20mm/day (Roubos, 1985; Geraerts et al., 1987). Pulse-label and pulse-chase experiments in the presence of blockers of axonal transport such as colchicine and vinblastine, show an accumulation in the somata of intermediates and end products, and a drastic delay in the appearance of end products in the axon terminals. More detailed electron microscopy observations show that vinblastine, vincristine, and colchicine cause a strong accumulation of secretory granules within the CDC somata, whereas no repletion of the neurohaemal area occurs after experimentally induced hormone release (Roubos et al., 1987b; unpublished data). Apparently, this block of axonal granule transport is caused by action upon the microtubular system; microtubules disappear after colchicine treatment and transform into paracristalline structures upon reaction with Vinca alkaloids (Müller et al., 1987). These results suggest that microtubules are involved in the transport of secretory granules in CDC axons (Fig. 4).

NEUROHAEMAL RELEASE

Egg-laying is a cyclic process that is reflected by three different states of electrical CDC activity, viz. the active state ("discharge", lasting about one hour), which is followed by the inhibited state (c. 5 hours) and the subsequent resting state, which has a variable length and may last up to several days (Kits, 1980). All CDC are in the same electrical state, due to the fact that they are electrotonically coupled (ter Maat et al., 1983). In vitro and in vivo (fine wire) studies (ter Maat et al., 1986) have clearly demonstrated that the CDC-discharge is a crucial event in egg laying (Kits, 1980; Geraerts et al., 1987). The auto-excitatory peptide plays an important role in the onset of the CDC-discharge (Moed et al., 1987). Probably, it acts via receptors upon adenylate cyclase (Fig. 4). This enzyme has been demonstrated cytochemically in the axolemma of the neurohaemal CDC axon terminals and is active particularly during the active state (Roubos et al., 1981a). cAMP-analogues and IBMX can induce the discharge, indicating that an intracellular rise of cAMP is responsible for the start of the discharge <u>in vivo</u> (Buma et al., 1986). Calcium ions are required for the release of CDCH (Roubos et al., 1981b). Recently it has been found that the cGMP-analogue 8-bromoGMP induces high exocytosis activity in the CDC (Roubos et al., 1987b). Consequently, a role of cGMP in the control of peptide release by the CDC seems feasible. In addition to calcium ions and cyclic nucleotides, still other factors (e.g. internal pH and inositoltriphosphate) appear to be involved in the induction of the discharge (Moed et al., 1987). Possibly, first the auto-excitatory peptide and then CDCH is released. This non-simultaneous release would correlate with the consecutive release of contents of two morphologically different types of secretory granule (Roubos et al., 1987b).

Shortly after the start of the discharge the titer of CDCH in the haemolymph starts to increase rapidly (quantitative bioassays of ovulation inducing activity in the haemolymph; Geraerts et al., 1984). This CDCH-release is reflected at the ultrastructural level by the frequent occurrence in the neurohaemal terminals of exocytosis of secretory granule contents. These contents can be selectively demonstrated with ultrastructural tannic acid methods (e.g. Roubos and van der Wal-Divendal, 1980). They are immunopositive with anti-CDCH (Fig. 7a). Exocytosis is concomitant with a depletion of secretory granules from the terminals (Buma et al., 1984; Roubos, 1984; Roubos et al., 1987a).

About 3h after the start of the discharge the CDCH-titer has decreased to zero (Geraerts et al., 1984). In contrast, the number of exocytosis profiles remains fairly high up to many hours after the end of the discharge. This indicates that the CDC release non-ovulation inducing secretory material during the inhibited state (when the cells are electrically inactive) (Roubos, 1984). Probably, exocytosis during the inhibited state depends on the action of calcium released from mitochondria (which take up excess calcium during the discharge; Buma and Roubos, 1983). The released material (other CDC peptides?) may play a role in the control of the activity of some components of egg-laying behaviour and/or in the regulation of the secretory activity of accessory sex glands.

Recent immunoelectron microscopy studies indicate that, in addition to CDCH, the CDCH-like peptide, αCDCP and calfluxin are released into the haemolymph. This means that the CDC release products into the haemolymph that are derived from both the CDCH- and the CDCH-like precursor (W.R.A. van Heumen and E.W. Roubos, unpublished results). The relations between the release of these peptides, the electrical state of the CDC and the physiological effects of these peptides have to be established.

NONSYNAPTIC COMMUNICATION

The cerebral commissure consists of two compartments that are separated by a continuous sheath of glial cells (Fig. 2). The outer compartment is formed by the neurohaemal area of the CDC, the inner consists of thousands of, mainly unidentified, axons. Ventral CDC send axons through the inner compartment. These give rise to collaterals, which divide into smaller collaterals that form an extensive network ("collateral system") throughout the inner compartment. Eventually, collaterals end blindly within the inner compartment. The collaterals never form synaptic contacts; exocytotic release of the contents of secretory granules takes place at nonsynaptic release sites, which lack the morphological specializations of the plasma membrane and the intercellular space characteristic of classical synapses (Fig. 7b; Schmidt and Roubos, 1987a). Preliminary immunoelectron microscopy studies (W.R.A. van Heumen and E.W. Roubos, unpublished results) suggest that not only CDCH but also other CDC peptides (αCDCP, CDCH-like peptide, calfluxin) are released from the collaterals into the interneuronal space. Interestingly, secretory granules in the collateral system appear to contain twice as less CDCH as compared to secretory granules in the neuro-haemal axon terminals.

Using the tannic acid-Ringer incubation-method (TARI-method; Buma et al., 1984) for the detection of exocytotic release of secretory granule contents <u>in vitro</u>, we found that elevation of the extracellular potassium concentration strongly stimulates exocytosis activity in the collaterals. No stimulation occurs in the absence of extracellular calcium ions.

Electron-dense material occurs apposed at the cytoplasmic side of the axolemma of collaterals (ethanolic phosphotungstic acid method). This material appears homologous with the presynaptic dense projections forming the "vesicular grid" in classical synapses. Such projections are also present in the neurohaemal axon terminals. Apparently, secretion from nonsynaptic release sites in CDC collaterals, shares fundamental characteristics with release from conventional neuronal release sites (neurohaemal axon terminals and classical synapses): exocytosis, association with a vesicular grid, induction by membrane depolarization, and dependence on extracellular calcium ions (Schmidt and Roubos, 1987b; E.D. Schmidt, unpublished data).

Nonsynaptic release of neuronal messengers has been implicated to be involved in nonsynaptic ("diffuse", "paracrine", "at a distance", "hormonal-like") communication within the central nervous system (e.g. Roubos et al., 1983; Schmitt, 1984; Schmidt and Roubos, 1987a,b). We suppose that the products released from the nonsynaptic release sites of the collateral system are involved in nonsynaptic communication of the CDC with remote targets in the central nervous system, thus controlling (some components of the) stereotyped phases of egg-laying behaviour. One possible target is the cerebral Ring neuron, which sends an axon branch through the inner compartment and, as was previously shown neurophysiologically, is controlled by the CDC in a non-synaptic fashion. The Ring neuron controls pedal motorneurons involved in locomotion. In addition, the CDC may control buccal and pedal motorneurons nonsynaptically (Jansen, 1984; Jansen and ter Maat, 1985).

The dynamics of collateral release appear to be completely different from those of neurohaemal release, as the collaterals show maximum exocytosis activity during the resting and inhibited state and low exocytosis activity during the discharge (Schmidt and Roubos, 1987b). This suggests that neurohaemal and collateral release are controlled independently. The glial sheath that separates the inner compartment from the outer may serve as a selective barrier preventing some substances (e.g. CDC peptides?) from passing between the two compartments (Schmidt and Roubos, 1987a).

INPUTS TO THE CDC

Field studies (Dogterom et al., 1985) indicate that the egg-laying activity of <u>L. stagnalis</u> is determined by the action of various different environmental factors, i.e. food quantity, photoperiod and water temperature. Laboratory studies (for review see Geraerts et al., 1987) have shown that these factors influence the rate of oviposition, suggesting modulatory effects on biosynthesis and release of CDCH. Furthermore, it has been shown that intrinsic (neuronal or neurohormonal) factors from the lateral lobes, small ganglionic appendices of the cerebral ganglia, stimulate the rate of oviposition (Geraerts, 1976). Subsequent ultrastructural morphometry studies have demonstrated that the lobes stimulate the protein synthetic machinery as well as the release activity of the CDC (Roubos et al., 1980). In addition, a number of other factors are known to affect CDC activities, including parasitical infection, tactile stimulation, copulation and clean water stimulation (for refs. see Geraerts et al., 1987).

The pathways, neural elements and chemical messengers that are involved in the relay of environmental stimuli to the CDC are only partly known (for reviews see Roubos, 1984; Geraerts et al., 1987). The clean water stimulus elicits the CDC discharge. It is transferred via the n. analis and the n. pallialis dexter externus, bilaterally to the CDC. The so called SWAP-neuron forms an extensive network in the pedal ganglia. It innervates the CDC as well as neurons controlling eating and copulation behaviour. Probably, under normal (non-egg laying) conditions, the network inhibits CDC activity. Tactile stimuli run via all peripheral nerves and project to the CDC via cholinergic synapses on the CDC axons. Copulation accelerates the onset of maturation of the female reproductive apparatus and of egg laying, probably via the transfer of substances from male accessory sex glands. The sensory cells involved may be located in the penis and the vagina.

Recent studies, involving light immunocytochemistry using an anti-CDCH antibody and <u>in situ</u> hybridization with a cDNA probe encoding CDCH, have indicated that CDCH is not only produced by the CDC but also by other central neurons (van Minnen et al., 1987).

Most of these CDC-like cells occur singly or in small groups in the pleural, pedal and visceral ganglia. They are considered as ectopic CDC, because they have the same size and ultrastructure as the CDC and have neurohaemal axon terminals in the periphery of nearby connectives. Possibly, they are involved in the control of egg laying and egg-laying behaviour as well. Furthermore, a group of CDC-like cells occurs laterally in each cerebral ganglion. These cells are much smaller than the CDC and their axons seem to make synaptic contacts with the CDC axons in the cerebral neuropiles. It has been suggested that they control CDC activity in cooperation with similar neurons in the accessory sex glands (see below).

ONTOGENY OF THE CDC

Light and electron microscopy studies (Roubos et al., 1987b); unpublished data) show that CDC are present already in very young snails 93 weeks old, 3mm shell height). During development the CDC-system increases in volume by cell growth as well as by cell multiplication. Ultrastructural evidence for synthesis, storage and release of CDC secretory material has been obtained in snails as small as 6mm. Secretory granules are remarkably small in juvenile snails but strongly increase in size up to adulthood. This granule "growth" occurs rather discontinuously, with a sudden strong increase from about 15mm shell height onwards. Possibly, this increase is related to the development of the reproductive system. In view of the supposed relationship between morphology and content of a secretory granule (Roubos and van de Ven, 1987), it seems that during the development of the CDC qualitative and/or quantitative changes occur in the granule contents, possibly caused by a change in the mode and/or rate of processing of precursor proteins into secretory peptides.

Recent light and electron microscopy studies show that the collateral system develops almost synchronously with the neurohaemal axon terminals. It first appears in snails of 10mm shell height. During development the collaterals strongly increase in size and, especially, in number. The diameter of the secretory granules increases, reaching a final size in sexually mature snails (>25mm). Release by exocytosis of secretory material into the intercellular space of the cerebral commissure was observed in collaterals from 10mm onwards. Immunoelectron microscopy shows that the released material in juveniles is CDCH(-like).

AGEING OF THE CDC

Under laboratory conditions <u>L. stagnalis</u> may live for about 2 years. By that time most snails have ceased laying eggs. We found that during ageing the CDC somata show a marked and progressive decrease in the number and volume of the Golgi apparatus, vacuolization and whirl formation of the RER, a dramatic increase in the number and size of lysosomal structures and a conspicuous increase in the number of microtubules. Secretory granules are positive with anti-CDCH, but no formation of new secretory granules takes place. These signs of degeneration do not occur in all CDC somata at the same time. During ageing some neurohaemal axon terminals show a strong depletion of secretory granules and reveal abundant signs of exocytosis, the released contents reacting positively with CDCH. However, other terminals do not show release activity and are studded with granules.

Apparently, the decrease of reproductive activity of <u>Lymnaea</u> during ageing is closely related to a successive degeneration of CDC. Cell by cell the CDC loose their capacity to synthesize and release secretory material. Neurophysiological and biochemical studies of ageing CDC are in progress. CDC degeneration may serve as a model for ageing studies of peptidergic neurons in general.

PRODUCTION OF CDCH(-LIKE) PEPTIDES OUTSIDE THE CNS

Using a combination of <u>in situ</u> hybridization, Northern blotting and immunocytochemistry, it has been shown that CDCH(-like) transcripts and peptides are also present in a variety of neural and non-neural tissues outside the CNS (van Minnen and Vreugdenhil 1987; van Minnen et al., 1987; E. Vreugdenhil, unpublished results). Peripheral neurons expressing CDCH(-like) material have been demonstrated in the oothecal gland, the muciparous gland and the pars contorta, which are female accessory sex glands. In these glands the processes of the neurons terminate on the secretory cells, suggesting that they control glandular secretory activity. CDCH-immunoreactive material has also been found in secretory cells of the prostate gland and sperm duct, as well as in the lumen of the male duct, suggesting that CDCH(-like) peptides are released and transported to the partner during copulation. The biological role of these peptides and their molecular structure, however, are as yet unclear. Possibly, the peptides are derived from different precursors that are encoded by a family of CDCH-genes. The possibility has been raised that all cells expressing one or more of these genes, within as well as outside the CNS, act together in reproduction (van Minnen et al., 1987).

ACKNOWLEDGEMENTS

The author is greatly indebted to Dr. W.P.M. Geraerts for reading the manuscript, and to Mrs. A.M.H. van de Ven, Mrs. C.M. Moorer-van Delft and Mrs. E. Veenstra for performing some of the experiments. Part of the studies have been made possible by grants to E.D. Schmidt and E. Vreugdenhil from the Foundation for Fundamental Biological Research (B.I.O.N.), which is subsidized by the Netherlands Organization for the Advancement of Pure Research (Z.W.O.).

REFERENCES

Buma, P., and Roubos, E.W., 1983, Calcium dynamics, exocytosis, and membrane turnover in the ovulation-hormone releasing caudo-dorsal cells of Lymnaea stagnalis, Cell Tiss. Res., 233:143.

Buma, P., Roubos, E.W., and Buijs, R.M., 1984, Ultrastructural demonstration of exocytosis of neural, neuroendocrine and endocrine secretions with an in vitro tannic acid (TARI-) method, Histochemistry, 80:247.

Buma, P., and Roubos, E.W., 1986, Ultrastructural demonstration of nonsynaptic release sites in the brain of the snail Lymnaea stagnalis, the insect Periplaneta americana, and the rat, Neuroscience, 17:867.

Buma, P., Roubos, E.W., and Brunekreef, K., 1986, Role of cAMP in electrical and secretory activity of the neuroendocrine caudo-dorsal cells of Lymnaea stagnalis, Brain Res., 380:26.

Chiu, A.Y., Hunkapillar, M.W., Heller, E., Stuart, D.K., Hood, L.E., and Strumwasser, F., 1979, Purification and primary structure of the neuropeptide egg-laying hormone of Aplysia californica, Proc. Natl. Acad. Sci. U.S.A., 76:6656.

Dictus, W.J.A.G., Jong-Brink, M. de, and Boer, H.H., 1987, A neuropeptide (calfluxin) is involved in the influx of calcium into mitochondria of the albumen gland of the freshwater snail Lymnaea stagnalis, Gen. Comp. Endocrinol., 65:439.

Dogterom, G.E., Bohlken, S., and Joosse, J., 1983, Effect of the photoperiod on the time schedule of egg mass production in Lymnaea stagnalis, as induced by ovulation hormone injections, Gen. Comp. Endocrinol., 49:255.

Dogterom, G.E., Thijssen, R., and Loenhout, H. van, 1985, Environmental and hormonal control of the seasonal egg-laying period in field specimens of Lymnaea stagnalis, Gen. Comp. Endocrinol., 57:37.

Ebberink, R.H.M., Loenhout, H. van, Geraerts, W.P.M., and Joosse, J. 1985, Purification and amino acid sequence of the ovulation neurohormone of Lymnaea stagnalis, Proc. Natl. Acad. Sci. U.S.A., 82:7767.

Geraerts, W.P.M., 1976, The role of the lateral lobes in the control of growth and reproduction in the hermaphrodite freshwater snail Lymnaea stagnalis, Gen. Comp. Endocrinol., 19:97.

Geraerts, W.P.M., Maat, A. ter, and Hogenes, T.M., 1984, Studies on release activities of the neurosecretory caudo-dorsal cells of Lymnaea stagnalis, pp 44-50, in: "Biosynthesis, Metabolism and Mode of Action of Invertebrate Hormones," J. Joffmann, and M. Porchet, eds., Springer, Berlin.

Geraerts, W.P.M., Vreugdenhil, E., Ebberink, R.H.M., and Hogenes, T.M., 1985, Synthesis of multiple peptides from a larger precursor in the neuroendocrine caudo-dorsal cells of Lymnaea stagnalis, Neurosci. Lett., 56:241.

Geraerts, W.P.M., Maat, A. ter, and Vreugdenhil, E., 1987, The peptidergic neuro-endocrine control of egg-laying behavior in Aplysia and Lynmaea, in: "Invertebrate Endocrinology," Vol. 2, H. Laufer, and R. Downer, eds., Liss, New York, (In Press).

Goldschmeding, J.T., Wilbrink, M., and Maat, A. ter, 1983, The role of the ovulation hormone in the control of egg-laying behaviour in Lymnaea stagnalis, pp 251-255, in: Molluscan neuro-endocrinology, J. Lever, and H.H. Boer, eds., North-Holland Publishing Company, Amsterdam.

Jansen, R.F., 1984, Neuronal and hormonal control of the egg-laying behavior in the pond snail Lymnaea stagnalis, Thesis, Vrije Universiteit, Amsterdam.

Jensen, R.F., and Maat, A. ter, 1985, Ring neuron control of columellar motor neurons during egg-laying behavior in the pond snail, J. Neurobiol., 16:1.

Joosse, J., 1964, Dorsal bodies and dorsal neurosecretory cells of the cerebral ganglia of Lymnaea stagnalis L., Arch. Neerl. de Zool., 15:1.

Kits, K.S., 1980, States of excitability in ovulation hormone producing neuroendocrine cells of Lymnaea stagnalis (Gastropoda) and their relation to the egg-laying cycle, J. Neurobiol., 11:397.

Maat, A. ter, Roubos, E.W., Lodder, J.C., and Buma, P., 1983, Integration of biphasic synaptic input by electrotonically coupled neuroendocrine caudo-dorsal cells in the pond snail, J. Neurophysiol., 49:1392.

Maat, A. ter, Dijcks, F.A., and Bos, N.P.A., 1986, In vivo recordings of neuroendocrine cells (caudo-dorsal cells) in the pond snail, J. Comp. Physiol. A, 158:853.

Mahon, A.C., Nambu, J.R., Taussig, R., Shyamala, M., Roach, A., and Scheller, R.H., 1985, Structure and expression of the egg-laying hormone gene family in Aplysia, J. Neurosci., 5:1872.

Minnen, J. van, and Vreugdenhil, E., 1987, The occurrence of gonadotropic hormones in the central nervous system and reproductive tract of Lymnaea stagnalis. An immunocytochemical and in situ hybridization study, pp. 62-67, in: Neurobiology. Molluscan models, H.H. Boer, W.P.M. Geraerts, and J. Joosse, eds., North-Holland Publishing Company, Amsterdam.

Minnen, J. van, Haar., C. van der, Raap, A.K., and Vreugdenhil, E., 1987, Localization of ovulation hormone-like neuropeptide in the central nervous system of the snail Lymnaea stagnalis with immunocytochemistry and in situ hybridization, Cell Tiss. Res., (In press).

Moed, P.J., Boss, N.P.A., Kits, K.S., Maat, A. ter, 1987, The role of release products and second messengers in the regulation of electrical activity of the neuroendocrine caudodorsal cells of Lymnaea stagnalis, pp. 194-199, in Neurobiology. Molluscan models, H.H. Boer, W.P.M. Geraerts, and J. Joosse, eds., North-Holland Publishing Company, Amsterdam.

Müller, L.J., Ven, A.H.M., van de, and Roubos, E.W., 1987, Neurons of the snail Lymnaea stagnalis as a model for the study of clinical neurotoxic side effects of anti-tumor agents, pp. 138-142, in: Neurobiology, Molluscan models, H.H. Boer, W.P.M. Geraerts, and J. Joosse, eds., North-Holland Publishing Company, Amsterdam.

Roubos, E.W., 1984, Cytobiology of the ovulation-neurohormone producing caudo-dorsal cells of the snail Lymnaea stagnalis, Int. Rev. Cytol., 89:295.

Roubos, E.W., 1985, Intracellular and extracellular control of neuroendocrine activity in the freshwater snail Lymnaea stagnalis, pp 47-49, in: Current trends in comparative endocrinology, B. Lofts, and W.N. Holmes, eds., Hong Kong.

Roubos, E.W., and Wal-Divendal. R.M. van der, 1980, Ultrastructural analysis of peptide-hormone release by exocytosis, Cell Tiss. Res., 207:267.

Roubos, E.W., and Ven, A.M.H. van de, 1987, Morphology of neurosecretory cells in Basommatophoran snails homologous with egg-laying and growth-hormone producing cells of Lymnaea stagnalis, Gen. Comp. Endocrinol., (In press).

Roubos, E.W., Geraerts, W.P.M., Boerrigter, G.H. and Kampen, G.P.J. van, 1980, Control of activities of the neurosecretory light green and caudodorsal cells and of the endocrine dorsal bodies by the lateral lobes in the freshwater snail Lymnaea stagnalis (L.), Gen. Comp. Endocrinol., 40:446.

Roubos, E.W., Keijzer, A.N. de, and Buma, P., 1981a, Adenylate cyclase activity in axon terminals of ovulation-hormone producing neuroendocrine cells in Lymnaea stagnalis, Cell Tiss. Res., 220:665.

Roubos, E.W., Schmidt, E.D., and Moorer-van Delft, C.M., 1981b, Ultrastructural dynamics of exocytosis in the ovulation neurohormone producing caudo-dorsal cells of the freshwater snail Lynmaea stagnalis (L.), Cell Tiss. Res., 215:63.

Roubos, E.W., Buma, P., and Roos, W.F. de, 1983, Ultrastructural correlates of electronic and neurochemical communication in Lymnaea stagnalis, with particular reference to nonsynaptic transmission and neuroendocrine cells, pp 78-81, in: Molluscan neuro-endocrinology. J. Lever, and H.H. Boer, eds., North-Holland Publishing Company, Amsterdam.

Roubos, E.W., Leeuwen, J.P.T.M. van, and Maijers, A., 1985, Ultrastructure of gap junctions in the central nervous system of Lymnaea stagnalis, with particular reference to electrotonic coupling between the neuroendocrine caudo-dorsal cells, Neuroscience, 14:711.

Roubos, E.W., Ven, A.M.H. van de, and Minnen, J. van, 1987a, Immunoelectron microscopy of formation, degradation and exocytosis of the ovulation neurohormone of Lymnaea stagnalis, Cell Tiss. Res., (In press).

Roubos, E.W., Ven, A.M.H. van der, Schmidt, E.D., and Heumen, W.R.A. van, 1987b, Structural aspects of biosynthesis, storage and release of neuropeptides in freshwater snails, pp. 82-88, in: Neurobiology. Molluscan models, H.H. Boer, W.P.M. Geraerts, and J. Joosse, eds., North-Holland Publishing Company, Amsterdam.

Schmidt, E.D., and Roubos, E.W., 1987a, Morphological basis for nonsynaptic communication within the central nervous system by exocytotic release of secretory material from the egg-laying stimulating neuroendocrine caudo-dorsal cells of Lymnaea stagnalis, Neuroscience, 20:247.

Schmidt, E.D., and Roubos, E.W., 1987b, Morphology and dynamics of nonsynaptic release sites of the caudo-dorsal cells of Lymnaea stagnalis, pp. 89-94, in: Neurobiology. Molluscan models, H.H. Boer, W.P.M. Geraerts, and J. Joosse, eds., North-Holland Publishing Company, Amsterdam.

Schmitt, F.O., 1984, Molecular regulators of brain function: a new view, Neuroscience, 13:991.

Vlieger, T.A. de, Kits, K.S., Maat, A. ter, and Lodder, J.C., 1980, Morphology and electrophysiology of the ovulation hormone producing neuro-endocrine cells of the freshwater snail Lymnaea stagnalis (L.), J. Exp. Biol., 84:259.

Vreugdenhil, E., Geraerts, W.P.M., Jackson, J.F., and Joosse, J., 1985, The molecular basis of the neuro-endocrine control of egg-laying behaviour in Lymnaea, Peptides, 6 (suppl. 3):465.

Wendelaar Bonga, S.E., 1971, Formation, storage, and release of neurosecretory material studied by quantitative electron microscopy in the freshwater snail Lymnaea stagnalis (L.), Z. Zellforsch., 113:490.

EMERGING IDENTITY IN CYTOPHYSIOLOGY OF SYNAPTIC AND NEUROSECRETORY TERMINALS

D.W. Golding, D.V. Pow*, Emine Bayraktaroglu+, Barbara A. May and R.M. Hewit

Department of Zoology
University of Newcastle upon Tyne
*Present address,
Department of Human Anatomy
University of Oxford

+Department of Biological
Sciences
Middle-East Technical University
Ankara

INTRODUCTION

Comparison of the organization of presynaptic terminals that typically release neurotransmitters at specialized junctions with other cells, and neurosecretory endings contributing to neurohaemal complexes and releasing hormones into the blood stream, has been a feature of interest since the commencement of their study (e.g., see Palay, 1958). Although terminals show considerable variation, the great majority, whether synaptic or neurohaemal, have in common a highly distinctive pattern of ultrastructure (Figs. 1, 2).

Possible secretory inclusions usually fall clearly into two categories. First, secretory granules are typically 70-200 nm in diameter and have electron dense contents. They show considerable variation in size and appearance, and can often be used as a basis for classifying the neurones involved. Second, synaptic vesicles are present in synaptic terminals, and comparable elements (Sharrer, 1968) are encountered in neurohaemal endings (hence their name 'synaptoid'). They typically show great homogeneity, measure 20-50 nm in diameter, and have lucent contents after conventional fixation. However, in contrast to the granules they have intensely dense cores following treatment with the Z10 reagent (Fig. 3). The inclusions normally exhibit a marked zonation. The vesicles are most densely concentrated adjacent to presumptive sites of release, which are marked, particularly in anthropods, by paramembranous thickenings or bars. In contrast, the granules 'stand back' and where they are abundant dominate the body of the terminal. There are doubtless many exceptions to this pattern. We have encountered two types of neurones (one endocrine and the other with terminals in the CNS) in the brain of <u>Lumbricus</u> that have secretory granules in abundance and show discharge of their contents by exocytosis, whereas vesicles are at most so poorly represented that they probably have no major role.

SYNAPTIC AND SYNAPTOID

The ultrastructural pattern described above is routinely encountered in components of most neurohaemal complexes. These range from the primitive complex of <u>Lumbricus</u> (Fig. 4), (Golding and Whittle, 1977), and the release sites formed by neurosecretory somata in <u>Helix</u> (Bayraktaroglu et al., 1988b) (Fig. 5), to the classic, well differentiated neurohaemal organs of anthropods (Figs. 3, 6) and vertebrates. Most progress in elucidating the roles of possible secretory inclusions in these terminals has been made with respect to the granules.

In contrast, our investigations have focused on the synaptoid vesicles. We have shown, first, that even features peculiar to synaptic structures in particular animals are mirrored by their synaptoid counterparts (May and Golding, 1982a). The cerebral ganglia of polychaete annelids are particularly amenable for such analysis since parallel observations can be made, in the same section or specimen, on synaptic contacts formed en passant and synaptoid endings of the same fibres (cf., Anwyl and Finlayson, 1973). Lucent vesicles at both sites form tight clusters with interstitial dense material that make contact with the plasmalemma adjacent to the synaptic cleft and neural lamella, respectively (Figs. 7-9). The vesicles show identical patterns of impregnation with Z10 (May and Golding, 1982b). Similarly, Binnington and Lane (1982) have shown that synaptoid vesicles in insects are clustered around the 'T-bars', and around the associated intramembranous particles. These particles are very characteristic of synaptic terminals in these animals. The identity of such configurations in the corpus cardiacum of the locust or cricket as 'synaptic' is unmistakeable (Fig. 6) (Golding and Pow, unpublished), just as, for example, cilia in sense organs can be confidently identified as such. A small minority of synaptic and neurohaemal terminals contain dense-cored vesicles after conventional fixation (Figs. 10, 11), but the duality of secretory inclusions based on size and distribution remains unmistakeable. In summary, we would categorically affirm that synaptic and synaptoid terminals are indistinguishable ultrastructurally.

A second line of research has concerned the involvement of synaptoid vesicles in secretory discharge. Such investigations are usually hampered by both the ephemeral nature of the process of exocytosis and the difficulties of distinguishing exocytotic omega profiles from their endocytotic counterparts. Consequently, we have exploited a system in <u>Helix</u> that is characterized by the presence of synaptoid vesicles that are dense cored after initial fixation in aldehydes (they are typically lucent when treated with OsO_4 alone). Furthermore, we have employed tannic acid which, when infused at neutral pH in Ringer solution (the TARI technique, Buma et al., 1984), 'freezes' the cores of secretory inclusions in position as they are discharged, without penetrating cell membranes or blocking further release. When the exocytotic profiles that have accumulated during the time of infusion are subsequently fixed, their discharged materials show an enhanced density due to a mordanting action of the acid.

Electron microscopy has revealed that both secretory granules and synaptoid vesicles discharge their contents by exocytosis (Figs. 12, 13) adjacent to the neural lamella (which is often bathed directly in the haemolymph) (Bayraktaroglu et al., 1988a). Release by granules and vesicles, respectively, is usually independent, although the possibility that both categories contribute to sites of compound exocytosis cannot be ruled out.

Crucially, some peptidergic neurosecretory cells have amines colocalized within them (Meister et al., 1985; Vieillemaringe et al., 1982). Such amines are stored within synaptoid vesicles in the median eminence (Richards and Tranzer, 1974). The same pattern probably applies to the many invertebrate aminergic neurosecretory cells that have now been described, as to elements that secrete GABA (Anderson and Mitchel, 1986). Further biochemical similarity of synaptic and synaptoid vesicles is shown by the common presence of a particular protein, synapsin I, in their bounding membranes. This protein is not present in granules (Navone et al., 1984).

The capacities of synaptoid vesicles to incorporate extracellular markers, and to sequester Ca^{2+} (review by Morris et al., 1981) are well known. In fact, such findings are perfectly consistent with the identification of these elements as, in effect, synaptic vesicles,

---

Figs. 1-6. (Opposite page) Fig. 1. Synaptic terminal - cerebral ganglion, <u>Lumbricus terrestris</u>; arrow, presynaptic thickening. Fig. 2. Neurosecretory terminal - sinus gland <u>Carcinus maenas</u>; arrows, synaptoid thickenings. Fig. 3. Neurosecretory terminal - Z10, <u>C. maenas</u>. Figs. 4-6. Neurosecretory terminals - cerebral ganglia, <u>L. terrestris</u> and <u>Helix pomnatia</u>, and corpus cardiacum, <u>Achaeta domesticus</u>, respectively. Latter shows dense bar or ribbon (arrow) with associated vesicles. g, secretory granules; v, synaptic/synaptoid vesicles; nl, neural lamella; s, stroma; vl, vascular lamella. Bars, 200nm. (Fig. 5 courtesy of Dr. A.C. Whittle.)

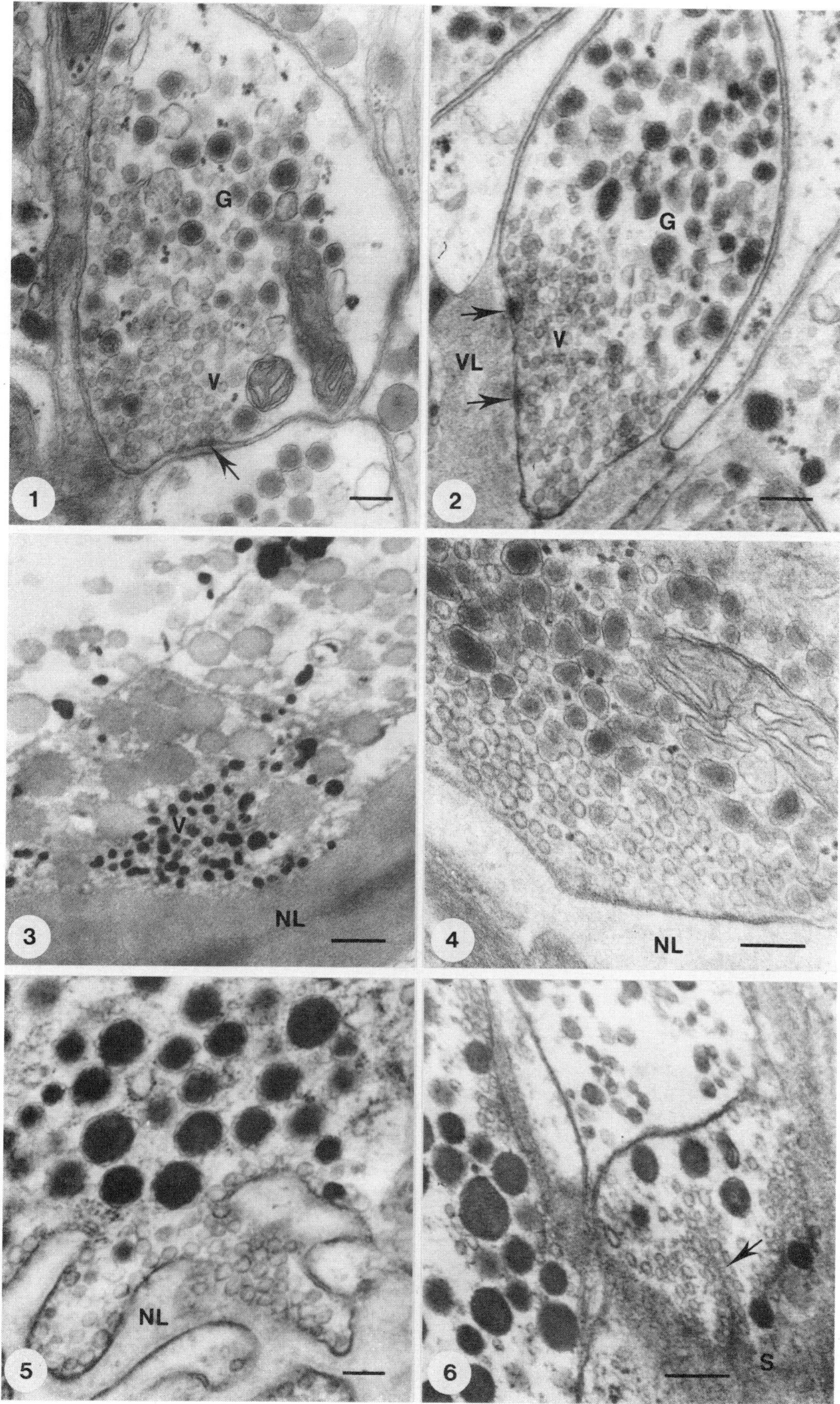

G
V
1
G
V
VL
2
V
NL
3
NL
4
NL
5
S
6

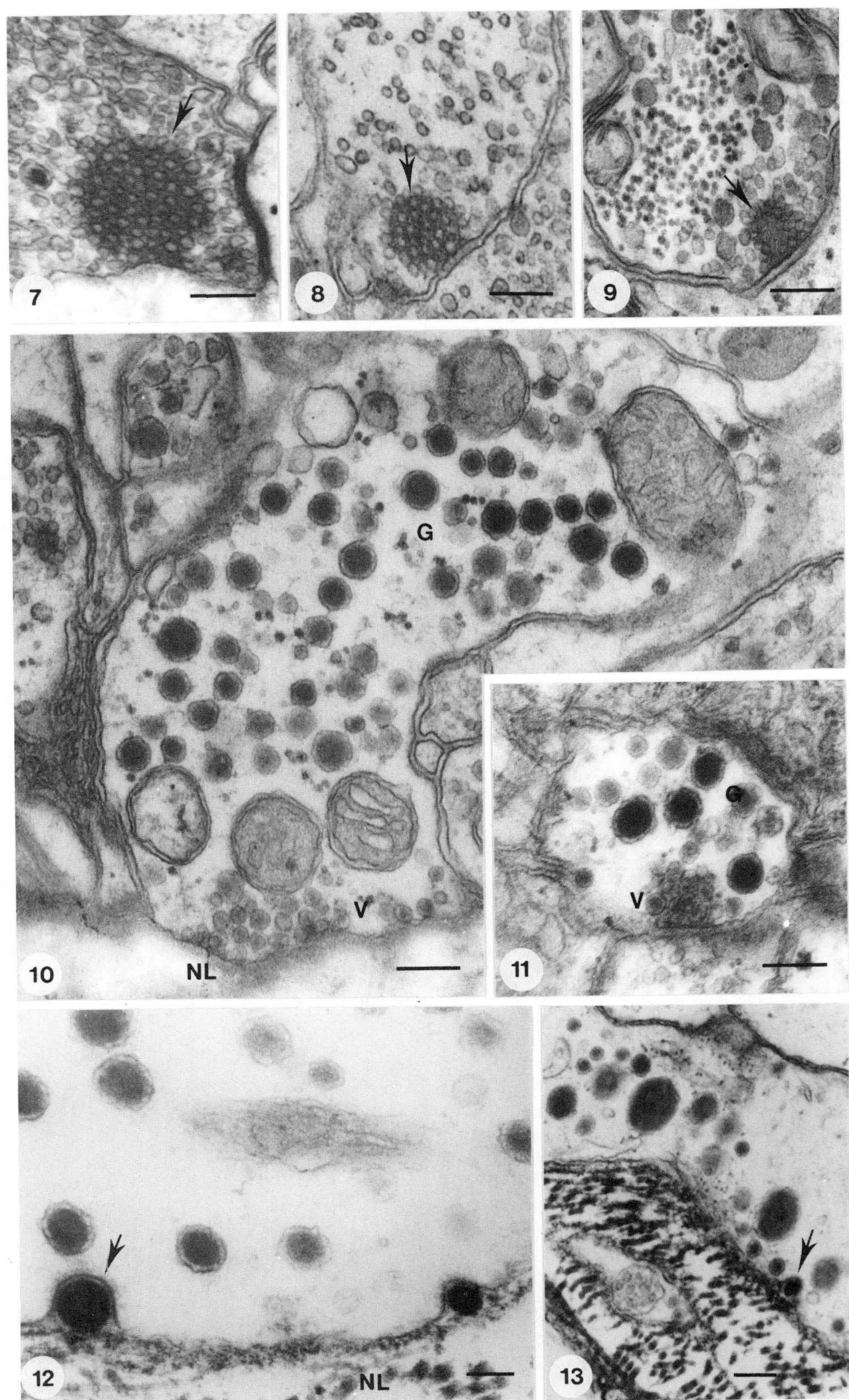

7
8
9
G
V
10
NL
11
V
12
NL
13

since the latter share these properties (Whittaker, 1982). Lastly, both synaptic and synaptoid vesicles are thought to be proliferated from SER (Rambourg and Droz, 1980; Zamora et al., 1984). These various lines of evidence support our contention that synaptic and synaptoid vesicles constitute a single category of inclusions.

SYNAPTIC AND PARASYNAPTIC

With respect to presynaptic endings, attention has until recently been focused almost exclusively on synaptic vesicles. However, whereas vesicle exocytosis is encountered extremely rarely, we found that definitive images of granule exocytosis, involving a multiplicity of neurone types and a number of species, are readily observed within the neuropiles of annelid ganglia (May, 1980; Golding and May, 1982; Golding and Bayraktaroglu, 1984). Comparable findings have now been made on groups ranging from platyhelminthes to vertebrates (Figs. 14-19) (see Buma and Roubos, 1986, and Golding and Pow, 1987a,b, for references).

Sites of granule discharge are marked by the presence of omega profiles, with dense material at various stages of dissolution being present within the indentations. Terminals may show multiple exocytosis involving the independent discharge of a number of granules (Fig. 14), or compound exocytosis in which one or more granules have apparently fused with another already engaged in discharge (Fig. 15). In some cases, dissolution of the material does not always keep pace with release, and pools of material then accumulate in the extracellular space.

Quantitative studies on the octopaminergic synapses in the corpus cardiacum of the locust have shown that granule exocytosis constitutes a correlate of secretory release resulting from neural activity. When the fibres are exposed to tannic acid _in vitro_ (see above), the incidence of omega profiles is elevated by high K+ and this effect is Ca^{2+}-dependent (Pow and Golding, unpublished). Furthermore, injection of tannic acid _in vivo_, followed by the initiation of flight, a condition known to be accompanied by activation of the terminals, results in the appearance of a 20-fold increase in the number of profiles (Fig. 16) in comparison with unflown controls (Pow and Golding, 1987).

Nerve terminals show a variety of morphological patterns of secretory release, (Fig. 20) (see Golding and Pow, 1987a,b, for references). At one extreme, both vesicles and granules show exocytosis (synaptic release) within the confines of the presynaptic thickening. At the other, membrane thickenings are absent, and both vesicles and granules discharge at apparently unspecialized sites (non-synaptic, or better, parasynaptic release, Schmitt, 1984). Finally, the two may be combined, with vesicle exocytosis at specialized junctions and granule discharge mainly at undifferentiated sites (Shkolnik and Schwartz, 1980; Golding and May, 1982). Our observations indicate that the latter pattern characterizes the great majority of terminals, including the archetypal synaptic endings that furnish the cholinergic innervation of the vertebrate adrenal medulla (Golding and Pow, 1987a) and endocrine pancreas (Pow, unpublished).

We have detected two principal variations upon this final theme. Quantitative studies on active terminals within the locust corpus cardiacum have established that granule discharge is untargeted - exocytosis is equally likely to be associated with regions of the terminal surface adjacent to other nerve fibres, glia, etc., as to regions abutting the postsynaptic cells (Pow and Golding, 1987). In contrast, granule discharge from within

---

Figs. 7-13.   (Opposite page) Fig. 7. Tight cluster of synaptoid vesicles - <u>Harmothoe imbricata</u>. Fig. 8. Synaptic vesicle cluster - cerebral ganglion, <u>Nereis diversicolor</u>. Fig. 9. Synaptic vesicle cluster in neuroeffector junction - nuchal organ, <u>N. diversicolor</u> . Fig. 10. Neurosecretory terminal with dense cored vesicles - <u>H. imbricata</u>. Fig. 11. Comparable synaptic terminal - <u>N. diversicolor</u>. Figs. 12, 13. Exocytosis (arrows) of secretory granules and synaptoid vesicle, respectively - <u>Helix aspersa</u> (tannic acid). g, secretory granules; v, synaptic/synaptoid vesicles; nl, neural lamella. Bars, 200nm. (Fig. 9. courtesy of Dr. A.C. Whittle.)

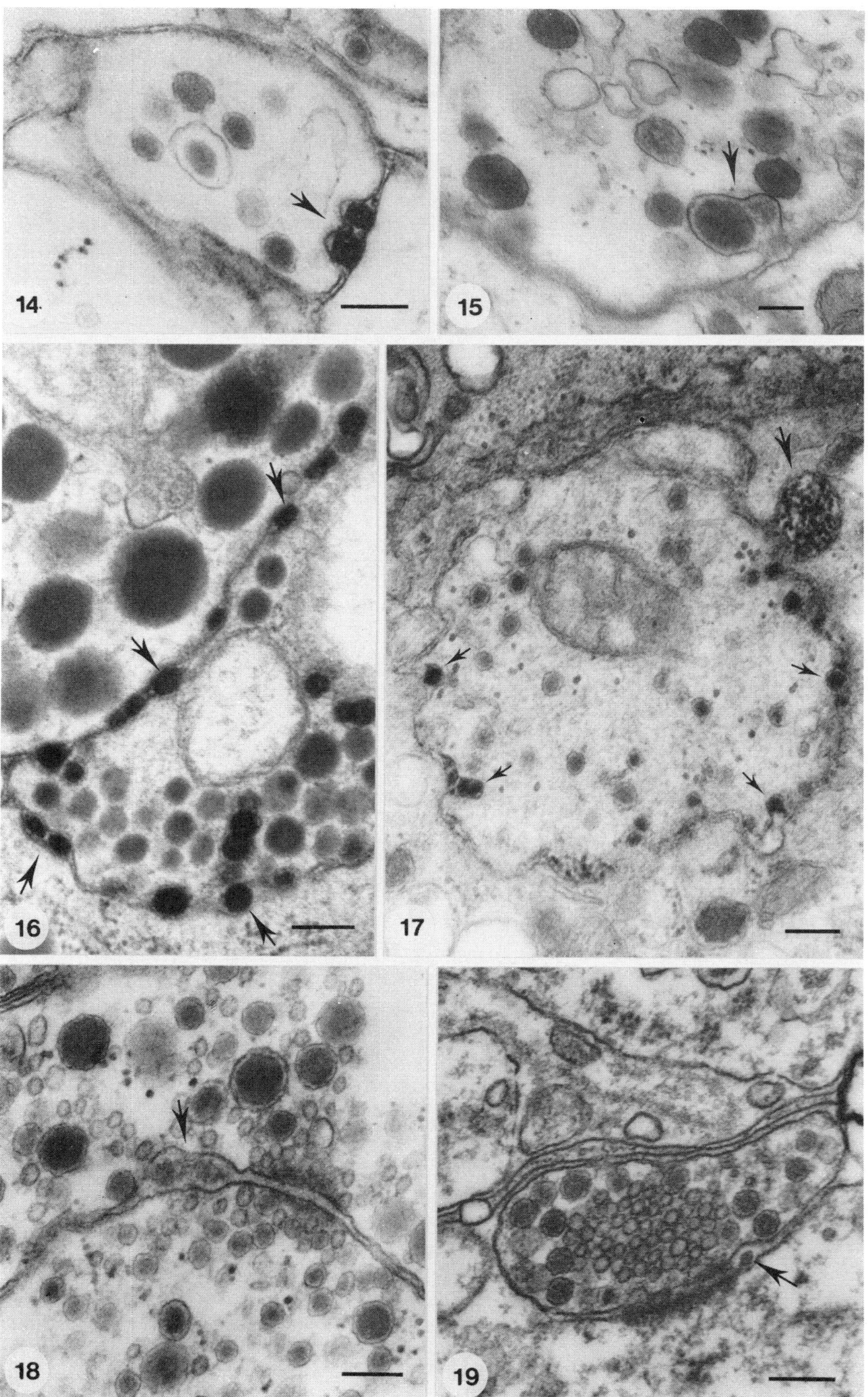

Figs. 14-19.   (Opposite page) Fig. 14. Multiple exocytosis - neuropile, <u>H. aspersa</u> (tannic acid). Fig. 15. Compound exocytosis - neuropile, <u>L. terrestris</u>. Fig. 16. Untargeted granule exocytosis - secretormotor terminal, corpus cardiacum, <u>Schistocerca gregaria</u> (tannic acid). Fig. 17. Targeted granule exocytosis, directed towards the postsynaptic cell (lower) and not towards the connective tissue (upper). Large arrow, exocytosis of chromaffin granule. Adrenal chromaffin gland, <u>Carassius auratus</u> (tannic acid). Figs. 18, 19. Sites of granule exocytosis (arrows) adjacent to synaptic thickenings and clefts - <u>L. terrestris</u> and rat adrenal medulla, respectively. Bars, 200 nm.

fibres synapsing upon chromaffin cells in the goldfish (but not in the frog or mammal) is targeted upon the postsynaptic cells (Fig. 17) (Golding, unpublished). Various intermediates between these two configuations are possible, and scrutiny of our own and published micrographs suggests that many terminals combine a generally untargeted pattern of granule exocytosis with an unusually high incidence of the phenomenon within regions of undifferentiated membrane immediately adjacent to membrane thckenings (Figs. 18, 19).

The pattern of secretory release shown by most neurones, as described above, has numerous implications. First we infer that these cells typically exhibit a three-fold differentiation of their surface with respect to adaptations for release: (i) The greater part of the plasmalemma, enveloping the cell body, axon, etc. allows low levels of discharge at most (ii) Presynaptic thickenings are specialized for the fusion of synaptic vesicles (although some granule discharge may also take place here) and (iii) other expansive regions of the terminal surface which are adapted as potential sites of granule exocytosis though these are apparently unspecialized morphologically.

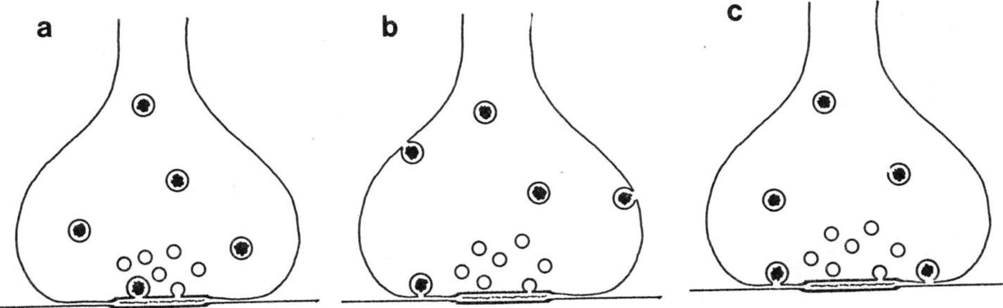

Fig. 20.   Patterns of discharge of secretory granules from typical synaptic terminals - a, synaptic; b, parasynaptic (untargeted); c, parasynaptic (targeted).

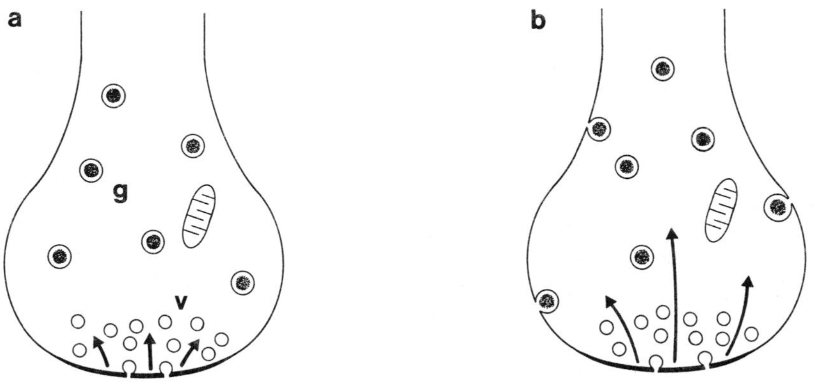

Fig. 21.   The contrasting topographical relationships of Ca^{2+} channels to vesicles and granules, respectively, may lead to mainly vesicle discharge at lower levels of stimulation and recruitment of granule release at higher levels.

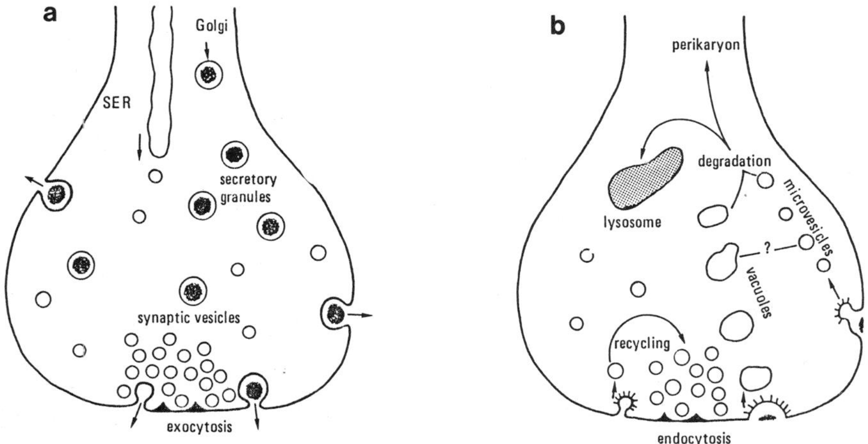

Fig. 22.   a, diverse origins, and patterns of discharge, of synaptic vesicles and secretory granules, respectively; b, diverse retrieval patterns and fates of vesicles and granules, respectively.

A second inference relates to the selective control of secretory release processes. A differential coupling of stimulus to release has been described with respect to acetylcholine (mainly from vesicles) and neuropeptide (from granules), with proportionally more peptide being released during high frequency stimulation (Lundberg, 1981). This relates to parasympathetic varicosities that lack synaptic thickenings, etc. and may possibly be due to different thresholds of cytosolic Ca^{2+} concentrations necessary for vesicle and granule fusion, respectively. Within more conventional terminals, the contrasting topographical relationships of vesicles and granules to the Ca^{2+} channels thought to be located within the synaptic thickening (Pumplin et al., 1987) may well have further implications for the dynamics of release from within the two types of inclusions (Fig. 21).

Third, the combination of vesicle exocytosis at specialized junctions, and granule discharge from sites more widely distributed across the terminal surface, suggests that a wider sphere of influence of peptide mediators, with effects beyond those on cells contacted directly, is a common characteristic.

TWIN-TRACK GLANDULAR UNITS

The emerging picture of the neurone, whether conventional or neurosecretory, is of a glandular unit possessing two distinct mechanisms for the elaboration and discharge of neurochemical mediators (see Golding and Pow, 1987b, for references). The secretory inclusions apparently differ in their respective origins (granules - Golgi apparatus; vesicles - SER), in the character of the mediators they contain (peptides may be stored exclusively within granules), and in the organization of their release sites (Fig. 22a). The patterns of endocytosis following discharge and the fate of the retrieved membranes is similarly dichotomous (Fig. 22b). The distinctive chemistry of the contents and membranes of the granules and vesicles. militates against the repeated suggestions concerning the origin of one category from the other. The mode of secretion involving the granules is apparently identical to that of conventional gland cells, whereas the vesicle mechanism with its rapidity of discharge, specialized sites of release and capacity for recycling, is one of the most distinctive in the living world.

REFERENCES

Anderson, R. and Mitchell, R., 1986, Uptake and autoreceptor-controlled release of ^{3}H-GABA by the hypothalamic median eminence and pituitary neurointermediate lobe, Neuroendocrinology, 42:277.

Anwyl, R. and Finlayson, L.H., 1973, The ultrastructure of neurons with both a motor and a neurosecretory function in the insect *Rhodnius prolixus*, Z. Zellforsch., 146:367.

Bayraktaroglu, E., Golding, D.W. and Whittle, A.C., 1988, Synaptic and synaptoid vesicles constitute a single category of inclusions: dense-cored synaptoid vesicles in Helix discharge their contents by exocytosis. (Submitted).

Bayraktaroglu, E., Whittle, A.C. and Golding, D.W., 1988, Neurosecretory cells with 'synaptoid perikarya' in Helix - a definitive description of secretory release from the somata of endocrine neurones. (Submitted).

Binnington, K.C. and Lane, N.J., 1982, Presence of T-bars, intramembrane particle arrays and exocytotic profiles in neuroendocrine terminals of an insect, Tissue & Cell, 14:463.

Buma, P. and Roubos, E.R., 1986, Ultrastructural demonstration of nonsynaptic release sites in the central nervous system of the snail *Lymnaea* stagnalis, the insect Periplaneta americana, and the rat, Neuroscience, 17:867.

Buma, P., Roubos, E.W. and Buijs, R.M., 1984, Ultrastructural demonstration of exocytosis of neural, neuroendocrine and endocrine secretions with an in vitro tannic acid (TARI-) method, Histochemistry, 80:247.

Golding, D.W. and Bayraktaroglu, E., 1984, Exocytosis of secretory granules - a probable mechanism for the release of neuromodulators in invertebrate neuropiles, Experientia, 40:1277.

Golding, D.W. and Pow, D.V., 1987a, 'Neurosecretion' by a classic cholinergic innervation apparatus; a comparative study of adrenal chromaffin glands in vertebrates, Cell Tissue Res., 249:421.

Golding, D.W. and Pow, D.V., 1987b, The new neurobiology - ultrastructural aspects of peptide release as revesled by studies of invertebrate nervous systems, in: "Neurohormones in Invertebrates", M.C. Thorndyke and G.J. Goldsworthy, eds., Cambridge University Press.

Golding, D.W. and Whittle, A.C., 1977, Neurosecretion and related phenomena in annelids, Int. Rev. Cytol., Suppl. 5:189.

Lundberg, J.M., 1981, Evidence for the coexistence of vasoactive intestinal polypeptide (VIP) and acetylcholine in neurons of cat exocrine glands. Morphological, biochemical and functional studies, Acta physiol. Scand., Suppl. 496:1.

May, B.A., 1980, Ultrastructural correlates of secretory release in invertebrate nervous systems, Gen. Comp. Endocrinol., 40:375.

May, B.A. and Golding D.W., 1982a, Synaptic and synaptoid vesicles constitute a single category of inclusions new evidence from invertebrate nervous systems, Acta Zool. (Stockh.), 63:111.

May, B.A. and Golding D.W., 1982b, Synaptic and synaptoid vesicles constitute a single category of inclusions: new evidence from Z10 impregnation, Acta Zool. (Stockh.), 63:171.

Meister, B., Hokfelt, T., Vale, W.W. and Goldstein, M., 1985, Growth hormone releasing factor (GRF) and dopamine coexist in hypothalamic arcuate neurons, Acta physiol. Scand., 124:133.

Morris, J.F., Nordmann, J.J. and Shaw, F.D., 1981, Granules, microvesicles, and vacuoles, in: "Neurosecretion: molecules, cells, systems," D.S. Farner and K. Lederis, eds., Plenum Press, New York.

Navone, F., Greengard, P. and De Camilli, P., 1984, Synapsin I in nerve terminals: selective association with small synaptic vesicles, Science, 226:1209.

Palay, S.L., 1958, The morphology of synapses in the central nervous system, Exptl. Cell Res., Suppl. 5:275.

Pow, D.V. and Golding D.W., 1987, 'Neurosecretion' by aminergic synaptic terminals in vivo - a study of secretory granule exocytosis in the corpus cardiacum of the flying locust, Neuroscience, (in press).

Pumplin, D.W., Reese, T.S. and Llinas, R., 1981, Are the presynaptic membrane particles the calcium channels? Proc. Nat. Acad. Sci. (Wash.), 78:7210.

Rambourg, A. and Droz, B., 1980, Smooth endoplasmic reticulum and axonal transport, J. Neurochem., 35:165.

Richards, J.G. and Tranzer, J.P., 1974, The characterization of monoaminergic nerve terminals in the brain by fine structural cytochemistry, in: "Neurosecretion - the final neuroendocrine pathway," F.G.W. Knowles and L. Vollrath, Eds., Springer-Verlag, Berlin.

Scharrer, B., 1968, Neurosecretion. XIV. Ultrastructural study of sites of release of neurosecretory materials in blatterian insects, Z. Zellforsch., 89:1.

Scholnik, N.J. and Schwartz, J.H., 1980, Genesis and maturation of serotonergic vesicles in identified giant cerebral neuron of Aplysia, J. Physiol., 43:945.

Schmitt, F.O., 1984, Molecular regulators of brain function: a new view, Neuroscience., 13:991.

Vieillemaringe, J., Duris, P., Bensch, C. and Girardie, J., 1982, Co-localization of amines and peptides in the same neurosecretory cells of locusts, Neurosci. Lett., 31:237.

Whittaker, V.P., 1982, Biophysical and biochemical studies of isolated cholinergic vesicles from Torpedo marmorata, Fed. Proc., 41:2759.

Zamora, J., Garosi, M. and Ramirez, V.D., 1984, Poststimulatory endocytosis, microvesicle repopulation and changes in the smooth endoplasmic reticulum in nerve endings of the median eminence superfused in vitro, Neuroscience, 13:105.

ISOLATED NEUROHYPOPHYSIAL NERVE ENDINGS, A PROMISING TOOL TO STUDY

THE MECHANISM OF STIMULUS-SECRETION COUPLING

Jean J. Nordmann, Govindan Dayanithi, Monique Cazalis[+], Marlyse Kretz-Zaepfel and Didier A. Colin

Centre de Neurochimie
5, rue Blaise Pascal
67084 Strasbourg Cedex
France

[+]INSERM U.176
rue Camille Saint-Saens
33077 Bordeaux Cedex
France

INTRODUCTION

Douglas and co-workers demonstrated in their pioneering work that depolarization of the plasma membrane and the presence of external calcium are two of the prerequisites in excitable cells for an increase in secretion to occur (Douglas and Rubin, 1963; Douglas and Poisner, 1964). They showed that activation of cholinergic receptors in the adrenal medulla or potassium-induced depolarization of the neurohypophysis gave rise to calcium uptake in these tissues. Consequently they hypothetised that catecholamine release and neurohypophysial hormone secretion are induced by an increase in the cytoplasmic calcium concentration. In the seventies, it was shown that the depolarization induced $^{45}Ca^{2+}$ uptake and the release of hormones from the neural lobe could be inhibited by agents known to block calcium channels (Dreifuss et al., 1973). The picture which emerged from these results was that in the neural lobe the arrival of action potentials promotes the opening of voltage-sensitive calcium channels and this leads, due to the large calcium gradient across the membrane, to the entry of calcium. The resulting increase in the free cytoplasmic calcium concentration would then, by an unknown process, trigger exocytosis. Although the neural lobe is the material of choice for studying secretion per se, it suffers from certain disadvantages such as the presence of non-neuronal cells, of a tortuous extracellular space making the diffusion into or out of the tissue very slow, and of the presence of molecules in the extracellular space which prevent, because of their high affinity for lectin-like material, the use of the latter as label for the cell membrane.

We have recently developed in our laboratory a preparation of isolated nerve endings which circumvents most of the problems described above (Cazalis et al., 1987a). We summarize here the work done with our colleagues during the last few years. Although the precise mechanism(s) by which Ca triggers secretion is not yet understood, the isolated nerve endings show promise as a potent tool for studying the coupling between electrical and neurosecretory activities.

PREPARATION OF THE ISOLATED NERVE ENDINGS

The neural lobes are obtained either from rats or cows. It is important to keep the preparation at a temperature around 37°C throughout the experiment. The neurohypophyses are impaled with insect pin to the surface of a Petri dish coated with dental wax or Sylgard Resin (Dow Corning) filed with normal saline. The pars intermedia is gently peeled away from the neural lobe under a dissecting microscope. The neurohypophyses are rinsed with normal saline and, before homogenization are dipped in a medium containing 270 mM sucrose, 10 mM HEPES, pH 7.2 and 0.01 mM EGTA. The presence of EGTA seems to be the key point in obtaining terminals which will release hormones during depolarization.

Indeed, although isolated neurosecretory nerve terminals were described more than twenty years ago (La Bella and Sanwal, 1965), it is only recently that a preparation has been obtained from which the hormones are released by exocytosis (Cazalis et al., 1987a). We thought at first that nerve endings (also termed neurosecretosomes, nerve terminals or kikis) isolated from bovine neural lobes did not release AVP in response to a potassium challenge because secretion had already occurred at the slaughter house during haemorrhage of the animals. In other words we thought that the mechanism leading to exocytosis was somehow inactivated. However, recent experiments in our laboratory show that isolated bovine nerve terminals do respond to depolarisation if prepared in a medium containing EGTA. In fact, these findings are not surprising. First, during homogenization the inside of some of the nerve terminals is exposed to the external medium as judged by the presence of Lucifer Yellow in some endings following homogenization in a medium containing the extracellular marker (Nordmann et al., 1987). Secondly, we found, using the fura-2 indicator, that a HEPES buffered sucrose medium without EGTA contains $1-10\mu M$ free calcium (Brethes et al., 1987), i.e. a cytoplasmic concentration which is large enough to induce hormone release (see below). Thus release might have occurred during homogenization due to the presence of calcium coming either from the homogenizing medium and/or from cellular calcium reservoirs disrupted during homogenization. In summary the control of the ionized calcium concentration in the homogenization medium is of extreme importance for maintaining the secretory machine.

After gentle homogenization the sample is centrifuged at 300g for 5 min in order to eliminate the cell debris (pituicytes, capillaries, red blood cells). The supernatant is centrifuged at 3400g for 20 min and the resulting pellet mostly represents peptidergic nerve endings as judged from their positive labelling with anti-neurophysin antibodies.

HORMONE RELEASE FROM NERVE ENDINGS AND NERVE SWELLINGS

The neuronal compartment of the neurohypophysis can be subdivided into three subcompartments (Morris, 1976). (a) The axons which contain microtubules and possibly neurosecretory granules (NSG), (b) the nerve swellings which contain large number of NSG and (c) the nerve endings which, beside eventual NSG and mitochondria, must contain microvesicles in order to belong to this compartment. In previous studies (Morris and Nordmann, 1980; Lescure and Nordmann, 1980; Nordmann, 1985) we have shown that the nerve endings were the first to show sign of NSG depletion during potassium-induced depolarization or dehydration. Furthermore, we showed that maximal hormone release, measured indirectly by calculating the decrease of the volumetric density of the NSG occurs, following K-depolarization, in the region of the nerve ending abutting the basal membrane. However, during prolonged stimulation of secretion such as that observed during dehydration, depletion of the nerve swellings also occurs but we did not know at the time if the granules could be released from the swellings or if they had to move to the endings before to release their content. Furthermore, because of the lack of membrane marker we did not know if the endocytotic vacuoles observed in the swellings during dehydration (Nordmann, 1985) originated from the endings or from the swellings.

The above-mentioned nerve endings preparation may help us to answer this question. Preliminary morphometric analysis of the two pellets has shown that the 300g pellet contain mostly swellings whereas the 3400g pellet contains large number of nerve endings. Our first observation (in preparation) was that nerve terminals isolated from animals dehydrated (5 days) did release only small amounts of hormone upon K-induced depolarization whereas release occurred at a larger extent from either the 300g pellet or from the neural lobe <u>in toto</u>. This could be explained by the release of hormone from the swellings (see also the paper by Morris et al in this volume). However are these results incompatible with those previously obtained?

The rat neural lobe contains a mean of 3.4×10^{7} nerve endings and 7.1×10^{6} nerve swellings. Their mean surface area are $7.1 \mu m^{2}$ and $25.6 \mu m^{2}$ respectively (Nordmann, 1977). Furthermore, whereas the swellings contain on average 2.2×10^{3} neurosecretory granules there are 2.6×10^{2} NSG in the nerve endings. We also know that after 10 min of K-depolarization the swellings (i.e. the 300g pellet) have release c. 1% of their hormone content and that the endings have secreted c. 3.3% of their content (in preparation). This corresponds to the loss of about 22 and 8.5 NSG per swellings and per endings respectively. Thus one can calculate that during 10 min of depolarization there was, in the swellings, about 1 granule released per μm^2 of plasma membrane whereas one μm^2 of nerve ending plasma membrane released 1.2 neurosecretory granule. These results are strengthened by the recent observation of Morris et al., (this volume). They showed, using the tannic acid technique, the released content of NSGs next to swellings. Thus one can conclude that the nerve swellings can release neurohypophysial hormones. The fact that we did not observe a significant decrease in the NSG volumetric density in the swellings during acute K^+-depolarisation (Morris and Nordmann, 1980) could be explained by the impossibility to detect, with stereology, variations corresponding to that described above (i.e. c. 1%). On the other hand the depletion of NSG in the swellings during dehydration (Nordmann, 1985) is consistent with the above-mentioned calculations.

CALCIUM CONCENTRATIONS IN THE NEUROSECRETOSOMES

Using the Ca^{2+} indicator fura-2 (Grynkievicz et al., 1985) we showed that under resting conditions the nerve endings have a mean internal calcium concentration of 350 nM (Brethes et al., 1987). Potassium- or veratridine-induced depolarization gave rise to an increase of c. 300-400 nM. However it has to be pointed out that these values represent the mean cytoplasmic concentration and, as we shall see below, might not represent the calcium concentration at the site of exocytosis.

We also measured the increase in calcium concentration as a result of electrical stimulation of the endings. Fig. 1 shows the increase in internal Ca^{+2} following stimulation at different frequencies. Note that at 1Hz there was no detectable increase in the free calcium concentration whereas at higher frequencies the internal calcium increased significantly. Similarly, no increase in hormone release was observed at 1Hz whereas at higher frequencies neuropeptide release was increased (unpublished). But one of the most

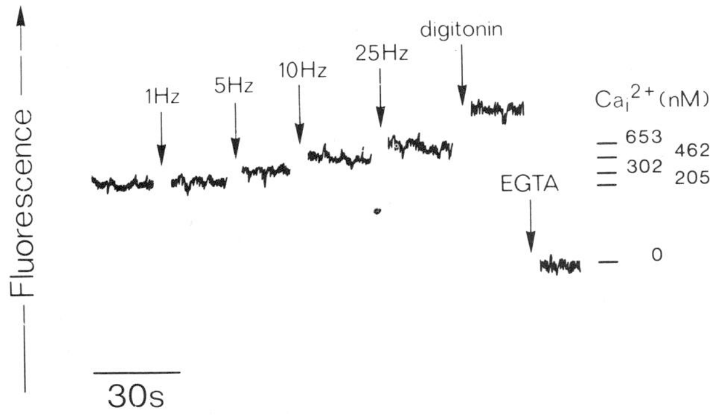

Fig 1. Effects of electrical stimulation on free Ca^{2+}_i. The isolated nerve endings were first incubated in 1 mM Ca-containing Locke solution. They were then stimulated at different frequencies as indicated. Note that the trains of stimuli were separated by 180-s silent intervals. The free internal calcium concentration was measured after addition of digitonin and EGTA (see Brethes et al., 1987).

exciting experiments was to study the role of the firing pattern of the magnocellular neurones on the intraterminal calcium concentration. One can record, <u>in vivo</u>, the activity of the hypothalamic neurones and, after digitization of the signal, use the pattern of firing to trigger a stimulator (Dutton and Dyball, 1979; Ingram et al., 1982; Cazalis et al., 1985). The neural lobe is impaled on a platinum electrode and stimulated <u>in vitro</u> with the same pattern as the one arriving at the nerve terminals under hormonal demand. Under these conditions one can analyse the importance of the bursting pattern in inducing hormone release. The main conclusions were that the pattern of firing within a burst and the interval duration between the bursts were the two major parameters for inducing maximal neuropeptide release. We found that four "AVP-like" bursts (i.e. burst recorded from an AVP-containing neuron) separated by 180s silent periods would release more hormone than when separated by intervals of smaller duration (Cazalis et al., 1985). Furthermore the importance of the silent intervals was demonstrated by stimulating neural lobes for a total of 40 min. Whereas some neurohypophyses were stimulated with bursts delivered without silent intervals (i.e. in this case a total of 89 bursts were delivered) others were stimulated with fifty bursts (i.e. 21s interval duration) or with twelve bursts only separated by 180s silent period. These experiments clearly showed that twelve bursts delivered during 40 min released much more hormone than more bursts given with shorter silent intervals (Nordmann and Stuenkel, 1986). Fig. 2 illustrates the above-mentioned results. The neural lobes stimulated with four "AVP-like" bursts showed a larger $^{45}Ca^{2+}$ uptake than when stimulated with four bursts delivered without intervals or than neurohypophyses stimulated with four bursts separated by silent intervals but in which the pulses were given at a constant frequency of 13Hz (Fig. 2a). Fig. 2b shows that four "AVP-like" bursts induced more AVP release when delivered with silent intervals than when given without silent periods. These results were strengthened when, with Ed Stuenkel, we found that the electrical properties of the axons and the nerve terminals could be correlated with the above-mentioned results. For instance we found that the axons could not follow for a prolonged period of time a train of stimuli (Nordmann and Stuenkel, 1986). Not only did the latency of conduction increase during a train of electrical pulses but the amplitude of the compound action potential decreased as a function of the length of the applied stimulus, suggesting the failure of some of the axons to conduct, for a

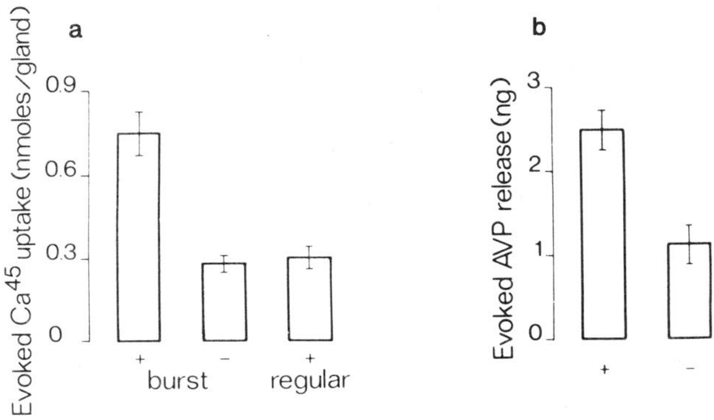

Fig.2.   Radioactive calcium uptake in (a) and AVP release from (b) neurohypophyses electrically stimulated with four AVP-like bursts separated (+) or not separated (−) by 21 s intervals or stimulated with four trains of stimuli given at a regular frequency of 13Hz which is also the mean frequency within the AVP-like burst (for Methods see Cazalis et al., 1985).

prolonged period of time, action potentials. We also found that even after a single "AVP-like" burst there was a decrease in the number of axons able to conduct action potentials (Nordmann and Stuenkel, 1986). However, the axons did recover their electrical properties if the bursts were delivered with 180s silent intervals. In summary, the results described by our and other laboratories could be explained by the failure of the axons of the hypothalamo-neurohypophysial tract to conduct, for a prolonged period of time, action potentials occurring at frequencies equal or superior to 1Hz. It is worthwhile to point out that these experiments were performed <u>in vitro</u> and that experiments performed <u>in vivo</u> are needed before any final conclusions could be reached. We found that these observations could be correlated with the increase in the intraterminal calcium concentration $(Ca^{2+})_i$; see Brethes et al., 1987). For instance, the $(Ca^{2+})_i$ rose more when the terminals were stimulated electrically with five "AVP" bursts separated by 180s interval periods than when the stimuli were given without silent periods between the bursts. These experiments show what had already been hypothetized for many years but, to our knowledge, never demonstrated, i.e. a correlation between the electrical pattern of a neuron, the release of its neuropeptide (or neurotransmitter) and the intraterminal free calcium concentration.

IONIC CHANNELS IN NERVE ENDINGS

Veratridine, which promotes the opening of Na^+ channels induces the release of vasopressin from isolated nerve endings confirming the presence of sodium channels in the membrane of the neurosecretosomes. The veratridine-induced depolarization, as measured with the cyanide dye $Di(S)-C_3-5$ (Nordmann et al., 1982), is blocked by Na^+-free medium suggesting that sodium is, in the presence of the alkaloid, the charge carrier for the depolarizing current. The plausible mechanism by which veratridine promotes hormone release is that alkaloid-induced depolarization promotes the activation of calcium channels and the resulting increase of the cytoplasmic calcium concentration triggers exocytosis (Nordmann and Dyball, 1978). This hypothesis is confirmed by the blocking effect of the calcium channel antagonist D600 on the veratridine-induced increase of the calcium concentration and of hormone release.

The first evidence that the neural lobe has voltage-dependent channels came from the results obtained with cations such as Mn^{2+} or Co^{2+} or with the verapamil derivative D600 (Dreifuss et al., 1973; Russell and Thorn, 1974). At the time these agents had just been found to block calcium channels in the squid axon (Baker et al., 1973) and it was Peter Baker who suggested to us their use in order to demonstrate, for the first time, to our knowledge, that the activation of voltage dependent calcium channel was indeed one of the steps leading to exocytosis. During the last years progress has been made concerning the characterization of these channels. First K^+-induced secretion is blocked by cations such as Cd^{2+}, Gd^{3+}, Co^{2+} and Mn^{2+}. The typical concentrations for half-maximal inhibition being about 10,80,1100 and 1100μM respectively. The Ca^{2+} channel antagonist D600 (methoxyverapamil) has an IC_{50} of c. 10μM and we recently found (in preparation) that a new antagonist, D888 (desmethoxyverapamil), has an IC_{50} of c. 1μM. Potassium- and electrically-induced neurohormone release can also be blocked (IC_{50} = 4μM) by the dihydropyridines (DHP) nitrendipine and nicardipine. Furthermore we found that BAY K 8644, which has a structure similar to the dihydropyridine nifedipline, potentiates AVP release when the membrane was not fully depolarized. However, as shown in Fig. 3, in the presence of 100 mM potassium, Bay K 8644 became an inhibitor of release.

As we shall see below, these results are interesting and will allow one to further characterize the calcium channels implicated in the secretory process. One point worth mentioning is that dihydropyridines do not affect the release of neurotransmitters from brain synaptosomes (Reynolds et al., 1986; Suskiw et al., 1986; Rampe et al., 1984) suggesting that in the neural lobe the type(s) of calcium channel(s) involved in the release process are different (or possibly the channels are in a different state) then those located on the neurohypophysial nerve terminals.

Until very recently no toxin had been found which will block calcium channel with the same specificity as, for example, tetrodotoxin or a-bungarotoxin bind to sodium channels or acethylcholine receptors. However the recently discovered peptide ω-conotoxin, isolated from the venom of the fish-hunting snail <u>Conus geographus</u>, seems to have the qualities required to be a blocker of some of calcium channels. In collaboration with Francois Couraud and Nicole Martin-Moutot (Dayanithi et al., submitted) we have observed that in the isolated neural lobe one of these toxins, ω-CgTx GVIA (hereafter abbreviated ω-CgTx) has a very high affinity for its receptor. Indeed its association kinetic constant has a value of $5.6 \times 10^{-6} M^{-1} s^{-1}$ and the binding was found to be irreversible. Furthermore, in the neurosecretosomes, the concentration of ω-CgTx necessary for half of its receptors to be occupied is 2×10^{-10} M. We also found that ω-CgTx blocked K^+- and electrically induced AVP release but that it had no effect on the release of hormone from permeabilized nerve endings (see below), suggesting that it only acts at a step before the intra-terminal rise of free calcium.

The findings that nicardipine, nitrendipine and ω-CgTx block neurohormone release and that BAY K 8644 potentiates secretion are relevant for the characterization of the type of Ca channel involved in the process of secretion. Electrophysiological studies on chick sensory neurones (Fox et al., 1987) showed that three types (known as T, N and L) of calcium channels can be observed. Whereas the L-type is sensitive to dihydropyridines, both the N- and T-type are insensitive. ω-CgTx persistently blocks the N- and the L-channels. Moreover the presence of L-type channels on the neurosecretosomes has been confirmed by recent electrophysiological techniques (Lemos and Nowicky, 1987). Using the "terminal-attached" patch technique Jose Lemos and Martha Nowycky found that, in the presence of BAY K 8644, the predominant divalent cation inward currents correspond to L-type calcium channel activity. Thus, the results obtained on the neural lobe strongly suggest that the activation of L-type channels is the first step in the chain of events leading to secretion.

The patch-clamp technique has also allowed analysis of the macroscopic and the single channel currents in isolated neurosecretosomes (Lemos and Nordmann, 1986). This technique has also been used on the whole neural lobe (Mason and Dyball, 1986) and on isolated nerve endings obtained from the sinus gland of the crab (Lemos et al., 1986). We found that inside-out patches have several types of channels. One is highly selective for K^+, it is voltage dependent, has a slope conductance of about 130 pS and in the presence of calcium increases its probability of opening. The nerve endings have at least two other types of channels. One "fast" and one "slow" channel. The slow channel is permeable to Na^+ and K^+ but not to chloride and openings are seen with calcium concentrations above $0.1 \mu M$. It is interesting to note that it has the same characteristics as the "s" channel first found in the nerve terminals of the sinus gland (Lemos et al., 1986).

One of the most intriguing findings made during our recent studies on the mechanism of secretion of AVP and OT release was to find that relaxin (for review see Bryant-Greenwood, 1982), a 6000-daltons peptide synthetized by the corpus luteum, was able to modulate neurohormone release under basal <u>and</u> stimulated conditions (Dayanithi et al., 1987). Whereas OT and AVP basal release (i.e. the amount of hormone secreted in a medium containing 5 mM K^+) were inhibited, secretion of both hormones was potentiated when it was induced with a 100 mM K^+-containing medium or when the neural lobes were electrically stimulated. Half-maximal potentiation was observed at a relaxin concentration of c. 1nM. Furthermore we also found that relaxin interacts with a site coupled or at least close to calcium channels and since it does not have any effect on the amount of hormone secreted from permeabilized nerve endings, i.e. acts at a step preceding exocytosis. We do not yet know if relaxin acts directly on the calcium channel, at a site distinct from the calcium channel, or at a site associated but distinct from the calcium channel. These findings might be of interest for it is the first time, to our knowledge, that a circulating hormone has been shown to affect, directly at the level of the nerve endings, the release of AVP and OT by possibly acting on the gating mechanism of calcium channels. One main question however remains. If relaxin does indeed modulate neurohypophysial hormone release <u>in vivo</u>, where does it come from? Although in some species it is synthesized in the corpus luteum, one does not yet know if it can cross the blood brain barrier. Another alternative, which has yet not been demonstrated is the synthesis of relaxin in the CNS. This would seem a reasonable hypothesis.

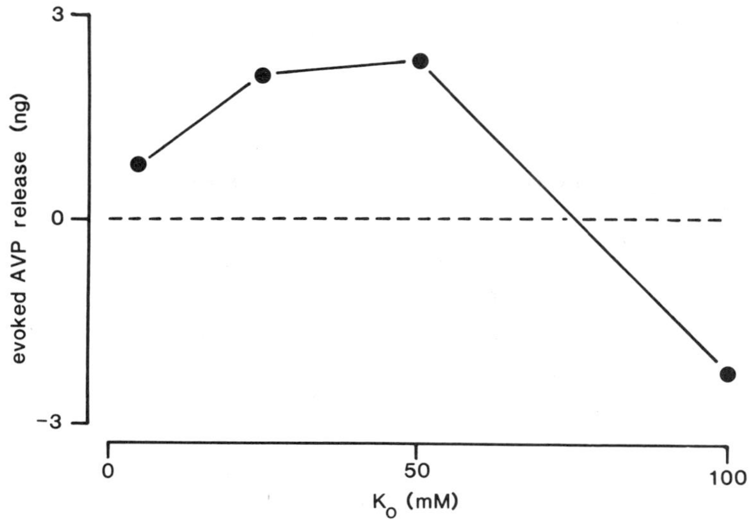

Fig.3.   Effects of the dihydropyridine derivative BAY K 8644 on the evoked AVP release from K^+-induced depolarized nerve terminals. The results are given as evoked amounts of AVP release compared to that determined in the absence of BAY K 8644. With depolarization obtained with K^+ concentrations up to 50 mM the dihydropyridine potentiates secretion whereas at higher concentrations release is inhibited.

MINIMAL REQUIREMENTS FOR HORMONE RELEASE

We have taken the advantages of the detergent effects of digitonin to permeabilize isolated nerve endings. This treatment allows the direct access to the cytoplasm, a way to bypass the depolarization step involved in the secretory process (Cazalis et al., 1987b; Nordmann et al., 1987). In contrast to the extracellular concentration of calcium necessary to activate half-maximal release (c. $300\mu M$), after permeabilization of the terminals with digitonin, an increased secretion is observed with calcium concentrations in the micromolar range. Although release is somewhat enhanced in the presence of ATP and to a lesser extent by GTP, it is not strictly speaking under the dependence of nucleotides. Anions such as Br^-, Cl^-, I^- and SCN^- inhibit AVP release at relatively high concentrations but we recently found that chloride ions (at concentrations in the range of the millimole, i.e. at concentrations similar to those encountered <u>in vivo</u>) are most certainly necessary for secretion to occur. This is exciting for neurosecretory granule membranes have been shown to have an anion channel (Lemos and Nordmann, 1986) and in permeabilized neurosecretosomes secretion can be blocked by SITS, a relatively specific anion channel blocker and also by the more specific chloride channel blocker N144 (unpublished). We still do not know the mechanism by which chloride ions allow exocytosis to occur but, as hypothetized by others (Pollard et al., 1977; Cohen et al., 1982), chemio-osmotic lysis might be among the final steps before secretion occurs. This still needs, however, to be fully demonstrated.

In conclusion we hope to have shown that the neurosecretosomes have the properties necessary to analyse the mechanism leading to the secretion of neuropeptides.

ACKNOWLEDGEMENTS

This paper is dedicated to Peter Baker who has greatly contributed to our understanding of the mechanisms regulating the cellular calcium concentration and the process of exocytosis. Peter Baker died prematurely on March 10, 1987. Peter was a very enthusiastic person and loved to communicate and discuss ideas. Not only will he be missed by Phillis, Lucy, Sarah, Alexandre and Charlotte, but also by the entire scientific community.

REFERENCES

Baker, P.F., Meves, H.S. and Ridgway, E.G., 1973, Effects of manganese and other agents on the calcium uptake that follows depolarization of squid axons, J. Physiol. Lond., 231:511.

Brethes, D., Dayanithi, G., Letellier, L. and Nordmann, J.J., 1987, Depolarization-induced Ca^{2+} increase in isolated neurosecretory nerve terminals measured with fura-2, Proc. Acad. Sci. USA, 84:1439.

Bryant-Greenwood, G.D., 1982, Relaxin as a new hormone, Endocr. Rev., 3:62.

Cazalis, M., Dayanithi, G. and Nordmann, J.J., 1985, The role of patterned burst and interburst interval on the excitation-coupling mechanism in the isolated rat neural lobe, J. Physiol. Lond., 369:45.

Cazalis, M., Dayanithi, G. and Nordmann, J.J., 1987a, Hormone release from isolated nerve endings of the rat neurohypophysis, J. Physiol. Lond., 390:55.

Cazalis, M., Dayanithi, G. and Nordmann, J.J., 1987b, Requirements for hormone release from permeabilized nerve endings isolated from the rat neurohypophysis, J. Physiol. Lond., 390:71.

Cohen, F.S., Akabas, M.H. and Finkelstein, A., 1982, Osmotic swelling of phospholipid vesicles causes them to fuse with a planar phospholipid bilayer membrane, Science, 217:458.

Dayanithi, G., Caxalis, M. and Nordmann, J.J., 1987, Relaxin affects the release of oxytocin and vasopressin from the neurohypophysis, Nature, 325:813.

Dayanithi, G., Martin-Moutot, N., Colin, D.A., Kretz-Zaepfel, M., Couraud, F. and Nordmann, J.J., 1987, The calcium channel antagonist ω-conotoxin blocks secretion from peptidergic nerve terminals, submitted.

Douglas, W.W. and Poisner, A.M., 1964, Stimulus-secretion coupling in a neurosecretory organ: the role of calcium in the release of vasopressin from the neurohypophysis, J. Physiol. Lond., 172:1.

Douglas, W.W. and Rubin, R.P., 1963, The mechanism of catecholamine release from the adrenal medulla and the role of calcium in stimulus-secretion coupling, J. Physiol. Lond., 67:388.

Dreifuss, J.J., Grau, J.D. and Nordmann, J.J., 1973, Effects on isolated neurohypophysis of agents which affect the membrane permeability to calcium, J. Physiol. Lond., 231:96.

Dutton, A. and Dyball, R.E.J., 1979, Phasic firing enhances vasopressin release from the rat neurohypophysis, J. Physiol. Lond., 290:433.

Fox, A.P., Nowycky, M.C. and Tsien, R.W., 1987, Kinetic and pharmacological properties distinguishing three types of calcium currents in chick sensory neurones, J. Physiol. Lond., in press.

Grynkiewicz, G., Poenie, M. and Tsien, R.Y, 1985, A new generation of Ca^{2+} indicators with greatly improved fluorescence properties, J. biol. Chem., 260:3440.

Ingram, C.D., Bicknell, R.J., Brown, D. and Leng, G., 1982, Rapid fatigue of neuropeptide secretion during continual electrical stimulation, Neuroendocrinol., 35:424.

La Bella, F.S. and Sanwal, M., 1965, Isolation of nerve endings from the posterior pituitary gland, J. Cell. Biol., 25:179.

Lemos, J.R. and Nordmann, J.J., 1986, Ionic channels and hormone release from peptidergic nerve terminals, J. exp. Biol., 124:53.

Lemos, J.R., Nordmann, J.J., cooke, I.M. and Stuenkel, E.L., 1986, Single channels and ionic currents in peptidergic nerve terminals, Nature, 319:410.

Lemos, J.R. and Nowycky, M.C., 1987, One type of channel in nerve terminals of the rat neurohypophysis is sensitive to dihydropyridines, Neurosc. Abstr., in press.

Lescure, H. and Nordmann, J.J., 1980, Neurosecretory granules release and endocytosis during prolonged stimulation of the rat neurohypophysis in vitro, Neuroscience, 5:651.

Mason, W.T. and Dyball, R.E.J., 1986, Single ion channel activity in peptidergic terminals of the isolated rat neurohypophysis related to stimulation of neural stalk axons, <u>Brain Res.</u>, 383:279.

Morris, J.F., 1976, Hormone storage in individual neurosecretory granules of the pituitary gland: a quantitative ultrastructural approach to hormone storage in the neural lobe, <u>J. Endocrinol.</u>, 68:209.

Morris, J.F. and Nordmann, J.J., 1980, Membrane recapture after hormone release from nerve endings in the neural lobe of the rat pituitary, <u>Neuroscience</u>, 5:639.

Morris, J.F., Pow, D.V. and Shaw, F.D., This volume.

Nordmann, J.J., 1976, Evidence for calcium inactivation during hormone release in the rat neurohypophysis, <u>J. exp. Biol.</u>, 65:669.

Nordmann, J.J., 1977, Ultrastructural morphometry of the rat neurohypophysis, <u>J. Anat.</u>, 123:213.

Nordmann, J.J., 1985, Hormone content and movement of neurosecretory granules in the neural lobe during and after dehydration, <u>Neuroendocrinol.</u>, 40:25.

Nordmann, J.J. and Dyball, R.E.J., 1978, Effects of veratridine on Ca fluxes and the release of oxytocin and vasopressin from the isolated rat neurohypophysis, <u>J. gen. Physiol.</u>, 72:297.

Nordmann, J.J. and Stuenkel, E.L., 1986, Electrical properties of axons and neurohypophysial nerve terminals and their relationship to secretion in the rat, <u>J. Physiol. Lond.</u>, 380:521.

Nordmann, J.J., Desmazes, J.P. and Georgescault, D., 1982, The relationship between the membrane potential of neurosecretory nerve endings, as measured by a voltage sensitive dye, and the release of neurohypophysial hormones, <u>Neuroscience</u>, 7:731.

Nordmann, J.J., Dayanithi, G. and Lemos, J.R., 1987, Isolated neurosecretory nerve endings as a tool for studying the mechanism of stimulus-secretion coupling, <u>Biosc. Rep.</u>, in press.

Pollard, H.B., Tack-Goldmann, K.M., Pazoles, C.J., Creutz, C.E. and Shulman, N.R., 1977, Evidence for control of serotonin secretion from human platelets by hydroxy ion transport and osmotic lysis, <u>Proc. Nat. Acad. Sci. USA</u>, 74:5295.

Rampe, D., Janis, R.A. and Triggle, D.J., 1984, BAY K 8644, a 1, 4-dihydropyridine Ca^{2+} Channel activator: dissociation of binding and functional effects in brain synaptosomes, <u>J. Neurochem.</u>, 43: 1688.

Reynolds, I.J., Wagner, J.A., Snyder, S.H., Thayer, S.A., Olivera, B.M. and Miller, R.J., 1986, Brain voltage-sensitive calcium channels subtypes differentiated by -conotoxin fraction GVIA, <u>Proc. Acad. Sci. USA</u>, 83:8804.

Russell, J.T. and Thorn, N.A., 1974, Calcium and stimulus-secretion coupling in the neurohypophysis. II. Effects of lanthanum, a verapamil analog (D600) and prenylamine on ^{45}Ca transport and vasopressin release in isolated rat neurohypophyses, <u>Acta endocrinol.</u>, 76:471.

Suskiw, J.B., O'Leary, M.E., Murawsky, M.M. and Wang, T., 1986, Presynaptic calcium channels in the rat cortical synaptosomes: fast-kinetics of phasic calcium influx, channel inactivation, and relationship to nitrendipine receptors, <u>J. Neurosci.</u>, 6:1349.

NEURONAL-GLIAL AND SYNAPTIC PLASTICITY IN THE ADULT OXYTOCINERGIC SYSTEM

D.T. Theodosis, D.A. Poulain, C.Montagnese and J.D. Vincent

I.N.S.E.R.M.
U.176 and Université de Bordeaux II
Bordeaux, France

The magnocellular neurones that secrete oxytocin occur in the same hypothalamic regions in which one also finds the neurones secreting vasopressin (mainly the supraoptic (NSO) and paraventricular nuclei (PVN)). Nevertheless, it is possible to differentiate the two populations anatomically (see, for example, Rhodes et al., 1981; Theodosis et al., 1986 a), because of their different peptide content, and physiologically (see Poulain and Wakerley, 1982), on the basis of their different behaviour in response to various stimuli to hormone release.

Oxytocinergic neurones are particularly suitable for structure- function analysis and can serve as excellent models to study the various aspects of neurosecretion, as attested to by the numerous studies reviewed in this volume. Upon activation by various stimuli (suckling, vaginal distension and chronic dehydration), the neurones significantly increase their electrical activity which then enhances hormone release from their terminals (see Poulain and Wakerley, 1982). The enhanced hormone demands are accompanied by an increased biosynthetic activity in their cell bodies and this is reflected morphologically in increased nuclear and cytoplasmic volumes (Morris, 1978; Chapman et al., 1986) and in the proliferation and hypertrophy of organelles such as the Golgi complex and Nissl substance, respectively (Morris, 1978). These cytological changes are hardly surprising since they occur in all actively secreting cells. More surprising is that the oxytocinergic system also undergoes reversible structural changes that modify the relationship of its neurones to adjacent glial cells and its synaptic circuitry, and consequently, its overall anatomy.

NEURONAL GLIAL CHANGES

The cell bodies and dendrites of magnocellular neurones occur quite closely packed together in the hypothalamic nuclei, but under basal conditions of hormone release, they usually remain separated one from the other by neuropile elements, and in particular, by fine astrocytic processes. As in most other structures in the mammalian CNS, neuronal surfaces are rarely seen to be directly juxtaposed, with no intervening glial elements. Quantitative ultrastructural analyses, on magnocellular nuclei in which the neuronal elements were immunostained for their content of either oxytocin or vasopressin, revealed that, under basal conditions of hormone release, in male and female animals, not only are neuronal surface juxtapositions rare, but they are also limited in extent and occur equally between oxytocinergic and vasopressinergic elements (Theodosis et al.,1986a; Chapman et al., 1986). Thus, in the SON of normally hydrated male and virgin female rats, about 20% of all oxytocinergic or vasopressinergic cell bodies are directly juxtaposed to an adjacent neurosecretory cell body or dendrite (see Fig. 3) but the neuronal juxtapositions involve no more than 2% of all the neuronal surface membrane in the nucleus.

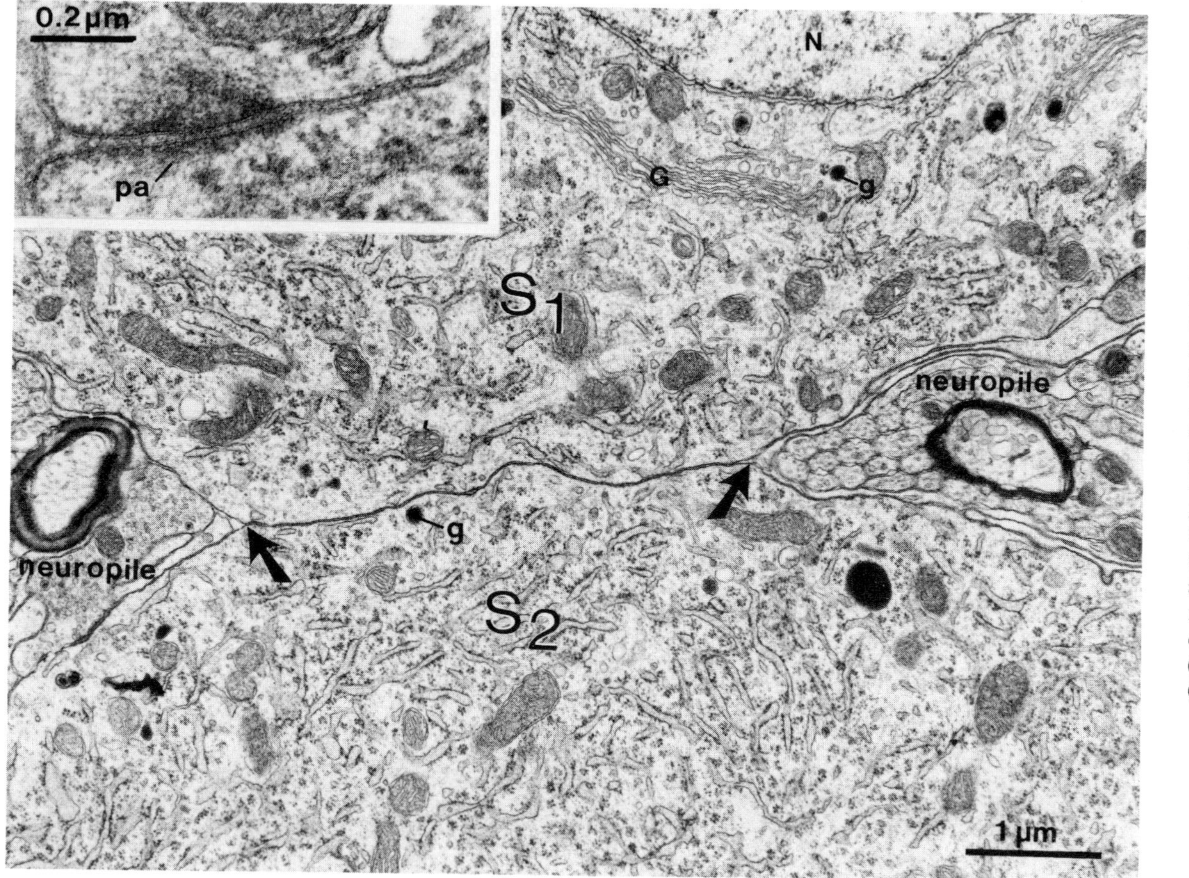

Fig 1. Electron micrograph of directly juxtaposed magnocellular neurones in the supraoptic nucleus of a lactating rat. The surface membranes of two neurosecretory soma profiles (S_1, S_2) are seen to be extensively juxtaposed, that is, they are not separated by any neuropile or glial elements (between arrows). Higher magnification of such juxtapositions (inset) shows that the extracellular space between the juxtaposed plasmalemmae appears normal; the only membrane specializations observed are attachment plates (p.a.). g : neurosecretory granules, G : Golgi apparatus, N : nucleus.

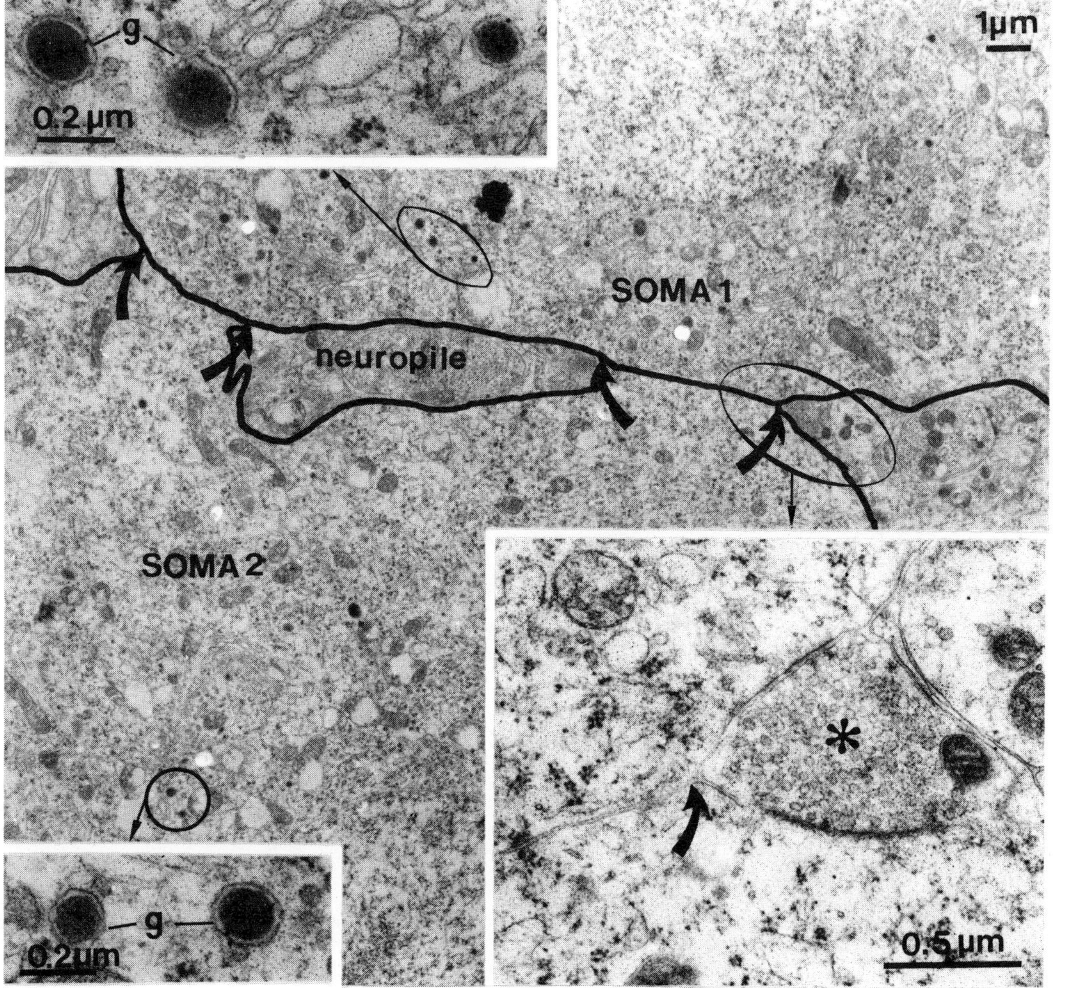

Fig 2. Electron micrograph of 2 oxytocinergic neurones in the SON of a primiparous rat that had received a continuous intracerebroventricular infusion of oxytocin for 8 days. The surface membranes of the two soma profiles (outlined in black) are extensively juxtaposed (between curved arrows). The oxytocinergic nature of the neurones is revealed by colloidal gold particles over their secretory granules (g, left insets) after immunostaining with specific anti-oxytocin serum and immunoglobulin-coupled colloidal gold (post-embedding method). The two somata are also contacted by the same presynaptic terminal (asterisk and right inset).

159

On the other hand, in the nuclei of animals stimulated to release oxytocin (for example, parturient, lactating and chronically dehydrated rats), glial coverage of oxytocinergic neurones markedly diminishes (Fig. 1 and 2) and almost every oxytocin-secreting neurone is found directly juxtaposed, at one point or another of its surface, to an adjacent cell body and/or dendrite, which usually is also oxytocinergic (Fig. 3) (Theodosis et al., 1986a; Chapman et al., 1986). These changes are specific to oxytocin neurones since plasmalemma juxtapositions between the neighbouring vasopressin neurones continue to be few and limited in extent, even during dehydration which stimulates particularly vasopressin release (Fig. 3). Interestingly, when vasopressin cells are directly juxtaposed to another neuronal element, the latter is usually oxytocinergic. To date, most of these analyses have been carried out in the SON, because of its cellular homogeneity, and in this nucleus, the majority of directly juxtaposed oxytocinergic elements were found in the dorsal confines of the nucleus, corroborating the well-known dorsal preponderance of oxytocin cells in this nucleus. However, oxytocinergic profiles exist everywhere in the SON, including the ventral glial lamina which contains many oxytocinergic dendritic profiles. What must be kept in mind, especially in terms of functional consequences, is that one often finds directly juxtaposed oxytocinergic profiles occurring as clusters of 3-5 elements. Moreover, such clusters may be found anywhere within the nuclei. Recently, we have noted that a similar anatomical reorganisation occurs in the more heterogeneous PVN, and here also, the changes affect exclusively the magnocellular oxytocinergic system.

The individual plasmalemma juxtapositions in the stimulated animals can be quite extensive (often over several micrometers in any plane of section) and may involve more than 10% of all the oxytocinergic surface membrane in the SON (Theodosis et al., 1986; Chapman et al., 1986). The extracellular space between the directly juxtaposed neuronal surfaces appears unmodified and is similar in width to that seen between other cellular elements in the nucleus (11-13 nm). The only visible plasmalemma specializations are punctae

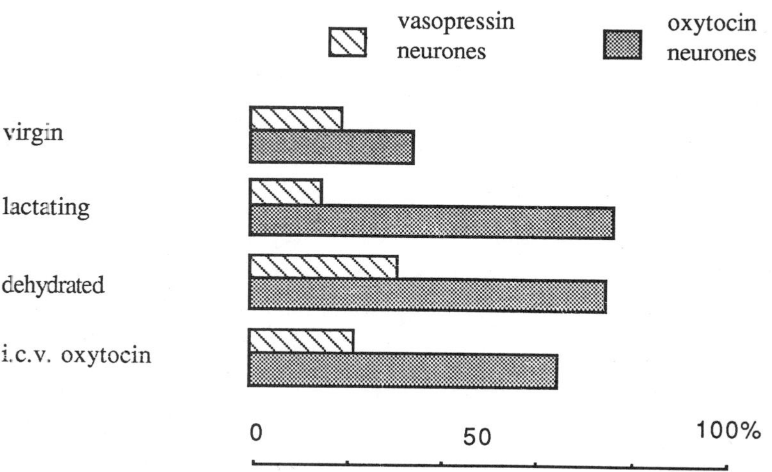

Fig. 3. Specificity of the neuronal-glial changes in the rat SON. Oxytocinergic neurones were identified by ultrastructural immunocytochemistry using anti-oxytocin serum and immunoglobulin-coupled colloidal gold. As the SON contains neurosecretory neurons which are either oxytocinergic or vasopressinergic, in the analyses illustrated here, all immunonegative profiles were assumed to be vasopressinergic. The proportion of oxytocin- and vasopressin-secreting neurones showing neuronal juxtapositions were thus determined on immunostained ultrathin sections of the SON of animals under different physiological and experimental conditions. Note that among the vasopressin neurones, the proportion of directly juxtaposed cells does not vary between groups, whereas there is a highly significant increase (p<0.001) in the proportion of juxtaposed oxytocin neurones during lactation, prolonged dehydration and oxytocin infusion in the 3rd ventricle. Adapted from Trends Neurosci., 1987.

adhaerentes or attachment plates (Fig. 1), a type of cell junction seen in all nervous tissue, and which is usually attributed an adhesive function. The appearance of attachment plates highlights further the plasticity of the system since their incidence increases significantly as the amount of juxtaposed neuronal surface increases (Theodosis et al., 1981, Theodosis and Poulain, 1984; Theodosis et al., 1986a; Chapman et al., 1986). Gap juctions, inferred from dye coupling in <u>in vitro</u> preparations of the nuclei (Andrew et al., 1981), have never been seen to couple any neurosecretory element within the magnocellular nuclei, either by conventional or freeze fracture electron microscopy. Gap juctions do exist in these nuclei, but they couple glial processes.

Once stimulation ceases, glial processes are again seen to separate the neurones. Unlike the disappearance of glial processes, which can take place relatively quickly, (for example, within a few hours prior to parturition (Montagnese et al., 1987)), the reappearance of glial processes between the neurones appears to depend on the length of time the system has remained stimulated. In rats that have undergone two consecutive pregnancies and lactations, the incidence of plasmalemma juxtapositions returns to 'virgin' values two months after weaning, whereas one month is sufficient after a first lactation (Theodosis and Poulain, 1984). On the other hand, if the young are taken away immediately after birth, glial reinsertion has taken place as early as two days post partum (Montagnese et al., 1987). A similar pattern is seen when animals are rehydrated after a period of dehydration: the number of juxtaposed neurones returns to normal values within five days of rehydration in normal rats dehydrated for 10 days by drinking saline (Tweedle and Hatton, 1984) whereas, in Brattleboro rats suffering from diabetes insipidus since birth, it still remains elevated after one month of vasopressin treatment (Chapman et al., 1986). Conversely, one can maintain the increase in juxtapositions by maintaining stimulation. Rats are usually weaned after 21-22 days of lactation but a perfectly normal lactation can be

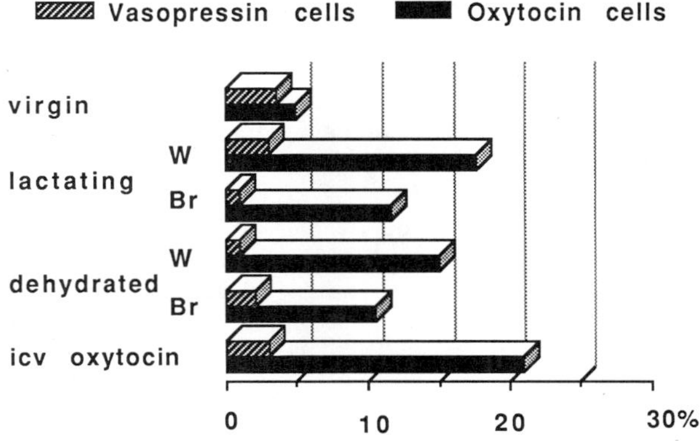

Fig. 4.   Shared synapses in the rat SON. Identification of the neurones was the same as that described in Figure 3. Note that the incidence of "double" synapses in the SON of unstimulated animals (virgin) is low and that they contact an equal proportion of oxytocin- and vasopressin- secreting cells. The proportion of vasopressin neurones receiving such synapses remains the same after stimulation. However, the proportion of oxytocin-neurones sharing the same synapse significantly increases (p<0.001) during lactation and prolonged dehydration, in Wistar (W) and Brattleboro homozygote (Bv) rats, and after 8 days of oxytocin infusion in the 3rd ventricle of normally hydrated, non-lactating Wistar rats.

maintained for much longer periods of time, providing the mothers are supplied with new litters; such a prolonged lactation (for over a month, for example) is accompanied by the same anatomical reorganisation of the oxytocinergic system as that seen during a normal lactation (Montagnese et al., 1987). It should be noted that there are no further increases in the incidence of juxtapositions after such prolonged stimulation, as there is not, if the system is doubly stimulated, as is the case, for example, in dehydrated lactating females (Theodosis et al., 1981). Perhaps this is not too surprising when one considers that almost every oxytocinergic neurone is directly juxtaposed to another neuronal element at the very beginning of stimulation, as, for example, within the few hours preceding parturition.

The cellular mechanisms underlying these changes are at present unknown but it is very probable that both neurones and glial cells are actively involved. The active participation of the neurones is obvious. Under stimulation, the neurones hypertrophy and it is not unlikely that they also change in shape, as noted in other adult neuronal systems (Purves and Hadley, 1985). In addition, the cell bodies of these neurones produce and destroy attachment plates in concert with the amount of their surface in juxtaposition, and also take part in synapse remodelling, as will be described below. The participation of astrocytes in these changes is less obvious. It could have been imagined that glial processes are squeezed away when the neurones hypertrophy. This is certainly not the case since both types of neurone hypertrophy when chronically stimulated, as for example during dehydration (Chapman et al., 1986) and glial coverage of both oxytocinergic and vasopressinergic cells actually increases (Theodosis et al., 1986a; Chapman et al., 1986). What is peculiar to oxytocinergic cells is that the proportion of their plasmalemma covered by glial elements diminishes after stimulation. Neuronal juxtapositions result, therefore, not so much from active glial retraction, but, rather, from an insufficient glial coverage. Another question raised by such observations is why should the glial environment become modified only around oxytocinergic neurones? One possibility is that glial cells react to varying concentrations of extracellular K+ consequent to changing neuronal activity (see Coles, 1986). Again, the specificity of the changes argues against such a possibility : although the electrical activity of both oxytocin and vasopressin cells is enhanced under chronic dehydration (see Poulain and Wakerley, 1982), it is only oxytocinergic cells that are left with less glial coverage (Chapman et al., 1986). A more specific signal must therefore be responsible.

SYNAPTIC CHANGES

Magnocellular cell bodies and dendrites receive a rich synaptic input and in animals in which hormone demands are minimal, most terminals impinge onto one single post-synaptic element in any plane of section examined. Upon stimulation, however, we often see the same terminal synapses onto 2 or even 3 postsynaptic elements, simultaneously, which in the main are oxytocinergic (Fig. 4). In the SON of lactating rats, for example, such shared synapses couple over 20% of all oxytocinergic cell bodies (relative to less than 3% of all vasopressinergic cells) (Theodosis et al., 1986a). This amplification of synaptic input coincides with the neuronal-glial changes and it is not uncommon to see a cluster of oxytocinergic profiles whose surface membranes are in extensive juxtaposition also sharing one or more synaptic terminal.

The shared synapses are highly plastic, since they disappear once the system is no longer stimulated and reappear upon new stimulation (Theodosis and Poulain, 1984; Tweedle and Hatton, 1984; Chapman et al., 1986; Montagnese et al., 1987). Like the plasmalemmal juxtapositions, the rapidity with which they disappear depends on the duration of stimulation: after a normal lactation, at least one month is necessary for their incidence to return to virgin values (Theodosis and Poulain, 1984); if the young are removed immediately at birth, they are no longer visible within 10 days (Montagnese et al., 1987).

There are now many examples of synaptic plasticity in the adult mammalian brain but little is known of the cellular processes involved (see Cotman and Nieto-Sampedro, 1984). From many of these studies, it is clear that synaptogenesis involves the active participation of both pre- and post-synaptic partners and that the rate of synapse formation depends on the intensity of stimulation. In addition, post-synaptic densities are highly plastic structures which can be formed within minutes (Cohen and Siekevitz, 1978) and

162

whose shape and form can be influenced significantly by the activity in the adjacent terminal (Greenough et al., 1978; Vrensen and Nunes-Cardoro, 1981). It may be possible then, that within the magnocellular nuclei, shared synapses are formed as extensions of hyperactive preexistent terminals that have induced the adjacent postsynaptic cell to create a postsynaptic density.

FACTORS INVOLVED IN THE STRUCTURAL PLASTICITY

From the variety of physiological and experimental conditions under which the magnocellular nuclei become structurally modified, it is evident that the factors responsible are complex and multifactorial. Since the oxytocinergic system is always reorganized at lactation, suckling may have been considered as a primary stimulus to bring on the changes. However, the nuclei are already reorganised at parturition (Hatton and Tweedle, 1982; Theodosis and Poulain, 1984), when the young have not yet begun to suckle. Gestation is also not indispensable since the system becomes reorganized in animals that have not undergone a normal pregnancy but have been induced to lactate by exposing them continuously to suckling litters (Montagnese et al., 1987). Finally, the nuclei can also change and to the same extent under prolonged dehydration, in male and female rats (Tweedle and Hatton, 1984; Chapman et al., 1986). Nevertheless, all these seemingly disparate situations may create similar conditions that in turn allow for the same structural changes to occur. More specifically, there are now several observations to suggest that increased levels of hormones, such as oxytocin and/or steroids, associated with all the conditions under which the oxytocinergic system is reorganised may take part, singly or in combination, in inducing such an anatomical reorganisation.

That oxytocin itself may play a key role in the restructuring of its own system is suggested by several observations. In non-pregnant, unsuckled rats, prolonged infusion of oxytocin, or its agonists, into the third ventricle is capable of inducing morphological changes in the SON, identical to those seen under more physiological conditions (Theodosis et al., 1986b). The intracerebroventricular infusions of oxytocin would thus mimic an enhanced central release of the peptide. There is now much evidence that oxytocin is indeed released locally within the magnocellular nuclei (Moos et al., 1984; Mason et al., 1986), perhaps from oxytocinergic terminals that synapse on oxytocinergic neurones (Theodosis, 1985). On the other hand, under all the conditions when the nuclei are structurally modified, there appears to be an increased peripheral release of oxytocin, as seen from the increased electrical activity of the neurones. This is most obvious in lactating rats, during the suckling-induced milk ejection reflex, and in parturient rats, when there is an abundant pulsatile secretion of oxytocin (Summerlee, 1981). As for late gestation and dehydration, they are associated with a tonic release of the hormone (see Poulain and Wakerley, 1982). It is tempting to speculate, then, that the increased activity of the oxytocinergic system causes not only an increased peripheral release of hormone, but also a local release, which would then affect, directly or indirectly, the anatomical arrangement of its neurones.

In addition to oxytocin, parturition and suckling-induced or post-partum lactation are characterized by increased plasma levels of ovarian and placental steroids. It can be argued that sexual steroids only play a minor role since during dehydration, the nucleus is as reorganised as during lactation. Nevertheless, dehydration is associated with an increased secretion of corticosteroids (Coover et al., 1979; Ezzarani et al., 1985) which could assume the same function as the sexual steroids, as happens, for example, when corticosteroids maintain lactation in ovariectomized rats (MacDonald and Matt, 1984). Steroid-accumulating neurones have been described in other areas of the CNS, and the hormones have been implicated in sexual dimorphism of certain nervous structures (see Bottjer and Arnold, 1984; Arnold and Gorski, 1984) and in compensatory collateral sprouting after lesion (Azmitia and Zhou, 1986). In comparison, therefore, it can be postulated that steroids may affect the structure of the oxytocinergic system by acting directly on the oxytocinergic neurones and/or their surrounding glia. Alternatively, as steroids are known to modulate the electrical activity of magnocellular neurones (Negoro et al., 1973; Akaishi and Sakuma, 1985), they could act by facilitating oxytocin release.

Although it is tempting to speculate that elevated levels of steroid hormones or other reproductive hormones, such as prolactin, are responsible for the structural plasticity of the SON, it is obvious that, in themselves, they cannot be sufficient, since the nuclei remain unmodified for quite long periods of time when their plasma levels are significantly enhanced, as for example, during pseudogestation and gestation (Montagnese et al., 1987). A more likely hypothesis is that these hormones act synergistically with oxytocin and/or other factors to produce the plastic changes.

STRUCTURAL PLASTICITY AND THE FUNCTION OF THE OXYTOCINERGIC SYSTEM

It is obvious from the foregoing that the oxytocinergic system possesses a remarkable capacity to modify itself in response to physiological stimuli. But is this plasticity of any consequence to the function of its neurones? During parturition and especially lactation, oxytocinergic neurones display a particular electrical activity, the high frequency discharge of action potentials, which is synchronized across the whole neuronal population in the hypothalamus (Poulain and Wakerley, 1982; Belin and Moos, 1986). Since the morphological reorganization affects primarily oxytocinergic neurones, an obvious hypothesis is that it contributes to this synchronization.

The possible impact of shared synapses on synchronization of neuronal firing appears evident, since they allow several postsynaptic elements to receive the same information simultaneously. GABA is a major inhibitory transmitter in the magnocellular nuclei (Randle et al., 1986) and we recently showed that the majority of "double" synapses in the rat SON are GABAergic (Theodosis et al., 1986c). It may be, then, that an increased inhibitory input to oxytocin neurones serves to coordinate their accelerated firing by permitting activity to occur only at particular intervals, as is the case during the milk ejection reflex induced by suckling. On the other hand, some shared synapses are dopaminergic (Buijs et al., 1984) and this would imply a shared excitatory input (see Poulain and Wakerley, 1982), which could further facilitate neuronal coordination.

Extensive neuronal surface juxtapositions may also contribute to synchronization of firing, by enhancing neuronal excitability through electrical field interactions and/or through changes in the concentration of extracellular ions, particularly K+. Electrical field effects imply increased flow of electrical current through the extracellular space without the participation of any specialized structural connections (see Dudek et al., 1986). The electrical activity in one neurone would thus modify the excitability of its neighbours, especially if no glial elements exist to buffer the interaction. Neuronal firing is also accompanied by important movements of water and ions across neuronal membranes, which can modify not only the composition of the extracellular fluid but also the size and resistance of the extracellular space (see Coles, 1986). Because of the membrane properties of glial cells, and because of their position between neurones, they are usually considered essential to the maintenance of the homeostatis of the extracellular space. Lack of glial elements would then result in accumulation of ions, particularly K^+, which could considerably modify the excitability of the extensively juxtaposed oxytocinergic somata and dendrites in the hypothalamus.

The potential contribution and relative importance of electrical field and ionic effects on the synchronization of oxytocin neurones has still to be determined. They certainly cannot be held totally responsible for the neuronal coordination of the oxytocinergic system, since synchronized electrical activity is recorded from neurones in four spatially separated nuclei (Belin and Moos, 1986). However, they may be important within each nucleus, by further facilitating the synchronized activity that has been set off by other factors, such as synaptic drive, synaptic coupling and perhaps electrotonic coupling.

REFERENCES

Akaishi T. & Sakuma Y., 1985a, Estrogen exites oxytocinergic but not vasopressinergic cells in the paraventricular nucleus of female rat hypothalamus, <u>Brain Res.</u>, 335:302.

Andrew R.D., MacVicar B.A., Dudek F.E. and Hatton G.I., 1981, Dye transfer through gap junctions between neuroendocrine cells of rat hypothalamus, Science, 211:1187.

Arnold A.P. & Gorski R.A., 1984, Gonadal steroid induction of structural sex differences in the central nervous system, Ann. Rev. Neurosci., 7:413.

Azmitia E.C. & Zhou F.C., 1986, Chemically induced homotypic collateral sprouting of hippocampal serotoninergic afferents, Exp. Brain Res. Suppl., 13:129.

Belin V. & Moos F., 1986, Paired recordings from supraoptic and paraventricular oxytocin cells in suckled rats : recruitment and synchronization, J. Physiol. (Lond.), 377:369.

Bottjer S.W. & Arnold A.P., 1984, Hormones and structural plasticity in the adult brain, Trends Neurosci., 7:168.

Buijs R.M., Geffard M., Pool C.W. & Hoorneman E.M.D., 1984, The dopaminergic innervation of the supraoptic and paraventricular nucleus. A light and electron microscopical study, Brain Res., 323:65.

Chapman D.B., Theodosis D.T., Montagnese C., Poulain D.A. & Morris J.F., 1986, Osmotic stimulation causes structural plasticity of the neurone-glia relationships of the oxytocin but not vasopressin secreting neurones in the hypothalamic supraoptic nucleus, Neuroscience, 17:679.

Cohen R.S. and Siekevitz P., 1978, Form of the postsynaptic density : a serial section study, J. Cell Biol., 78:36.

Coles J.A., 1986, Homeostasis of extracellular fluid in retinas of invertebrates and vertebrates, Prog. Sensory Physiol., 6:105.

Coover G.D., Heybach J.P., Lenz J. & Miller J.F., 1979, Corticosterone "basal levels" and response to ether anesthesia in rats on a water deprivation regimen. Physiol. Behav., 22:653.

Cotman C.W. & Nieto-Sampedro M., 1984, Cell biology of synaptic plasticity, Science, 225:1287.

Dudek F.E., Snow R.W. & Taylor C.P., 1986, Role of electrical interactions in synchronization of epileptiform bursts, In: Advances in Neurology (eds. Delgado-Escueta A.V., Ward A.A. Jr., Woodbury D.M. and Porter R.J.), Raven Press, New York, vol. 44, pp. 593-617.

Ezzarani E.-A., Laulin J.-P. & Brudieux R., 1985, Effets de la privation d'eau sur la production de corticostérone chez le rat mâle Brattleboro génétiquement privé de vasopressine, J. Physiol. (Paris), 80:157.

Greenough W.T., West R.W. & De Voogt T.J., 1979, Subsynaptic plate perforations : changes with age and experience in the rat, Science, 202:1096.

Hatton G.I. & Tweedle C.D., 1982, Magnocellular neuropeptidergic neurons in the hypothalamus : increases in membrane apposition and number of specialized synapses from pregnancy to lactation, Brain Res. Bull., 8:197.

MacDonald G.J. & Matt D.W., 1984, Adrenal and placental steroid secretion during pregnancy in the rat, Endocrinol., 114:2068.

Mason W.T., Hatton G.I., Ho Y.W., Chapman C. & Robinson I.C.A.F., 1986, Central release of oxytocin, vasopressin and neurophysin by magnocellular neurones depolarization : evidence in slices of guinea pig and rat hypothalamus, Neuroendocrinol., 42:311.

Montagnese C.M., Poulain D.A., Vincent J.D. and Theodosis D.T., 1987, Structural plasticity in the supraoptic nucleus during gestation, post-partum lactation and suckling-induced pseudogestation and lactation, J. Endocrinol., 115 (in press).

Moos F., Freund-Mercier M.J., Guerné Y., Guerné J.M., Stoeckel M.E. & Richard Ph., 1984, Release of oxytocin and vasopressin by magnocellular nuclei in vitro : specific facilitatory effect of oxytocin on its own release, J. Endocrinol., 102:63.

Morris J.F., 1978, Structural aspects of hormone production and storage. pp. 601-618, in: Cell Biology of Hypothalamic Neurosecretion (Vincent J.D. and Kordon C., eds), CNRS, Paris.

Negoro H., Vissussewan S. & Holland R.C., 1973a, Unit activity in the paraventricular nucleus of female rats at different stages of the reproductive cycle and after ovariectomy with or without strogen or progesterone treatment, J. of Endocrinol., 59:545.

Poulain D.A. & Wakerley J.B., 1982, Electrophysiology of hypothalamic magnocellular neurones secreting oxytocin and vasopressin, Neuroscience, 7:773.

Purves D. & Hadley R.D., 1985, Changes in the dendritic branching of adult mammalian neurones revealed by repeated imaging in situ., Nature, 315:404.

Randle J.C., Bourque C.W. & Renauld L.P., 1986, Characterization of spontaneous and evoked inhibitory postsynaptic potentials in rat supraoptic neurosecretory neurones in vitro., J. Neurophysiol., 56:1703.

Rhodes C.H., Morrell J.I. & Pfaff D.W., 1981, Immunohistochemical analysis of magnocellular elements in rat hypothalamus : distribution and number of cells containing neurophysin, oxytocin and vasopressin, J. of Comp. Neurol., 198:45.

Summerlee A.J.S., 1981, Extracellular recordings from oxytocin neurones during the expulsive phase of birth in unanaesthetized rats, J. Physiol. (Lond.), 321:1.

Theodosis D.T., 1985, Oxytocin-immunoreactive terminals synapse on oxytocin neurones in the supraoptic nucleus, Nature, 313:682.

Theodosis D.T., Chapman D.B., Montagnese C., Poulain D.A. & Morris J.F., 1986a, Structural plasticity in the hypothalamic supraoptic nucleus at lactation affects oxytocin-, but not vasopressin- secreting neurones, Neuroscience, 17:661.

Theodosis D.T., Montagnese C., Rodriguez F., Vincent J.D. & Poulain D.A., 1986b,. Oxytocin induces morphological plasticity in the adult hypothalamo-neurohypophysial system, Nature, 322:738.

Theodosis D.T., Paut L. & Tappaz M.L., 1986c, Immunocytochemical analysis of the GABAergic innervation of the oxytocin and vasopressin-secreting neurones in the rat supraoptic nucleus, Neuroscience, 19:207.

Theodosis D.T. & Poulain D.A., 1984a, Evidence for structural plasticity in the supraoptic nucleus of the rat hypothalamus in relation to gestation and lactation, Neuroscience, 11:183.

Theodosis D.T., Poulain D.A. & Vincent J.D., 1981, Possible morphological bases for synchronization of neuronal firing in the rat supraoptic nucleus during lactation, Neuroscience, 6:919.

Tweedle C.D. & Hatton G.I., 1984, Synapse formation and disappearance in adult rat supraoptic during different hydration states, Brain Res. 309:373.

Vrensen G. & Nunes-Cardozo J., 1981, Changes in size and shape of synaptic connections after visual training : an ultrastructural approach of synaptic plasticity, Brain Res., 218:79.

CONTRIBUTIONS OF ELECTROPHYSIOLOGY TO THE STUDY OF NEUROSECRETION

Dennis W. Lincoln and John A. Russell*

MRC Reproductive Biology Unit
Centre for Reproductive Biology
37 Chalmers Street, Edinburgh, UK
*Department of Physiology, University of Edinburgh
Teviot Place, Edinburgh, UK

INTRODUCTION

This manuscript celebrates the contributions of our chairman, Dr Barry A. Cross, and those who worked with him or under his guidance in the Department of Veterinary Anatomy of the University of Cambridge, then in the Medical School of the University of Bristol and now in the AFRC Institute of Animal Physiology and Genetics Research at Cambridge and elsewhere.

THE CAMBRIDGE ERA, pre-1966

Geoffrey W. Harris, a father figure in modern neuroendocrinology, used electrical stimulation to elicit endocrine responses, most notably in his studies on the neuroendocrine control of ovulation and water balance (Harris, 1947) and subsequently, with Barry Cross, then a young veterinary graduate, on studies of the milk ejection reflex of the rabbit (Cross and Harris, 1952). These investigations were landmarks in the study of the neural control of the pituitary gland, though they provided little direct evidence about the electrophysiological events involved. Cross made the first recordings of the electrical activity of hypothalamic neurones while on a working visit to the Department of Anatomy of the University of California, Los Angeles (Cross and Green, 1959), and on returning to Cambridge sought to exploit this recording technology in an attempt to establish the electrophysiological determinants of hormone secretion. He and his new graduate students struggled against many difficulties that, unspoken, it was preferred not to recognise (Cross and Silver, 1965; Lincoln and Cross, 1967). In retrospect, the recordings made at that time of the electrical activity of single hypothalamic neurones lacked identification in terms of functional specificity. Likewise any attempt to correlate changes in electrical activity and hormone secretion from the pituitary gland were confounded by the vastly different time scales of the two events and by the necessity to use anaesthetics that disrupted the endocrine functions one wished to study. Times were hard in that period and Harris's reputed scepticism of electrical recording, as well as lesioning techniques, was perhaps justified.

THE BRISTOL ERA, 1967-1974

In 1967, Cross moved his team to the Department of Anatomy in the University of Bristol, and there the group was joined by Dr Richard E.J. Dyball, Dr Richard G. Dyer and Dr Jonathan B. Wakerley. Progressively the difficulties outlined above were addressed. Antidromic activation was introduced for the characterisation of recordings from magnocellular neurones (Dyball and Koizumi, 1969), and radio-telemetry of electrical activity and decerebrate hypothalamic islands were developed to avoid the complications of anaesthesia (Beechey and Lincoln, 1969; Cross and Dyer, 1971). Pursuing Cross's earlier work, further studies of the dramatic release of oxytocin associated with labour and milk ejection in the rabbit revealed a potentially useful model within which to study the neural control of the posterior pituitary (Lincoln 1971; 1974a). The location of the oxytocinergic neurones in the hypothalamic magnocellular nuclei had been established, procedures existed for the electrophysiological identification of these magnocellular neurones, and situations were known where hormone release appeared to be time-locked to specific afferent stimuli. However, it was technically not possible in 1970 to combine the radio-telemetry of electrical activity and intramammary pressure with the antidromic identification of putative oxytocinergic neurones in conscious lactating animals during reflex milk ejection.

The breakthrough occurred serendipitously during studies in which Jonathan Wakerley was attempting to elicit oxytocin release in the rat by dilatation of the vagina/cervix. To everyone's amazement, the pups that had been placed on the nipples of the anaesthetised rat, to monitor milk ejection from their weight gain, rose synchronously to their feet (stretch response) every 5-10 min and sucked with increased vigour for about 10 s, in the absence of any experimental stimulus (Wakerley and Lincoln, 1971a). Thus the milk-ejection reflex of the rat was discovered. Unlike the rabbit, discrete milk ejections were observed at regular intervals during the prolonged period of pup attachment and continued during anaesthesia. Oxytocin release was pulsatile and confined to periods in which the mother exhibited an electroencephalographic (EEG) state akin to slow-wave sleep (Lincoln et al., 1973; 1980). This last observation, whilst fascinating in its own right, provided an important key because it allowed one for the first time to study the electrophysiological events associated with suckling-induced oxytocin release in an anaesthetised animal. Within a year, the first recordings were made of putative oxytocinergic neurones in the paraventricular nucleus of the lactating rat during milk ejection. In a population of continuously active neurones (as distinct from the phasically firing neurones discussed below), an extraordinary, stereotyped and explosive acceleration in electrical activity, lasting 2-4 s, was observed to precede by approximately 10 s each rise in intramammary pressure that signalled milk ejection (Fig. 1) (Lincoln and Wakerley 1971; Wakerley and Lincoln 1973). Later similar changes were observed in the supraoptic nucleus (Lincoln and Wakerley, 1974). After several years of endeavour identical electrical recordings were eventually made in unanaesthetised lactating rats (Summerlee and Lincoln, 1981), quieting sceptics who considered the results obtained under anaesthesia to be artefactual. The measurement of intramammary pressure in the anaesthetised rat provided a second-by-second bioassay of oxytocin release without the need to remove blood, and strongly indicated that each milk ejection was associated with a transient period of oxytocin secretion (10 s or less). Recently, measurement of oxytocin by radioimmunoassay has confirmed the predicted pulsatile pattern of oxytocin secretion during suckling in the unanaesthetised rat (Fig. 1) (Higuchi et al., 1986). These electrical recordings, made in the early 1970's, were therefore the first to provide a direct and definite correlation between hypothalamic activity and hormone secretion in any mammalian system.

Among the very first neurones recorded from the paraventricular nuclei of the rat were some observed to fire in phases, with 20-40 s periods of activity of 5-15 spikes/s alternating with corresponding periods of silence (Fig. 2) (Wakerley and Lincoln, 1971b). It was obvious from a number of double recordings that these cells did not fire in synchrony with each other, and thus one deduced that each must be controlled by an intrinsic pacemaker. The activity of these neurones was not, with the occasional exception, correlated with the release of oxytocin at milk ejection, and so these cells were designated for the control of vasopressin secretion. Such a phasic pattern of discharge had an important implication for hormone secretion. The asynchronous firing of the putative vasopressinergic neurones was presumed to lead to continuous vasopressin secretion because at any time some terminals, at least, would be involved in secretion. Thus, the instantaneous rate of secretion would depend on the mean period of cell firing in relation to the mean period of silence, and the mean rate of firing during the burst. By analogy, the continuous background firing of the oxytocin cells should likewise result in the continuous secretion of small amounts of oxytocin. So by 1974 both oxytocin and vasopressin neurones had been extensively characterised in terms of their firing patterns and responses to various stimuli.

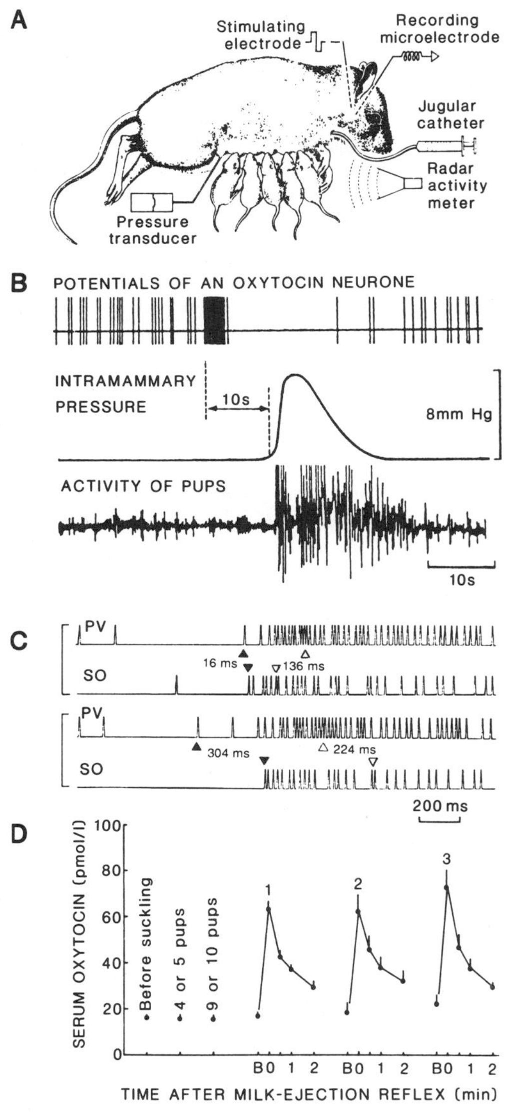

Fig. 1. The anaesthetised lactating rat: a model for the study of the electrical activity of putative oxytocinergic neurones. A. Identifies the procedures involved. An intramammary pressure recording from one nipple monitors milk ejection, and an i.v. catheter for oxytocin (and drug) allows calibration of the pressure recording. A stimulating electrode is placed in the posterior pituitary for antidromic identification of magnocellular neurones; its placement is confirmed by electrically evoking a pulse of oxytocin secretion. A litter of hungry pups is then applied to provide natural afferent stimulus; their activity in response to milk ejection is monitored by a doppler device. B. Illustrates the explosive acceleration in the activity of an oxytocin neurone about 10 s before the rise in intramammary pressure that signals milk ejection, and to which the pups respond by sucking with increased vigour. Illustration adapted from Lincoln, 1984. C. Two examples of dual recordings of the electrical activity of oxytocin neurones monitored simultaneously in separate magnocellular nuclei of a lactating rat. The recordings are projected on a fast time base to illustrate the relative high degree of synchrony involved. Data adapted from Belin et al., 1984. D. Serum oxytocin concentrations in nine lactating rats. Data are synchronised around the occurrence of the first three milk ejections observed after the application of the pups to the nipples. Note the attachment of the pups per se did not promote a release of oxytocin, and following each milk ejection the levels declined in parallel to the predicted half-life of the hormone in the circulation. Data adapted from Higuchi et al., 1986.

GENE EXPRESSION: LINKAGE OF ELECTROPHYSIOLOGICAL AND BIOSYNTHETIC PROPERTIES

Oxytocin and vasopressin neurones are remarkable, in the rat at least, for their electrophysiological dichotomy which, as will be discussed later, must be associated with fundamental differences in membrane kinetics, chemosensitivity and connectivity. The ancestral arginine vasotocin gene appears to have been duplicated early in vertebrate evolution, with point mutations following in the two genes thereafter (Acher, 1980). Interestingly, only one of these two genes appears to be expressed in a single magnocellular neurone (see Lincoln and Russell, 1986), though nothing is known of how the expression of the oxytocin gene, for example, suppresses the vasopressin gene and vice versa. However, many other genes must be co-expressed in a controlled sequence in order to confer on the oxytocin cell its specific electrophysiological properties, and the same integrated organisation must apply to the vasopressin cell. Indeed, the physiology of the magnocellular neurones has evolved to generate a pattern of hormone secretion that matches beautifully the physiology of the target tissues. Both the mammary gland in milk ejection and the uterus in the expulsion of the fetus respond most effectively to pulses of oxytocin and, as will be discussed later, subtle electrophysiological mechanisms have been evolved to generate pulsatile secretion, that are highly efficient in the use of electrical activity, function to conserve peptide stores and effectively separate signals from noise (see frequency facilitation). In its continuous mode of discharge, a steady secretion of oxytocin probably suits the renal mechanisms concerned with sodium secretion on which it acts. The kidney (and blood vessels) appears to require a sustained vasopressin signal that alters in a homeostatic manner, minute-by-minute, to maintain water balance. The asynchronous phasic pattern of discharge of the vasopressin cell is transformed into changes in rate of secretion by alterations in the duration of the burst and the firing rate within the burst, and each spike is used most economically by capitalising on the process of frequency facilitation. The distinction between oxytocin and vasopressin neurones is even more complex and subtle because abundant evidence has now accumulated to indicate co-synthesis and secretion of oxytocin or vasopressin with other regulatory peptides.

It is possible to manipulate the firing patterns of oxytocin and vasopressin neurones so that they imitate one another to an extent, but this is far short of converting the firing activity of an oxytocin cell into that normally associated with a vasopressin cell and _vice versa_. Hyperosmolarity and severe haemorrhage, for example, cause oxytocin cells to fire fast and continuously, though never in bursts or phases (Wakerley et al, 1975; Harris et al., 1975; Poulain et al., 1977; Wakerley et al., 1978). Bilateral carotid occlusion, on the other hand, causes a synchronised and marked acceleration in the firing of putative vasopressin cells, and that leads presumably to a surge in vasopressin secretion (Dreifuss et al., 1976a).

Nothing is known, however, of how the genes of either the oxytocin or vasopressin cells are controlled so as to generate their co-ordinated patterns of expression, but the message is clear - specialisation involves a radical, irreversible decision about the expression of a whole co-ordinated group of genes.

CENTRAL PROCESSING OF INFORMATION

<u>Afferent inputs and the synchronisation of the oxytocinergic neurones</u>

In the 1950's and 1960's it was supposed that milk ejection was a simple neuroendocrine reflex in which the magnocellular neurones were modulated, second-by-second, by the sucking activities of the young, as indicated by studies on the cat (Brooks et al., 1966). However, subsequent studies on the rat revealed no changes in the background discharge of neurones that could be related to the sucking of the pups. Yet, there is a positive relationship between the number of young sucking on the nipples and, for any individual oxytocin neurone, the magnitude of the neurosecretory burst and the size of the oxytocin pulse secreted (Lincoln and Wakerley, 1975; Meyer et al., 1987). It is not clear how this amplitude modulation is achieved, what the mechanisms are that cause the magnocellular neurones to fire in synchrony, nor how the interval between one milk ejection and the next is governed. These issues have been extensively evaluated in previous reviews (Poulain and Wakerley, 1982; Lincoln and Russell, 1985); thus discussion will be limited to a number of the more recent and seminal observations.

Lesioning, electrical stimulation and histochemical tracing techniques have been used to identify the pathways transmitting the stimulus applied by the sucking of the pups. The input lacks somatotopy, it is multi-segmental (Poulain and Wakerley, 1986) and projects rostrally through the ipsilateral dorsal-lateral funiculus of the spinal cord to relay in the lateral cervical nucleus of the brain stem (Dubois-Dauphin et al., 1985a,b). From there the pathway passes mainly contralaterally through the lateral tegmentum of the midbrain, possibly involving the peripeduncular nucleus (Juss and Wakerley, 1981; Hanson and Kohler, 1984). Interestingly, bilateral lesions in the ventral tegmentum disrupt the pattern of milk ejection, with occasional milk ejections following each other at unusually short intervals (Juss and Wakerley, 1981). The occurrence of such twin milk ejections suggests that the mechanisms determining the milk ejection interval involve either the midbrain or pathways traversing this region. The pathway rostral from the mesencephalon is unclear, but it is certainly indirect to the magnocellular nuclei and is probably both ipsilateral and contralateral (Dubois-Dauphin et al., 1985a). The reflex can be suppressed in many species by cognitive stress (Cross, 1955; Clarke and Lincoln, 1976) and in the rat by arousal (wakefulness) (Lincoln et al., 1980). Inputs from higher neural centres are not however essential for the reflex: neither cortical depression nor lesions of the anteroventral third ventricular region (AV3V) or septum block milk ejection (Lebrun et al., 1983; Blackburn et al., 1987). Interestingly, the association in the rat between EEG synchronisation and milk ejection does not seem to be causal, since lesions of the dorsal raphe nuclei cause desynchronisation of the EEG yet milk ejections continue, albeit less frequently (Clarke et al., 1985).

Remarkable dual recordings have recently been taken from the magnocellular nuclei by Richard's group in Strasbourg. These confirm, as the constant interval between electrophysiological activation and milk ejection had already indicated, that during suckling all magnocellular oxytocin neurones burst simultaneously, with a mean firing time lag of 122 msec to the peak firing rate between random and remote pairs of neurones in the supraoptic and paraventricular nuclei (Fig. 1) (Belin et al., 1984; Belin and Moos, 1986). The larger bursts began marginally before the shorter ones, and some cells are recruited only after the first milk ejection, though there is no evidence that either magnocellular nucleus consistently leads the other (Belin et al., 1984). There is accumulating evidence to indicate that oxytocin itself acts as an autoregulator upon oxytocinergic neurones to 'open the gate' that synchronises the bursting of these cells. Extensive projections link the four magnocellular nuclei (Silverman et al., 1981), the area of apposition between adjacent oxytocin neurones increases during lactation (Theodosis and Poulain, 1984; 1987; Hatton, 1986) and recent studies have identified oxytocin containing terminals impinging upon the perikarya of oxytocinergic neurones (Theodosis, 1985). Oxytocin is released centrally within the vicinity of the magnocellular nuclei (Moos et al., 1984; Mason et al., 1986), in vitro oxytocin excites continuous and not phasic cells in the supraoptic nucleus (Yamashita et al., 1987) and oxytocin injected into the third cerebral ventricle augments the neurosecretory bursts associated with milk ejection in the rat and reduces the interval between them (Fig. 2) (Freund-Mercier and Richard, 1984). An oxytocin antagonist given into the third cerebral ventricle has the opposite effect. Peripherally administered oxytocin does not augment or inhibit the reflex (Lincoln, 1974b), presumably because it cannot cross the blood brain barrier in effective amounts. Constant collision antidromic stimulation of oxytocin neurones produces some (orthodromic) increase in activity, but no bursts (Leng, 1981), and, whilst short trains of antidromic pulses (that presumably release oxytocin from central processes) do not produce immediate milk ejection bursts, they reduce the interval between them (Dyball and Leng, 1986). It is possible that the electrophysiological synchronisation of the oxytocin cells and their response to oxytocin but not vasopressin could relate to a specialisation at the receptor level for which there may be little or no visible ultrastructural correlate. Emerging, therefore, is evidence of a dual control system. The sucking stimulus remains the essential input and acts to unlock the 'gate' simultaneously in all four magnocellular nuclei, with a key generated in the brain stem. Once unlocked, the expression of the neurosecretory burst is probably determined by local mechanisms.

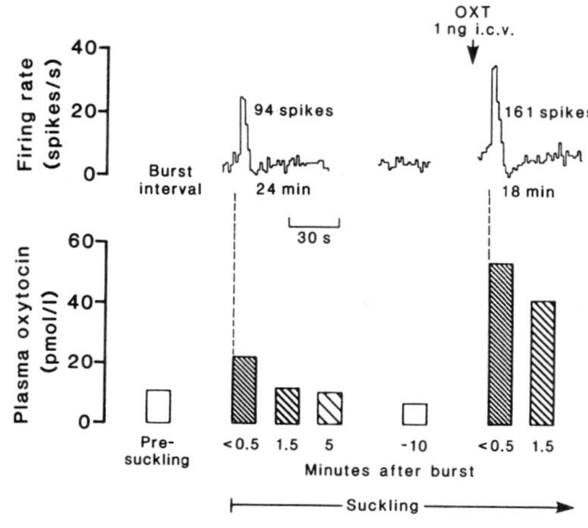

Fig. 2. Regulation of oxytocin neurones by oxytocin during suckling. The upper part records the firing rate of a putative oxytocin neurone in the paraventricular nucleus of a lactating rat. The number of spikes in the explosive period of neurosecretory activation was increased and the interburst interval reduced by the administration of oxytocin (1 ng, i.c.v). The lower part records the level of oxytocin in the peripheral plasma of the rat before and after the two milk ejections shown above. The level of oxytocin was increased presumably due to an increase in electrical activity within all (or many) of the oxytocin neurones in the four magnocellular nuclei. Data adapted from Meyer et al., 1987.

<u>Location of the central osmoreceptors and the control of phasic firing</u>

The initial supposition that phasic firing of the vasopressinergic neurone is an intrinsic property of this type of neurone has been extensively confirmed, though such phasic discharge can be initiated or augmented by synaptic activation, changes in the osmolarity of the extracellular fluid (Mason, 1980; Leng et al., 1982), stimulation by antidromic potentials (Dreifuss et al., 1976b) and by current injection (Andrew and Dudek, 1984).

Magnocellular vasopressin cells generate action potentials displaying a number of characteristics that are important both in the context of phasic firing and hormone secretion. These have been analysed with intracellular recording techniques <u>in vitro</u>. The depolarising phase consists of two components. Tetrodotoxin, the Na+ channel blocker, eliminates the

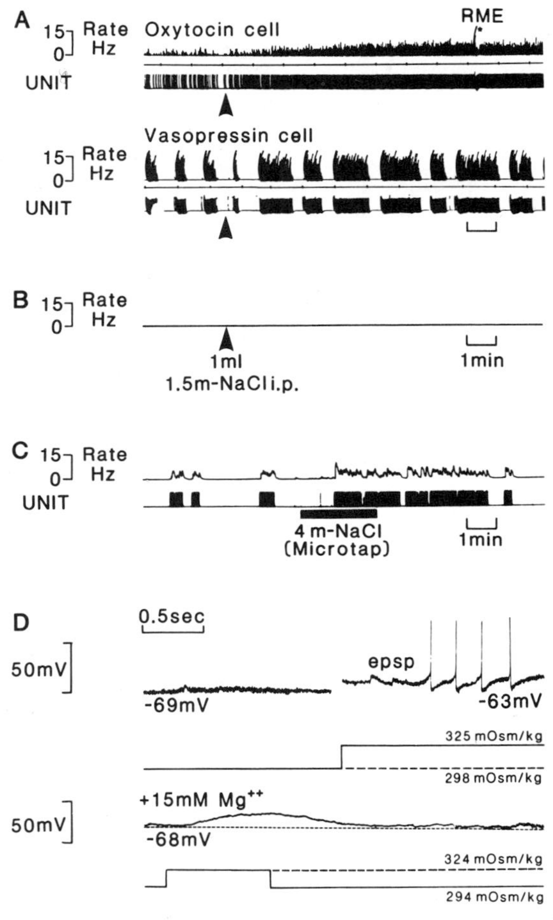

Fig. 3. Osmoresponsiveness of magnocellular neurones. A. Spike activity of putative oxytocin and vasopressin neurones, antidromically-identified, in the supraoptic nucleus of a urethane-anaesthetised lactating rat during suckling, before and after the administration of a hyperosmotic stimulus (1.5M NaCl, i.p.). Note the firing rate of the oxytocinergic neurone was progressively increased by the osmotic stimulus, whilst the effect on the vasopressinergic neurone was to increase the duration of the bursts and intraburst firing rate. Data adapted from Brimble and Dyball, 1977. B. A recording from an antidromically-identified neurone in the supraoptic nucleus, using an identical preparation to that above, following the placement of an acute electrolytic lesion in the AV3V region. Note the total lack of activity and response to an osmotic challenge. Data from Russell and Leng, unpublished. C. Activity of an antidromically-identified phasic neurone in the supraoptic nucleus of an intact rat. An extracellular electrode with a microtap was used to deliver minute quantities of 4M NaCl in the vicinity of the recorded cell. The increased frequency of firing and increased duration of bursts suggests osmosensitivity of either the recorded neurone or adjacent cells. Data adapted from Leng, 1980. D. Intracellular recordings from magnocellular neurones in the supraoptic nucleus of an <u>in vitro</u> hypothalamic slice. The upper trace illustrates a quiet neurone excited by an increase in the osmolarity of the superfusing medium (increased Na+). The cell is depolarised by the osmotic challenge, a number of epsp's are observed and the neurone fires a number of spikes when the threshold is reached (c -50 mV). In the lower trace synaptic input has been blocked by increased Mg++ in the superfusate. A similar osmotic challenge deloparizes the membrane of the cell but it now fails to generate spikes in the absence of a synaptic input. Data adapted from Mason, 1980.

fast phase of the spike and leaves a high threshold spike of reduced amplitude and increased duration. Exposure to low Ca++ has the reverse effect and eliminates the second component (Andrew and Dudek, 1984; Bourque and Renaud, 1985). Increasing the rate of firing from say 3 to 10 Hz can double the duration of the spike, and that is accounted for by a prolongation of the calcium conductance (Mason and Leng, 1984). The repolarising phase is a more complex event and involves a combination of currents. These include an inactivation of the fast Na+ current and the activation of a delayed rectifier. Two K+ currents also appear to be involved, one expressed as a fast transient current and the other as a much slower Ca++ dependent current. The latter results in the post-burst hyperpolarisation and is responsible for the period of silence observed in both oxytocin and vasopressin cells after a period of rapid spiking. This intrinsic hyperpolarisation probably also accounts for anumber of observations previously attributed to recurrent synaptic inhibition (Andrew, 1986).

Phasic firing depends on the prolonged hyperpolarisation phase discussed above for the period of silence and a depolarisation plateau to maintain firing during the burst. The slow depolarisation, involving an inward Ca++ current, serves to lower the membrane potential towards the spike threshold, after which the depolarising afterpotentials of the spikes summate to produce the depolarisation plateau potential which sustains activity (Bourque, 1986; Andrew, 1987a). Phasic bursting does not therefore appear to be driven by a patterned synaptic input. However, the onset of bursting depends on the presence of synaptic or non-synaptic depolarising events to trigger the initial spikes for the production of a regenerative depolarisation plateau. Such insight provides an explantation for why it is possible _in vivo_ to both promote and terminate phasic discharge through the application of antidromic potentials (Dreifuss et al., 1976b).

In addition, some magnocellular neurones display a sinusoidal oscillation in membrane potential which is attributed to a regular oscillation in synaptic input (Andrew, 1987b). The source of this input has not been determined, it could arise from within the magnocellular nuclei via a presynaptic cell or network of cells, or alternatively the input could reach the magnocellular nuclei from elsewhere. The supraoptic nuclei receive an excitatory cholinergic input from a region located a short distance dorsally (Hatton et al., 1983), and some cells in this region are driven orthodromically by stimulation of the neural stalk (Blount and Leng, 1985).

Phasic firing therefore appears to involve two interactive systems, the principal one intrinsic and the other synaptically driven. Mason (1980) demonstrated a slow depolarisation in magnocellular neurones in response to increased osmolarity during synaptic blockade, and thus provided the first clear evidence that magnocellular neurones had the potential to function as osmoreceptors (Fig. 3). However, normal osmosensitivity depends on the functional integrity of a synaptic input. The AV3V region contains osmosensitive neurones in the organum vasculosum of the lamina terminalis (OVLT) (Sayer et al., 1984) and connects with or receives fibres en route from the osmosensitive subfornical organ (Sibbald et al., 1984); and mediates the action of circulating angiotensin II on both oxytocin and vasopressin neurones (Ferguson and Renaud, 1986) possibly by using angiotensin II as a neurotransmitter (Lind et al., 1984; Lincoln and Russell, 1985; Okuya et al., 1987). Lesions of the AV3V region quieten or silence magnocellular neurones and abolish their response to osmotic stimulation (Fig. 3) (Dyball and Prilusky, 1981; Leng and Russell, 1986), indicating that this region provides an essential tonic excitatory input to the magnocellular neurones. Despite these results, the direct osmosensitivity of magnocellular neurones (demonstrated _in vitro_) does underlie their osmo-responsiveness since restoring tonic activity with locally applied glutamate after AV3V lesioning permits the excitation of magnocellular neurones by hyperosmotic stimuli (Leng et al., 1988). In phasic neurones, osmotically-induced depolarisation, possibly evoked by the suppression of a K+ conductance (Abe and Ogata, 1982), interacts with epsp's from the excitatory synaptic input to trigger and sustain the periods of firing. It is remarkable that oxytocinergic neurones continue to function as normal in the context of reflex milk ejection and parturition after the placement of AV3V lesions, though their osmosensitivity is lost (Blackburn et al., 1987; Russel et al., 1987).

TERMINAL MODULATION OF HORMONE SECRETION

Frequency facilitation

Harris established, before the era of electrical recording, that approximately 50 Hz was the optimal frequency with which to stimulate the posterior pituitary in order to evoke hormone secretion (Harris et al., 1969). This exceeds by more than an order of magnitude the background firing of most hypothalamic magnocellular neurones. Indeed, it was not until the electrophysiological determinants of oxytocin secretion were established in the rat that such high frequencies of firing were observed. It has been calculated that these high frequencies of firing increase by 2 or 3 orders of magnitude the amount of hormone secreted in response to each spike (Lincoln, 1974c; Dreifuss et al., 1981). This has profound implications, because it allows a pulse of oxytocin to be secreted over a short period of time whilst limiting the drain on the hormone stores imposed by the continuous background firing of the neurones. Put another way, it greatly improves the signal to noise ratio.

No full explanation exists for this phenomenon. The spike broadening associated with high frequency firing at the level of the cell body could, if similar membrane properties operate at the terminals, result in a much higher influx of Ca++ (Andrew, 1986). Furthermore the level of Ca++ within the cell might increase during rapid firing if the rate of influx exceeds the rate of extrusion or sequestration, and facilitate exocytosis of neurosecretory granules. An additional possibility has recently come to light. Leng and Shibuki (1987) have recorded a dramatic rise in extracellular K+ in the posterior pituitary in association with the secretion of the oxytocin pulse that promotes milk ejection in the rat (Fig. 4). Such a rise in the level of extracellular K+ would reduce the potassium gradient across the cell membrane, and that might, by reducing the K+ current, promote spike duration and the entry of Ca++ (Cazalis et al., 1987). Alternatively the very high extracellular K+ levels might serve to facilitate the spike invasion of the fine terminal arborisations of the magnocellular axons, thereby augmenting secretion.

Thus action potentials, whilst effectively equal as units of conductance, become orders of magnitude more effective at promoting hormone secretion when the interspike interval is reduced by an increase in firing rate or a change in the pattern of spike discharge.

Terminal modulation of secretion by extrinsic factors

The terminals of the posterior pituitary are outside the blood-brain barrier and are exposed to circulating peptides that cannot so effectively reach the magnocellular nuclei. The first evidence to indicate that the terminals could be subject to local regulation came from a single, simple experiment. Naloxone was given to an anaesthetised lactating rat that had failed to release oxytocin in response to electrical stimulation of the posterior pituitary. Minutes later the same burst of electrical stimulation released the normal, predicted amount of oxytocin, as judged from the intramammary pressure response. This single observation launched a new era because it established that secretion could be inhibited at the level of the magnocellular nerve terminals by endogenous opioid peptides (Clarke et al., 1979). Such a conclusion broke with the established dogma of 'stimulus-secretion coupling', because it implied that electrical activity transmitted from the magnocellular nuclei need not release oxytocin at the terminals.

The terminal inhibition of secretion was confirmed in studies on the posterior pituitary _in vitro_ and, strikingly, was shown to affect electrically-induced oxytocin secretion much more than vasopressin secretion (Fig. 5) (Bicknell and Leng, 1982). Now the question arose as to the source and nature of the endogenous opioid peptide. Considerable evidence was produced for both an innervation of the posterior pituitary by enkephalin-containing nerve fibres (possibly terminating on pituicytes) and for the co-existence of opioid peptides within the oxytocin- and vasopressin-containing neurones (reviewed by Bicknell, 1985; Lincoln and Russell, 1986). Of signal importance was the discovery of dynorphin and related peptides in the vasopressin- but not the oxytocin-containing neurones (Watson et al., 1982; Weber and Barchas, 1983), and of evidence indicating a release of dynorphin with vasopressin (Millan and Herz, 1985). It is now clear that the principal (or only) type of opioid receptor in the posterior pituitary is of the kappa sub-type (Herkenham et al., 1986),

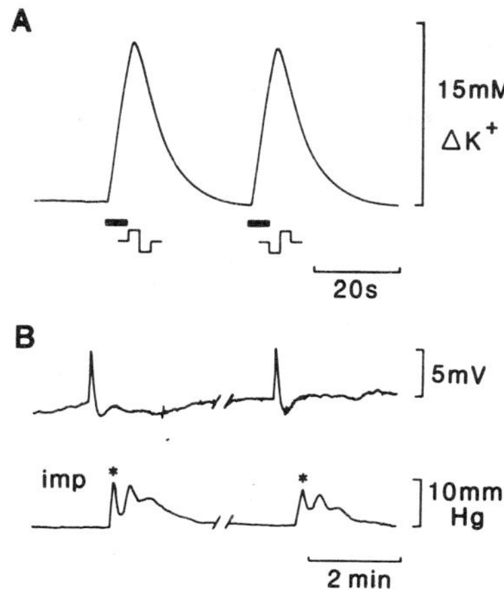

Fig. 4. Extracellular K+ changes in the posterior pituitary of the rat during A) activation of the neural stalk _in vitro_ and B) reflex milk ejection. Stimulation in A was applied at 20 Hz for 5s, and the K+ changes recorded through a K+ sensitive electrode. Milk ejection in B was evoked by 10 hungry pups and the milk ejections recorded from the changes in intramammary pressure (imp), the K+ flux being recorded in mV. Both recordings are adapted from Leng and Shibuki, 1987.

and many of these are located on oxytocin-containing terminals (Falke and Martin, 1985). These findings lead to the conclusion that the vasopressin-terminals in the posterior pituitary, by co-releasing dynorphin, act to inhibit the release of oxytocin: an example of heterotypic interaction between magnocellular neurones. This conclusion is supported by studies with various kappa-receptor specific ligands (Bicknell et al., 1985; 1987). The function of such an interaction is uncertain: it may suspend secretion of oxytocin during parturition at times when it would be inappropriate to give birth (Leng et al., 1987) or it may conserve oxytocin for reproductive purposes when both vasopressin and oxytocin

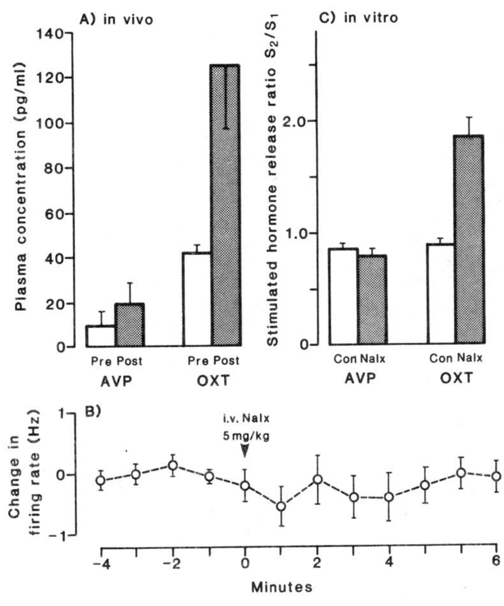

Fig. 5. Heterotypic interactions between nerve terminals in the posterior pituitary: inhibition of oxytocin release by opioid peptides.

A. Oxytocin and vasopressin levels in peripheral plasma of six urethane anaesthetised lactating rats: naloxone (5mg/kg i.v.) significantly increased plasma oxytocin levels after 6 min but did not alter the level of vasopressin. B. Changes in firing rate, recorded electrophysiologically, of oxytocinergic neurones in the supraoptic nucleus; the blood levels of oxytocin and vasopressin of these rats are shown in Part A. Naloxone had no effect on the firing rate of these cells. Data adapted from Bicknell et al., 1988. C. Effect of naloxone on the electrically-stimulated release of oxytocin and vasopressin from the isolated posterior pituitary _in vitro_. Data expressed as the ratio of release between the two periods of stimulation, with and without naloxone at 5 uM. Data adapted from Leng and Bicknell (1987).

neurones are stimulated by changes in salt and water balance (Hartman et al., 1987). Whether sufficient met-enkephalin is secreted from oxytocin terminals to affect oxytocin or vasopressin terminals is unclear (Gaymann and Martin, 1987).

In addition to the effects of opioid peptides on the secretory terminals there is substantial evidence that opioids have central effects, especially on oxytocinergic neurones. Acutely administered, opioid peptides (and opiates) inhibit oxytocin neurones both _in vitro_ (Wakerley et al., 1983; Pumford et al., 1987) and _in vivo_ (Wright and Clarke, 1983). More strikingly, after chronic intracerebroventricular morphine treatment, during which tolerance develops to the inhibitory effects, the administration of naloxone greatly excites the electrical activity of oxytocin neurones and produces a massive release of oxytocin (Bicknell et al., 1988; Rayner et al., 1988). Naloxone has no effect on the electrical activity of oxytocin neurones in normal anaesthetised rats indicating that these neurones are not under tonic opioid inhibition.

CONCLUDING REMARKS

Thirty years have now elapsed since Barry Cross first promoted our interest in the electrophysiological study of magnocellular neurones. The studies that followed, perhaps more than any others, have underwritten the neuroendocrine doctrine first elaborated by his mentor Geoffrey Harris. In doing so, they have also brought to our attention new phenomena. This list would include such events as pulsatile hormone secretion, phasic and synchronised burst firing, frequency facilitation and terminal modulation of stimulus secretion coupling. Thirty years ago it was anatomy and histochemistry that led the way, presenting the electrophysiologist with the challenge. In the past fifteen years, however, the position has been reversed. Today it is physiological observations, largely based on electrical recording, that present the structural anatomist and the molecular biologist with the leading questions. And finally, we have witnessed the total destruction of at least one sectarian barrier. A decade ago those who studied the peptidergic neurones of the magnocellular nuclei were outcasts in traditional neurophysiological circles that thought only of aminergic and cholinergic transmission. Today, thanks to the discovery of the widespread involvement of peptide messengers in the regulation of the brain, those who work with the electrophysiological study of magnocellular neurones find themselves at one of the cutting edges in the field of neuroscience, and with horizons that stretch far beyond the original concept of neurosecretion.

REFERENCES

Abe, H. and Ogata, N., 1982, Ionic mechanism for the osmotically-induced depolarisation in neurones of the guinea-pig supraoptic nucleus _in vitro_, J. Physiol., Lond., 327:157.

Acher, R., 1980, Molecular evolution of biologically active polypeptides, Proc. R. Soc., Lond., B., 210:21.

Andrew, R.D., 1986, Intrinsic membrane properties of magnocellular neurosecretory neurons recorded _in vitro_, Fed. Proc., 45:2306.

Andrew, R.D., 1987a, Endogenous bursting by rat supraoptic neuroendocrine cells is calcium dependent, J. Physiol., Lond., 384:451.

Andrew, R.D., 1987b, Isoperiodic bursting by magnocellular neuroendocrine cells in the rat hypothalamic slice, J. Physiol., Lond., 384:467.

Andrew, R.D. and Dudek, F.E., 1984, Analysis of intracellularly recorded phasic bursting by mammalian neuroendocrine cells, J. Neurophysiol., 51:552.

Beechey, P and Lincoln, D.W., 1969, A miniature FM transmitter for the radiotelemetry of unit activity, J. Physiol., Lond., 203:5P.

Belin, V. and Moos, F., 1986, Paired recordings from supraoptic and paraventricular oxytocin cells in suckled rats: recruitment and synchronization, J. Physiol., Lond., 377:369.

Belin, V., Moos, F. and Richard, Ph., 1984, Synchronization of oxytocin cells in the hypothalamic paraventricular and supraoptic nuclei in suckled rats: direct proof with paired extracellular recordings, Exp. Brain Res., 57:201.

Bicknell, R.J., 1985, Endogenous opioid peptides and hypothalamic neuroendocrine neurones, J. Endocr., 107:437.

Bicknell, R. J., Chapman, C. and Leng, G., 1985, Effects of opioid agonists and antagonists on oxytocin and vasopressin release _in vitro_, Neuroendocrinology, 41:142.

Bicknell, R.J., Chapman, C. and Zhao, B-G., 1987, Opioid receptor subtypes mediating inhibition of oxytocin, vasopressin and noradrenaline release from the rat neurohypophysis in vitro, J. Physiol., Lond., 388:9P.

Bicknell, R.J. and Leng, G., 1982, Endogenous opiates regulate oxytocin but not vasopressin secretion from the neurohypophysis, Nature, Lond., 298:161.

Bicknell, R.J., Leng, G., Lincoln, D.W. and Russell, J.A., 1988, Naloxone excites oxytocin neurones in the supraoptic nucleus of lactating rats after chronic morphine treatment, J. Physiol., Lond., (In press).

Blackburn, R.E., Leng, G. and Russell, J.A., 1987, Control of magnocellular oxytocin neurones by the region anterior and ventral to the third ventricle (AV3V region) in rats, J. Endocr., 114:253.

Blount, C.A. and Leng. G., 1985, Synaptic activation and inhibition in the lateral hypothalamus following stimulation of the neural stalk in the rat, J. Physiol., Lond., 361:30P.

Bourque, C.W., 1986, Calcium-dependent spike after-current induces burst firing in magnocellular neurosecretory cells, Neurosc. Lett., 70:204.

Bourque, C.W. and Renaud, L.P., 1985, Calcium-dependent action potentials in rat supraoptic neurosecretory neurones recorded in vitro, J. Physiol., Lond., 363:419.

Brimble, M.J. and Dyball, R.E.J., 1977, Characterization of the responses of oxytocin- and vasopressin-secreting neurones in the supraoptic nucleus to osmotic stimulation, J. Physiol., Lond., 271:253.

Brooks, C. McC., Ishikawa, T., Koizumi, K. and Lu, H-H., 1966, Activity of neurones in the paraventricular nucleus and its control, J. Physiol., Lond., 182:217.

Cazalis, M., Dayanithi, G. and Nordmann, J.J., 1987, Requirements for hormone release from permeabilized nerve endings isolated from the rat neurohypophysis, J. Physiol., Lond., 390:71.

Clarke, G., Juss, T.S. and Wakerley, J.B., 1985, Effect of lesions of the medial raphe on EEG patterns amd reflex milk ejection in the urethane-anaesthetized rat, J. Physiol., Lond., 364:57P.

Clarke, G. and Lincoln, D.W., 1976, Effects of catecholamine antagonists on the milk-ejection reflex of the anaesthetised rat, Br. J. Pharmac., 57: 458.

Clarke, G, Wood, P, Merrick, L. and Lincoln, D.W., 1979, Opiate inhibition of peptide release from the neurohumoral terminals of hypothalamic neurones, Nature, Lond., 282:746.

Cross, B.A., 1955, The hypothalamus and the mechanism of sympathetico-adrenal inhibition of milk ejection, J. Endocr., 12:15.

Cross, B.A. and Dyer, R.G., 1971, Unit activity in rat diencephalic islands: the effects of anaesthetics, J. Physiol., Lond., 212:467.

Cross, B.A. and Green, J.D., 1959, Activity of single neurones in the hypothalamus: effect of osmotic and other stimuli, J. Physiol., Lond., 148:554.

Cross, B.A. and Harris, G.W., 1952, The role of the neurohypophysis in the milk-ejection reflex, J. Endocr., 8:148.

Cross, B.A. and Silver, I.A., 1965, Effect of luteal hormone on the behaviour of hypothalamic neurones in pseudopregnant rats, J. Endocr., 31:251.

Dreifuss, J.J., Harris, M.C. and Tribollet, E., 1976a, Excitation of phasically firing hypothalamic supraoptic neurones by carotid occlusion in rats, J. Physiol., Lond., 285:171.

Dreifuss, J.J., Tribollet, E., Baertschi, A.J. and Lincoln, D.W., 1976b, Mammalian endocrine neurones: control of phasic activity by antidromic action potentials, Neurosc. Lett., 3:281.

Dreifuss, J.J., Tribollet, E. and Muhlethaler, M., 1981, Temporal patterns of neural activity and their relation to the secretion of posterior pituitary hormones, Biol. Reprod., 24:51.

Dubois-Dauphin, M., Armstrong, W.E., Tribollet, E. and Dreifuss, J.J., 1985a, Somatosensory systems and the milk-ejection reflex in the rat. I. Lesions of the mesencephalic lateral tegmentum disrupt the reflex and damage mesencephalic somatosensory connections, Neuroscience, 15:1111.

Dubois-Dauphin, M., Armstrong, W.E., Tribollet, E. and Dreifuss, J.J., 1985b, Somatosensory systems and the milk ejection reflex in the rat. II. The effects of lesions in the ventroposterior thalamic complex, dorsal columns and the lateral cervical nucleus-dorsolateral funiculus, Neuroscience, 15:1131.

Dyball, R.E.J. and Koizumi, K., 1969, Electrical activity in the supraoptic and paraventricular nuclei associated with neurohypophysial hormone release, J. Physiol., Lond., 201:711.

Dyball, R.E.J. and Leng, G., 1986, Regulation of the milk ejection reflex in the rat, J. Physiol., Lond., 380:239.

Dyball, R.E.J. and Prilusky, J., 1981, Responses of supraoptic neurones in the intact and deafferented rat hypothalamus to injections of hypertonic sodium chloride, J. Physiol., Lond., 311:443.

Falke, N. and Martin, R., 1985, Opioid binding in a rat neurohypophysial fraction enriched in oxytocin and vasopressin nerve endings, Neurosci. Lett., 61:37.

Ferguson, A.V. and Renaud, L.P., 1986, Systemic angiotensin acts at subfornical organ to facilitate activity of neurohypophysial neurons. Amer. J. Physiol., 251:R712.

Freund-Mercier, M.J. and Richard, Ph., 1984, Electrophysiological evidence for facilitatory control of oxytocin neurones by oxytocin during suckling in the rat, J. Physiol., Lond., 352:447.

Gaymann, W. and Martin, R., 1987, A re-examination of the localisation of immunoreactive dynorphin (1-8), [Leu] enkephalin and [Met] enkephalin in the rat neurohypophysis, Neuroscience, 20:1069.

Hanson, S. and Kohler, C., 1984, The importance of the peripeduncular nucleus in the neuroendocrine control of sexual behaviour and milk ejection in the rat, Neuroendocrinology, 39:563.

Harris, G.W., 1947, The innervation and actions of the neurohypophysis: an investigation using the method of remote-control stimulation, Phil. Trans. R. Soc., B., 232:385.

Harris, G.W., Manabe, Y. and Ruf, K.B., 1969, A study of the parameters of electrical stimulation of unmyelinated fibres in the pituitary stalk, J. Physiol., Lond., 203:67.

Harris, M.C., Dreifuss, J.J. and Legros, J.J., 1975, Excitation of phasically firing supraoptic neurones during vasopressin release, Nature, Lond., 258:82.

Hartman, R.D., Rosella-Dampman, L. and Summy-Long, J.Y., 1987, Endogenous opioid peptides inhibit oxytocin release in the lactating rat after dehydration and urethane, Endocrinology, 121:536.

Hatton, G.I., 1986, Placticity in the hypothalamic magnocellular neurosecretory system, Fed. Proc., 45:2328

Hatton, G.I., Ho, Y.W. and Mason, W.T., 1983, Synaptic activation of phasic bursting in rat supraoptic nucleus neurones recorded in hypothalamic slices, J. Physiol., Lond., 345:297.

Herkenham, M., Rice, K.C., Jacobson, A.E. and Rothman, R.B., 1986, Opiate receptors in the rat pituitary are confined to the neural lobe and are exclusively kappa, Brain Res., 382:365.

Higuchi, T., Tadokoro, Y., Honda, K. and Negoro, H., 1986, Detailed analysis of blood oxytocin levels during suckling and parturition in the rat. J. Endocr., 110:251.

Juss, T.S. and Wakerley, J.B., 1981, Mesencephalic areas controlling pulsatile oxytocin release in the suckled rat, J. Endocr., 91:233.

Lebrum, C.J., Poulain, D.A. and Theodosis, D.T., 1983, The role of the septum in the control of the milk ejection reflex in the rat: effects of lesions and electrical stimulation, J. Physiol., Lond., 339:17.

Leng, G., 1980, Rat supraoptic neurones: the effects of locally applied hypertonic saline, J. Physiol., Lond., 304:405.

Leng, G., 1981, The effects of neural stalk stimulation upon firing patterns in rat supraoptic neurones, Exp. Brain Res., 41:135.

Leng, G. and Bicknell, R.J., 1987, Hormone release from the neural lobe, in: "The Circumventricular Organs,", P. Gross, ed., CRC, Boca Raton.

Leng, G., Dyball, R.E.J. and Russell, J.A., 1988, Neurophysiological aspects of body fluid homeostasis, Comp. Biochem. Physiol., in press.

Leng, G., Mansfield, S., Bicknell, R.J., Brown, D., Chapman, C., Hollingsworth, S., Ingram, C.D., Marsh, M.I.C., Yates, J.O. and Dyer, R.G., 1987, Stress-induced disruption of parturition in the rat may be mediated by endogenous opioids, J. Endocr., 114:247.

Leng, G., Mason, W.T. and Dyer, R.G., 1982, The supraoptic nucleus as an osmoreceptor, Neuroendocrinology, 34:75.

Leng, G. and Russell, J.A., 1986, Role of the region asnterior and ventral to the third ventricle (AV3V) in the control of oxytocin secretion, Proc. Int. Union Physiol. Sci. XVI, P518.04.

Leng, G. and Shibuki, K., 1987, Extracellular potassium changes in the rat neurohypophysis during activation of the magnocellular neurosecretory system, J. Physiol., Lond., 392:97.

Lincoln, D.W., 1971, Labour in the rabbit: effect of electrical stimulation applied to the infundibulum and median eminence, J. Endocr., 50:607.

Lincoln, D.W., 1974a, Suckling: a time constant in the nursing behaviour of the rabbit, Physiol. Behav., 13:711.

Lincoln, D.W., 1974b, Does a mechanism of negative feedback determine the intermittent release of oxytocin during suckling?, J. Endocr., 60:193.

Lincoln, D.W., 1974c, Dynamics of oxytocin secretion, in: "Neurosecretion: the Final Neuroendocrine Pathway," F. Knowles & L. Vollrath, eds., Springer-Verlag, Heidelberg.

Lincoln, D.W., 1984, The posterior pituitary, in: "Hormonal control of reproduction," C. R. Austin and R. V. Short, eds., Cambridge University Press, Cambridge.

Lincoln, D.W. and Cross, B.A., 1967, Effect of oestrogen on the responsiveness of units in the hypothalamus, septum and preoptic area of rats with light-induced persistent oestrus, J. Endocr., 37:191.

Lincoln, D.W. and Russell, J.A., 1985, The electrophysiology of magnocellular oxytocin neurons, in: "Oxytocin: clinical and laboratory studies," J. A. Amico and A. G. Robinson, eds., Elsevier, New York.

Lincoln, D.W. and Russell, J.A., 1986, Oxytocin and vasopressin: new perspectives, in: "Neuroendocrine Molecular Biology," G. Fink, A. J. Harmar and K. W. McKerns, eds., Plenum Press, London.

Lincoln, D.W. and Wakerley, J.B., 1971, Accelerated discharge of paraventricular neurosecretory cells correlated with reflex release of oxytocin during suckling, J. Physiol., Lond., 222:23P.

Lincoln, D.W. and Wakerley, J.B., 1974, Electrophysiological evidence for the activation of supraoptic neurones during the release of oxytocin, J. Physiol., Lond., 242:553.

Lincoln, D.W. and Wakerley, J.B., 1975, Factors governing the periodic activation of supraoptic and paraventricular neurosecretory cells during suckling in the rat, J. Physiol., Lond., 250:443.

Lincoln, D.W., Hentzen, K., Hin, T., van der Schoot, P., Clarke, G., and Summerlee, A.J.S., 1980, Sleep: a prerequisite for reflex milk ejection in the rat, Exp. Brain Res., 38:151.

Lincoln, D.W., Hill, A. and Wakerley, J.B., 1973, The milk ejection reflex of the rat: an intermittent function not abolished by surgical levels of anaesthesia, J. Endocr., 57:459.

Lind, R.W., Swanson, L.W. and Ganten, D., 1984, Angiotensin II immunoreactivity in the neural afferents and efferents of the subfornical organ of the rat, Brain Res., 321:209.

Mason, W.T., 1980, Supraoptic neurones of rat hypothalamus are osmosensitive, Nature, Lond., 287:154.

Mason, W.T., Hatton, G.I., Ho, Y.W., Chapman, C. and Robinson, I.C.A.F., 1986, Central release of oxytocin, vasopressin and neurophysin by magnocellular neurone depolarization: evidence in slices of guinea pig and rat hypothalamus, Neuroendocrinology, 42:311.

Mason, W.T. and Leng. G., 1984, Complex action potential waveform recorded from supraoptic and paraventricular neurones in the rat: evidence for sodium and calcium spike components at different membrane sites, Exp. Brain Res., 56:135.

Meyer, C., Freund-Mercier, M.J., Guerne, Y. and Richard, Ph., 1987, Relationship between oxytocin release and amplitude of oxytocin cell neurosecretory bursts during suckling in the rat, J. Endocr., 114: 263.

Millan, M.J. and Herz, A., 1985, The endocrinology of the opioids, Int. Rev. Neurobiol., 26:1.

Moos, F., Freund-Mercier, M.J., Guerne, Y., Guerne, J.M., Stoeckel, M.E., and Richard, Ph., 1984, Release of oxytocin and vasopressin by magnocellular nuclei in vitro: specific facilitatory effect of oxytocin on its own release, J. Endocr., 102:63.

Okuya, S., Inenaga, K., Kaneko, T., and Yamashita, H., 1987, Angiotensin II sensitive neurons in the supraoptic nucleus, subfornical organ and anteroventral third ventricle of rats in vitro, Brain Res., 402:58.

Poulain, D.A. and Wakerley, J.B., 1982, Electrophysiology of hypothalamic magnocellular neurones secreting oxytocin and vasopressin. Neuroscience, 7:773.

Poulain, D.A. and Wakerley, J.B., 1986, Sensory projections from the mammary glands to the spinal cord in the lactating rat. II. Electrophysiological responses of dorsal horn neurones during stimulation of the nipples, including suckling, Neuroscience, 19:511.

Poulain, D.A., Wakerley, J.B. and Dyball, R.E.J., 1977, Electrophysiological differentiation of oxytocin and vasopressin secreting neurones, Proc. R. Soc., Lond., B., 196:367.

Pumford, K., Hunter, M.J. and Russell, J.A., 1987, Paradoxical effects of naloxone and morphine on the electrical activity of magnocellular neurones in the supraoptic nucleus (SON) in rat hypothalamic slices, Neurosc. Lett., Suppl., 29:544.

Rayner, V.C., Robinson, I.C.A.F. and Russell, J.A., 1988, Chronic intracerebroventricular morphine and lactation in rats: dependence and tolerance in relation to oxytocin neurones, J. Physiol., Lond., in press.

Russell, J.A., Blackburn, R., Mansfield, S. and Leng., 1987, Lesions of the region antero-ventral to the third ventricle (AV3V region) in the rat do not prolong parturition, J. Endocr., Suppl., 112:183.

Sayer, R.J., Hubbard, J.I. and Sirett, N.E., 1984, Rat organum vasculosum terminalis in vitro: responses to transmitters, Amer. J. Physiol., 247:R374.

Sibbald, J.R., Sirett, N.E. and Hubbard, J.I., 1984, Osmosensitive neurons in the rat subfornical organ, Proc. Univ. Otago Med. Sch., 62:93.

Silverman, A.J., Hoffman, D.L. and Zimmerman, E.A., 1981, The descending afferent connections of the paraventricular nucleus of the hypothalamus, Brain Res. Bull., 6:47.

Summerlee, A.J.S. and Lincoln, D.W., 1981, Electrophysiological recordings from oxytocinergic neurones during suckling in the unanaesthetized lactating rat, J. Endocr., 90:255.

Theodosis, D.T., 1985, Oxytocin-immunoreactive terminals synapse on oxytocin neurones in the supraoptic nucleus, Nature, Lond., 313:682.

Theodosis, D.T. and Poulain, D.A., 1984, Evidence that oxytocin-secreting neurones are involved in the ultrastructural reorganisation of the rat supraoptic nucleus apparent at lactation, Cell Tiss. Res., 235:217.

Theodosis, D.T. and Poulain, D.A., 1987, Oxytocin-secreting neurones: a physiological model for structural plasticity in the adult mammalian brain, Trends Neurosc., 10:426.

Wakerley, J.B. and Lincoln, D.W., 1971a, Intermittent release of oxytocin during suckling in the rat, Nature New Biol., 233:180.

Wakerley, J.B. and Lincoln, D.W., 1971b, Phasic discharge of antidromically-identified units in the paraventricular nucleus of the hypothalamus, Brain Res., 25:192.

Wakerley, J.B. and Lincoln, D.W., 1973, The milk-ejection reflex of the rat: a 20- to 40-fold acceleration in the firing of paraventricular neurones during oxytocin release, J. Endocr., 57:477.

Wakerley, J.B. and Lincoln, D.W., 1974, Electrophysiological evidence for the activation of supraoptic neurones during the release of oxytocin, J. Physiol., Lond, 242:533.

Wakerley, J.B., Noble, R. and Clarke, G., 1983, Effects of morphine and d-Ala, d-Leu enkephalin on the electrical activity of supraoptic neurosecretory cells in vitro., Neuroscience, 10:73.

Wakerley, J.B., Poulain, D.A. and Brown, D., 1978, Comparison of firing patterns in oxytocin- and vasopressin-releasing neurones during progressive dehydration, Brain Res., 148:425.

Wakerley, J.B., Poulain, D.A., Dyball, R.E.J. and Cross, B.A., 1975, Activity of phasic neurosecretory cells during haemorrhage, Nature, Lond., 258:82.

Watson, S.J., Akil, H., Fischli, W., Goldstein, A., Zimmermann, E.A., Nilaver, G. and van Wimersma Greidanus, Tj. B., 1982, Dynorphin and vasopressin: common localization in magnocellular neurons, Science, N.Y., 216:85.

Weber, E. and Barchas, J.D., 1983, Immunohistochemical distribution of dynorphin-B in rat brain: relation to dynorphin-A and alpha-neoendorphin systems, Proc. Natl Acad. Sci., USA, 80:1125.

Wright, D.M. and Clarke, G., 1983, Opiates, neurohypophysial hormones and water balance, Alcohol alcohol., 18:337.

Yamashita. H., Okuya, H., Inenaga, K., Kasai, M., Uesugi, S., Kannan, H. and Kaneuko, T., 1987, Oxytocin predominantly excites putative oxytocin neurones in the rat supraoptic nucleus in vitro, Brain Res., 416:364.

ELECTRICAL ACTIVITY OF PEPTIDERGIC NEURONES AND ITS RELATION TO HORMONE RELEASE

D.A. Poulain and D.T. Theodosis

INSERM U.176, Université de Bordeaux II
1 rue Camille Saint-Saëns
33077 Bordeaux Cédex (France)

As all other neurones, the magnocellular neurones that secrete vasopressin and oxytocin generate action potentials and an obvious question is whether the electrical activity displayed by their cell bodies is necessary to bring about hormone release from their terminals. Another question quickly arises when one considers the temporal organization of these action potentials. As we will see, vasopressinergic and oxytocinergic neurones display several distinct patterns of electrical activity and one of the outcomes of recent research has been to show that such patterns of firing are particularly adapted to generate temporal patterns of hormone release necessary to meet the requirements of the target organ.

I. ACTION POTENTIALS IN MAGNOCELLULAR NEUROSECRETORY CELLS

Oxytocinergic and vasopressinergic neurones represent the final common pathway of numerous neuroendocrine reflexes. Their cell bodies in the hypothalamus thus constitute the primary site of integration of an important afferent input, reflected electrophysiologically in post- synaptic excitatory and inhibitory potentials, to which the cells react by generating action potentials (Dudek et al., 1980; Mason, 1983), which are then conducted to axonal terminals in the neurohypophysis. This has been demonstrated by antidromic stimulation : action potentials can be triggered in the hypothalamic cell bodies when the neurohypophysis is stimulated electrically (Yagi et al., 1966). Alternatively, electrical stimulation of the pituitary tract evokes orthodromically a compound action potential in the neurohypophysis, both _in vitro_ (Ishida, 1970; Dreifuss et al., 1971) and _in vivo_ (Dyball and Leng, 1987). Action potentials along the axons appear to be of the classical, sodium-dependent type, and can be abolished _in vitro_ by reducing external sodium or by adding tetrodotoxin (TTX) to the medium (Dreifuss et al., 1971). The conduction velocity of neurosecretory axons, estimated from the latency of the antidromic spike, is about 0.5m/s. The refractory period of the antidromic potential being 3-5ms, the axons can thus conduct at rates as fast as 300 Hz. However, during long trains of high frequency stimulation, above 50 Hz, transmission slows down or even fails (Dreifuss et al., 1971; Pittman, 1983 ; Nordmann & Stuenkel ,1986), which may be due to the small caliber of the unmyelinated neurosecretory axons (0.2-0.5μm). We will see later that this phenomenon may constitute a rate-limiting factor in hormone release.

As neurosecretory terminals in the neurohypophysis have never been recorded intracellularly, their electrophysiological properties remain obscure. Nevertheless, biochemical experiments _in vitro_ on isolated neural lobes have revealed that an essential step for stimulus-secretion coupling is the entry of calcium into the terminals (for review, see Douglas, 1974; Nordmann, 1983). The calcium current is voltage dependent, as shown by depolarizing the glands, either by using high concentrations of K$^+$ in the bathing medium, or by stimulating electrically the neurohypophysis en masse, while blocking Na$^+$ channels in the terminals with tetrodotoxin (Douglas & Poisner, 1964; Dreifuss et al., 1971). However, a sodium current in the terminals may contribute to hormone release since electrical stimulation releases more hormone in normal than in TTX-containing media (for review and further discussion, see Gainer, 1980).

II. PATTERNS OF ELECTRICAL ACTIVITY

Action potentials in the magnocellular cell bodies are generated in succession, according to various temporal patterns. Such patterns are either specific to the type of cell and/or dependent on the physiological or experimental conditions. We will here discuss only those features of their organization which appear to be of relevance to hormone release (see Lincoln and Russel, this volume; see also Poulain & Wakerley, 1982; for a more detailed analysis, see Poulain et al., 1988).

Under basal conditions, vasopressinergic neurones exhibit a slow irregular pattern of action potential discharge, characterized by a low mean overall firing rate (<3 spikes/s). Upon steady stimulation (osmotic, cardiovascular), the neurones evolve a bursting or 'phasic' pattern of electrical activity. This consists of bursts of action potentials (range 5-100s, mode 20-25s) separated by periods with no action potentials (interburst intervals or 'silences' : range 5-100s, mode 20-25s), occuring more or less regularly. The mean intraburst firing rate varies from 5 to 15 spikes/s, and the overall mean firing rate is above 2.5 spikes/s. During a period of increasing stimulation, the neurones may evolve a fast continuous pattern of activity, but only temporarily. In this case, the mean overall firing rate is also above 3 spikes/s (Fig.1).

Under basal conditions, oxytocinergic neurones also display a slow irregular pattern

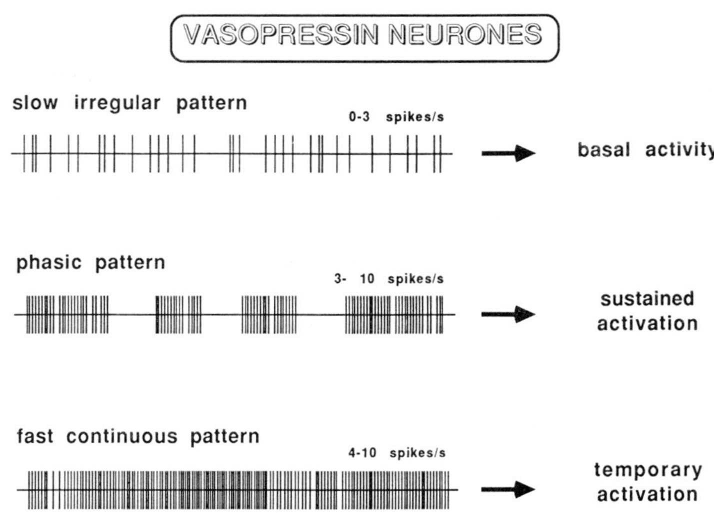

Fig. 1.   Patterns of electrical activity of vasopressin-secreting neurones. In these diagrams, each pattern is represented for a period about 5 min; each vertical line corresponds to an action potential. Numbers give the range of mean firing rates observed within the patterns. Note that the same neurone can exhibit different patterns according to its state of activation by various stimuli.

of discharge, indistinguishable from that of vasopressinergic neurones. However, when the cells are stimulated, their patterns of electrical activity depend on the type of stimulation. For example, during sustained activation (osmotic, cardiovascular), the neurones evolve a pattern of fast continuous activity but during suckling in lactating animals, they exhibit a highly specific activity, superimposed on the slow irregular or fast continuous background activity. This consists of high frequency discharges of action potentials, occuring periodically every 3-5 min, and lasting 3-4s; they are characterized by an initial peak at about 30-100 spikes/s which then decreases quite rapidly.

Several features of interest appear when we consider not only the sequence of firing rates, but also the interspike intervals in these patterns. Magnocellular cell bodies rarely exhibit very short interspike intervals (below 20ms), that is 'instantaneous' firing rates above 50 Hz, rates at which, as said above, axons often fail to conduct action potentials. Nevertheless, during a slow irregular pattern of spike discharge, the distribution histograms of interspike intervals always exhibit a certain proportion of short interspike intervals

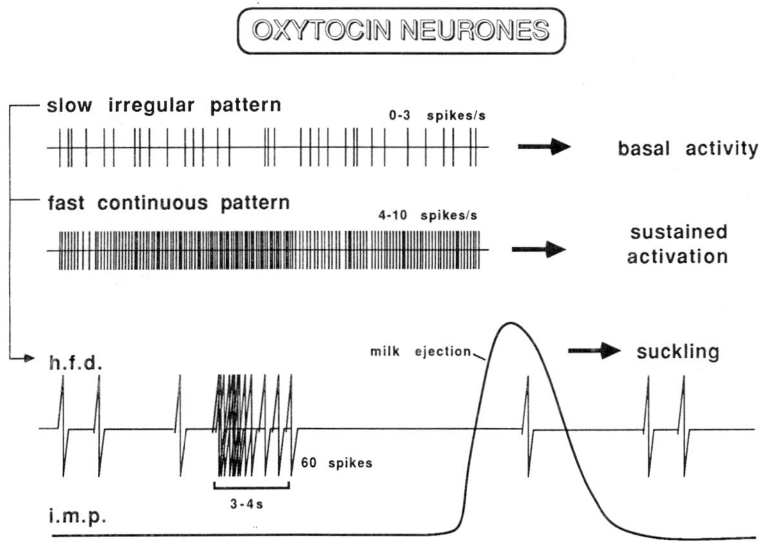

Fig. 2.   Patterns of electrical activity of oxytocin-secreting neurones. Same conventions as in Figure 1. Like vasopressin cells, the same oxytocin neurone can display slow irregular or fast continuous activities depending on the state of activation. In addition, it specifically reacts to suckling by intermittent high frequency discharges (h.f.d), synchronous in the whole population of oxytocin cells, superimposed on a background pattern of slow irregular or fast continuous activity. This leads to the release of a large pulse of oxytocin which causes a milk ejection, shown here by a transient increase in the intramammary pressure (i.m.p.). The high frequency discharge diagram is represented on an expanded time scale in comparison to the two background patterns.

(< 50ms), a feature which, as we will see, may be of consequence for basal hormone release. Interspike interval histograms for fast continuous activity or phasic activity look very similar, reflecting the increase in the proportion of short interspike intervals. However, detailed analyses of successive interspike intervals indicate that action potentials in magnocellular neurones tend to be more clustered than would be expected by a simple increase in firing rates given a random assortment of intervals. This phenomenon, particularly conspicuous during phasic activity, is also apparent during slow irregular and continuous activities (see Poulain et al., 1988).

In addition to the patterns observed at the individual neuronal level, it is also important to consider the activity of each neuronal population as a whole. In relation to hormone release, at least two general processes must be taken into account : first, that of synchronization of neuronal firing and second, that of recruitement of neuronal activity.

Simultaneous recordings of adjacent vasopressinergic neurones show that they have different levels of firing and often widely different patterns. In other words, under sustained stimulation, there is no synchronization throughout the vasopressinergic population, although this does not preclude a possible correlation in the electrical activity between adjacent neurones (Wakerley et al., 1978). As the intensity of stimulation increases progressively, more and more cells evolve a phasic pattern of electrical activity, that is vasopressin neurones are progressively recruited into phasic activity. Temporary synchronization may be brought about in vasopressinergic neurones, but only for brief periods of time, by carotid occlusion, a powerful vasopressin releasing stimulus (Dreifuss et al., 1976a), or by vaginal distension, a stimulus for the release of both vasopressin and oxytocin during labour (Dreifuss et al., 1976b).

Under normal conditions, oxytocinergic neurones also show no synchronization. When stimulation increases progressively, as during dehydration, there is a progressive recruitment of an increasing number of cells into fast continuous activity, but still without synchronization (Wakerley et al., 1978). On the other hand, during the milk ejection reflex induced by suckling, the whole population of oxytocinergic cells throughout the hypothalamus start to fire their high frequency discharges synchronously : there is still no spike-to-spike synchronization between cells, but the recruitment of the oxytocinergic population is immediate and complete within about 400ms (Belin et al., 1984).

Magnocellular neurones thus offer different models of population activity, either asynchronous with progressive recruitment, or synchronous with total recruitment. Are such complex organizations of electrical activity limited to their cell bodies or are they propagated down to their terminals? Although the ability of neurosecretory axons to conduct action potentials during prolonged stimulation, even at slow rates, has been questioned (Nordmann & Stuenkel, 1986), recent _in vivo_ studies showed that orthodromic action potentials in the neurohypophysis can follow high rates of stimulation, and that action potentials in the neurohypophysis can exhibit a phasic pattern indistinguishable from that seen at the level of the cell bodies (Dyball and Leng, 1987). In any case, it should be noted that firing rates in the cell bodies are rarely very high under physiological conditions, and when they are, it is only for brief periods of a few seconds, so that such a limitation would not really matter for hormone release.

III. FACILITATION, FATIGUE, AND HORMONES RELEASE

When action potentials reach the neurohypophysial terminals, they set up a complex series of electrochemical and morphological events that eventually result in the exocytotic release of hormone. This is not to say that each action potential is equipotent, releasing a fixed amount of hormone. On the contrary, current evidence clearly demonstrates that the effect of action potentials on hormone release depends on several parameters of the organization of spike discharge, namely its frequency, duration and periodicity.

Frequency facilitation of hormone release has been recognized _in vivo_ using electrical stimulation of the neurohypophysis : for example, in the rabbit, Harris et al. (1969) showed that milk ejections recorded from a cannulated mammary gland were best at stimulation frequencies of about 50 Hz. A better appraisal of the correlation between stimulus frequency and hormone release can be obtained _in vitro_, where the amount of hormone released can be measured directly in the medium bathing the neurohypophysis, rather than from the response of a target organ (see Douglas, 1974). Thus, when the neurohypophysis is stimulated with a fixed number of electrical stimuli, there is more hormone release at frequencies around 30-50 Hz; that is to say, the amount of hormone released per pulse increases with the frequency of stimulation (Dreifuss et al., 1971).

Together with the process of facilitation, however, there appears to be a process of fatigue, whereby the amount of hormone released per pulse declines with the duration of the train of pulses. When neurohypophyses are submitted to prolonged periods of stimulation, either by increasing external K^+ concentrations in the medium, or by applying prolonged trains of electrical pulses, there is a progressive decline in hormone release (see Ingram et al., 1982). This does not result from a depletion of hormone in the neurosecretory terminals but, rather, from a progressive failure of the processes related to stimulus-secretion coupling. The stores of hormone within the terminals are hardly depleted by more than 10-20% after strong stimulation (Thorn, 1966; Sachs et al, 1967), and the release of hormone can be re-activated by veratridine, a substance which opens calcium channels (Nordmann and Dyball, 1978).

The electrical, biochemical and morphological bases of facilitation and fatigue at the level of neurosecretory terminals have been investigated by many workers (see Douglas, 1974; Nordmann, 1983). Central to the notion of stimulus-secretion coupling as developed by Douglas is that action potentials arriving in the terminals depolarize the terminals and permit the entry of calcium, which will lead eventually to the fusion of secretory granules to the plasma membrane and exocytosis. It is clear then that facilitation results from a greater influx of free calcium at fast frequency stimulation. On the other hand, the bases for the process of fatigue are not fully understood. They involve various mechanisms leading to a decrease in free cytosolic calcium, but may also involve other mechanisms, such as the mobility of secretory granules within the terminals.

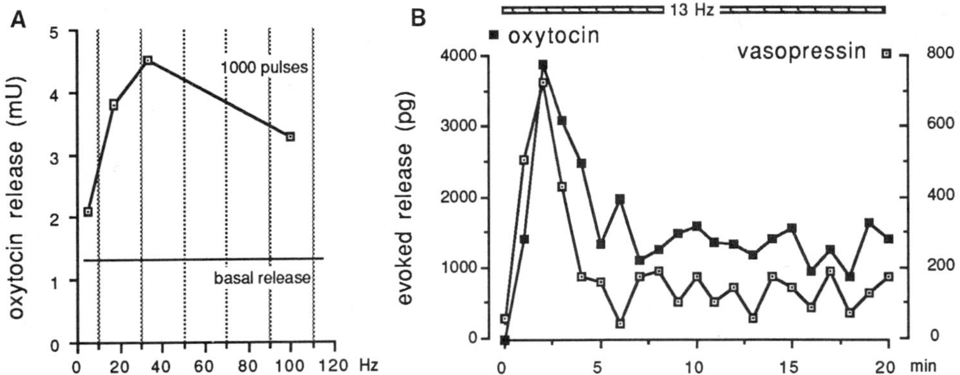

Fig. 3.   Facilitation and fatigue of hormone release. A. Facilitation : the same number (1000) of electrical pulses was applied to neurohypophyses _in vitro_ in a normal medium, at different frequencies. At 5 Hz (during 200s), there was hardly any release evoked by stimulation; at 35 Hz (during 29s), release was maximal (adapted from Dreifuss et al., 1971). B. Fatigue : failure of the glands to maintain a high level of hormone output during prolonged stimulation. The maximal rate of release was observed within 2 min of the onset of a 20 min period of stimulation at 13 Hz (from Bicknell et al., 1984)

IV. NATURAL PATTERNS OF ELECTRICAL ACTIVITY AND HORMONE RELEASE

It has become increasingly evident that the particular temporal patterns of electrical activity displayed by neurosecretory cells are of great significance to hormone release since they permit the process facilitation and reduce that of fatigue.

The high frequency discharge of action potentials in oxytocin cells at milk ejection constitutes the most conspicuous example of a pattern of discharge organized to facilitate hormone release. As pointed out by Lincoln (1974), about 9000 oxytocin cells produce a discharge of 60-100 spikes per cell within 3-4s, resulting in the release of 0.5-1mU oxytocin; the same neurones under normal conditions have a mean firing rate of 1.0 to 2.5 spikes/s, achieving a daily turn-over of oxytocin of 25-30mU (Jones & Pickering, 1972). During a high frequency discharge therefore, there is about 50-100 times more hormone released per action potential than under basal conditions.

The phasic pattern of electrical activity of vasopressin cells exemplifies not only the process of facilitation, but also how this pattern overcomes the process of fatigue. In _in vitro_ experiments, Dutton and Dyball (1978) used tape recordings of phasic patterns from various vasopressinergic cells to drive a stimulator to stimulate electrically the neurohypophysis; they then compared the effects of these natural patterns with the effects obtained from regular trains of stimuli generated by the stimulator, taking care that the natural and regular trains produced the same number of stimuli. Such experiments clearly demonstrated that the natural trains of stimuli were considerably more efficient for hormone

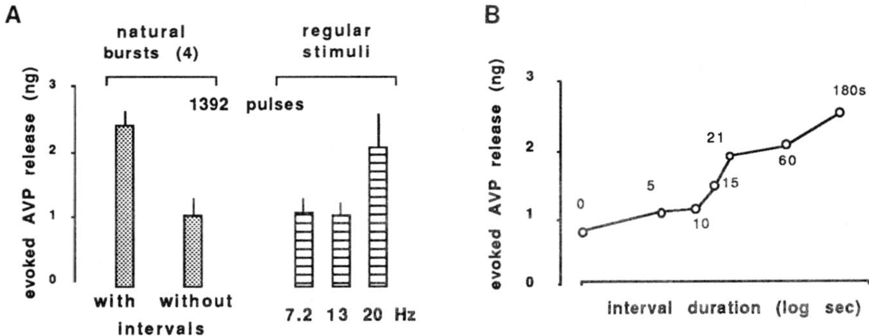

Fig. 4.   Efficiency of the phasic pattern for hormone release.
A burst of action potentials (previously recorded on magnetic tape from a rat vasopressin cell with phasic discharge) was used as a trigger for stimulating electrically neurohypophyses _in vitro_ (burst duration : 27s; number of spikes within the burst : 348, corresponding to a mean intraburst firing rate of 13 Hz). The mean interburst interval was here 21s, so that the mean overall firing rate was estimated to be 348/(27+21) = 7.2 Hz.
A. Neurohypophyses stimulated with four bursts separated by a normal 21s interval released twice as much hormone as those stimulated with 4 consecutive bursts with no resting interval. Stimulating with a regular sequence of pulses generated by the stimulator itself, either at 7.2 Hz (corresponding to the mean overall firing rate for the neurone), or 13 Hz (corresponding to the intraburst firing rate) was also less efficient than the natural pattern. To obtain a comparable level of hormone output as that obtained by a natural pattern, the regular stimulation had to be 20 Hz.
B. Effect of the interburst intervals on recovery from fatigue. Neurohypophyses _in vitro_ were stimulated with 4 bursts separated by intervals of 0, 5, 10, 15, 21, 60 or 180s. Note that hormone output is much higher when intervals exceed 10s (adapted from Cazalis et al., 1985).

release, particularly at low mean frequencies of action potential discharge.

The most obvious consequence of the patterning of electrical activity in bursts is that, for a given overall mean firing rate over a long period of time (that is the total number of spikes during the period of analysis divided by the duration of that period), intraburst firing rates are much higher than the mean firing rate, which could suffice to explain the better effect on hormone release. However, in the phasic pattern, the intraburst firing rate is not the sole parameter to consider. Burst duration, interburst intervals and the pattern of discharge within the bursts all contribute to maximalise hormone output. From _in vitro_ experiments using bursts and/or interburst intervals of various duration, it appears that hormone release is particularly enhanced for burst durations between 10-30s and interval durations of about 15-25s, a range of values which closely parallels those seen in normally phasically firing neurones (Shaw et al., 1984; Cazalis et al., 1985).

High frequency discharges in oxytocin cells and phasic patterns in vasopressin cells occur when the neurones are strongly activated and release a large amount of hormone. It must be kept in mind, however, that under basal conditions, most oxytocin and vasopressin cells display a slow irregular activity at a time when the levels of hormones are low (Wakerley et al., 1978). That such low levels of electrical activity contribute to basal hormone release has been difficult to demonstrate, mainly for technical reasons, such as the sensitivity of hormone assays or the difficulties of sampling methods. Nevertheless, transections of the pituitary stalk induce diabetus insipidus or stop reflex release of oxytocin. Moreover, recent _in vivo_ and _in vitro_ experiments have shown that low frequency stimulation can indeed elicit hormone release (Boer et al., 1980; Poulain & Tasker, 1985), and even more important, that natural slow irregular patterns of stimulation _in vitro_ are more efficient for hormone release than regular patterns of the same mean frequencies (Fig. 5).

This latter observation emphasizes the importance of interspike intervals in the processes of hormone release. When considering the natural patterns of electrical activity, it is probable that the the variability between successive interspike intervals is such as to, not only facilitate hormone release (during the clusters of action potentials), but also allow for recovery from the fatigue processes, during the long interspike intervals. A crucial issue, therefore, concerns the kinetics of the electrophysiological and biochemical events set off by the arrival of impulses at the terminals. A major drawback of studies on stimulus-secretion coupling in the neurohypophysis, is that, until now, the system has not been amenable to the

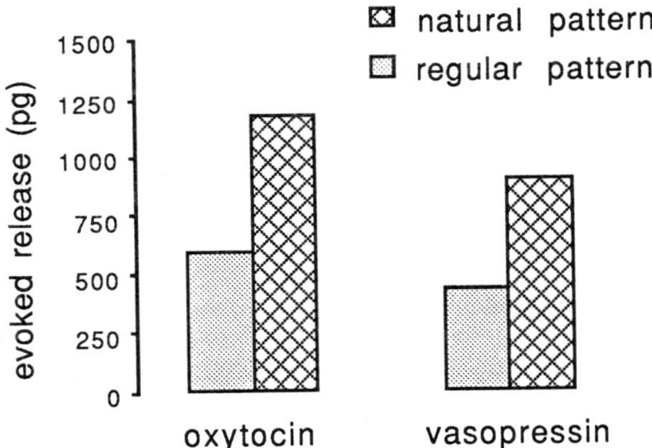

Fig. 5.   Hormone release and slow irregular patterns of firing. Neurohypophyses _in vitro_ were stimulated during 20 min with the same number of pulses, either with a 'natural' slow irregular pattern of action potential discharge previously recorded from a supraoptic neurones in the rat, or with a regular pattern generated by the stimulator (from Dayanithi and Poulain, unpublished observations).

kind of refined analysis performed for example on the squid stellate ganglion synapse or on the crab sinus gland, where direct evidence of the ionic currents involved were obtained by intracellular recordings. In studies on the neurohypophysis, it must be kept in mind that what is observed is a global effect of electrical activity on hormone output from thousands of terminals. Moreover, experimental stimulation, either by K+ induced depolarization or by synchronous electrical volleys applied to the pituitary stalk, bears little resemblance to the natural electrical activity which in this sytem is never synchronous from one neurone to another. Also, _in vitro_, evaluation of hormone dynamics is biased by a high basal output (which, in the absence of stimulation, is about 20 times higher than the normal turn-over _in vivo_), and measurements have a time resolution of at best 1min, whereas currents in terminals probably last a few hundreds milliseconds at most.

REFERENCES

Belin V., Moos F. and Richard Ph., 1984, Synchronization of oxytocin cellshypothalamic paraventricular and supraoptic nuclei in suckled rats: direct proof with paired extracellular recordings, Exp. Brain Res., 57:201.

Bicknell R.J., Brown D., Chapman C., Hancock P.D. and Leng G., 1984, Reversible fatigue of stimulus-secretion coupling in the rat neurohypophysis, J. Physiol. Lond., 348:601.

Boer K., Cransberg K. and Dogterom J., 1980, Effect of low-frequency stimulation of the pituitary stalk on neurohypophysial hormone release in vivo, Neuroendocr., 30:313.

Cazalis M., Dayanithi G. and Nordmann J.J., 1985, The role of patterned burst and interburst interval on the excitation-coupling mechanism in the isolated rat neural lobe, J. Physiol. Lond., 369:45.

Douglas W.W., 1974, Mechanism of release of neurohypophysial hormones: stimulus-secretion coupling, in: Handbook of Physiology,Endocrinology IV, part 1 (eds. Knobil E. & Sawyer W.H.) pp 191-224, Am.Physiol.Soc., Washington D.C.

Douglas W.W. and Poisner A.M., 1964, Stimulus-secretion coupling in a neurosecretory organ : the role of calcium in the release of vasopressin from the neurohypophysis, J. Physiol. Lond., 172:1.

Dreifuss J.J., Harris M.C. and Tribollet E., 1976a, Excitation of phasically firing hypothalamic supraoptic neurones by carotid occlusion in rats, J. Physiol. Lond., 257:337.

Dreifuss J.J., Kalnins I., Kelly J.S. and Ruf K.B., 1971, Action potentials and release of neurohypophysial hormones in vitro, J. Physiol. Lond., 215:805.

Dreifuss J.J., Tribollet E. and Baertschi A.J., 1976b, Excitation of supraoptic neurones by vaginal distension in lactating rats; correlation with neurohypophysial hormone release, Brain Res., 113:600.

Dudek F.E., Hatton G.I and MacVicar B.A., 1980, Intracellular recordings from the paraventricular nucleus in slices of rat hypothalamus, J. Physiol. Lond., 301:101.

Dutton A. and Dyball R.E.J., 1979, Phasic firing enhances vasopressin release from the rat neurohypophysis, J. Physiol. Lond., 290:433.

Dyball R.E.J. and Leng G., 1987, Single unit recordings from the rat neurohypophysis in vivo, J. Physiol. Lond., (in press).

Gainer H., 1978, Input-output relations of neurosecretory cells, in: Comparative Endocrinology (eds. Gaillard P.J. & Boer H.H.) pp 293-304. Elsevier, Amsterdam.

Harris G.W., Manabe Y. and Ruf K.B., 1969, A study of the parameters of electrical stimulation of unmyelinated fibres in the pituitary stalk, J. Physiol. Lond., 203:67.

Ingram C.D., Bicknell R.J., Brown D. and Leng G., 1982, Rapid fatigue of neuropeptide secretion during continual electrical stimulation, Neuroendocrinology, 35:424.

Ishida A., 1970, The oxytocin release and the compound action potential evoked by electrical stimulation on the isolated neurohypophysis of the rat, Jap. J. Physiol., 20:84.

Jones C.W. and Pickering B.T., 1972, Intra-axonal transport and turnover of neurohypophysial hormones in the rat, J. Physiol. Lond., 227:553.

Lincoln D.W., 1974, Dynamics of oxytocin secretion, in: Neurosecretion, "The final neuroendocrine pathway" (eds. Knowles F.G.W. & Vollrath L.) pp 129-133, Springer-Verlag, Berlin.

Mason W.T., 1983, Electrical properties of neurons recorded from the rat supraoptic nucleus in vitro, Proc. R. Soc. Lond., B 217:141.

Nordmann J.J., 1983, Stimulus-secretion coupling, <u>Prog. Brain Res</u>., 60:281.

Nordmann J.J. and Dyball R.E.J., 1978, Effect of veratridine on Ca fluxes and the release of oxytocin and vasopressin from the isolated rat neurohypophysis, <u>J. Gen. Physiol</u>., 72:297.

Nordmann J.J. and Stuenkel E.L., 1986, Electrical properties of axons and neurohypophysial nerve terminals and their relationship to secretion in the rat, <u>J. Physiol. Lond</u>., 380:521.

Pittman Q., 1983, Increases in antidromic latency of neurohypophyseal neurons during sustained activation, <u>Neurosci. Lett</u>., 37:239.

Poulain D.A., Brown D. and Wakerley J.B., 1988, Statistical analysis of patterns of electrical activity in vasopressin- and oxytocin-secreting neurones, <u>in</u> : Pulsatility in neuroendocrine systems (ed. Leng G.), CRC Press, New York (in press).

Poulain D.A. and Tasker J.G., 1985, Recurrent mammary gland contractions induced by a low tonic release of oxytocin in rats, <u>J. Endocr</u>., 107:89.

Poulain D.A. and Wakerley J.B., 1982, Electrophysiology of hypothalamic magnocellular neurones secreting oxytocin and vasopressin, <u>Neuroscience</u>, 7:773.

Sachs H., Share L., Osinchak J. and Carpi A., 1967, Capacity of the neurohypophysis to release vasopressin, <u>Endocrinology</u>, 81:755.

Shaw F.D., Bicknell R.J. and Dyball R.E.J., 1984, Facilitation of vasopressin release from the neurohypophysis by application of electrical stimuli in bursts. Relevant stimulation parameters, <u>Neuroendocrinol</u>., 39:371.

Shaw F.D. and Dyball R.E.J., 1984, The relationship between calcium uptake and hormone release in the isolated neurohypophysis. A reassessment, <u>Neuroendocrinology</u>, 38:504.

Thorn N.A., 1966, <u>In vitro</u> studies of the release mechanism for vasopressin in rats, <u>Acta Endocrinol</u>., 53:644.

Wakerley J.B., Poulain D.A. and Brown D., 1978, Comparison of firing patterns in oxytocin- and vasopressin-releasing neurones during progressive dehydration, <u>Brain Res</u>. 148:425.

Yagi K., Azuma T. and Matsuda K., 1966, Neurosecretory cell : capable of conducting impulse in rats, <u>Science</u>, 154:778.

RHYTHMIC PATTERNS OF DISCHARGE FROM PREOPTIC NEURONES:

RELATIONSHIP TO LH RELEASE IN CONSCIOUS RABBITS

A.J.S. Summerlee

Department of Anatomy
School of Veterinary Science
Park Row
Bristol BS1 5LS UK

INTRODUCTION

Neurones that stain immunohistochemically for gonadotrophin releasing hormone (GnRH) are scattered throughout the hypothalamus. They are found more abundantly in the medial preoptic area of the rat and rabbit brain, with smaller aggregations in the accessory olfactory bulbs, the rostral hippocampus and the septum (Bennett-Clark and Joseph, 1982; King and Anthony, 1984; Weindl and Sofroniew, 1981). There are few GnRH-cell bodies in the mediobasal hypothalamus (Silverman et al., 1982) and no discrete aggregations into classic hypothalamic nuclei equivalent to the magnocellular nuclei. The axons from the GnRH-cell bodies project to many parts of the brain which implies that GnRH may be involved in mediating/controlling central functions other than the well established roles controlling the release of luteinizing hormone (LH) and follicle-stimulating hormone (FSH) from the adenohypophysis. The most obvious convergence of GnRH-containing axons is found in the median eminence and the organum-vasculosum of the lamina terminalis (OVLT). The fibres that converge on the median eminence originate from the medial preoptic area and to a lesser extent from the paraventricular region (King and Anthony, 1984; Silverman et al., 1982; Witkin et al., 1982). In primates there are also cell bodies in the arcuate region that project to the median eminence.

In most experiments, the release of GnRH has been deduced from the sequential measurement of LH levels in peripheral plasma (Crowley et al., 1985; Karsch et al., 1984; Lincoln and Short, 1980; Lincoln et al., 1985; Summerlee, 1986), or from changes in the secretion of LH following administration of GnRH analogues or the removal of endogenous GnRH by active and passive immunization (Clarke et al., 1978; Fraser and McNeilly, 1983; Lincoln and Fraser, 1979). This has been possible because LH is released from the gonadotrophs within seconds of GnRH reaching the pituitary, and the half-life of LH is substantially shorter than the LH pulse interval under most conditions. In contrast, the measurement of FSH release is of little value because the FSH plasma profile is slower, more sustained and the half-life of FSH is much greater than the half-life of LH.

Pulsatile release of LH has now been observed in a wide range of species, at different stages of the reproductive cycle and in both sexes (see review Lincoln et al., 1985) and there is now convincing evidence that each LH pulse is driven by a GnRH pulse in the portal circulation. In the male rat (Ellis et al., 1983) and the ram (Lincoln and Fraser, 1979) passive immunization with GnRH antibodies halts the episodic release of LH. Conversely, the sheep pituitary can be driven to release LH during the non-breeding season by pulsatile administration of GnRH (Lincoln, 1979). Clarke et al., (1984) showed that sheep also responded to exogenous GnRH after hypothalamic-pituitary disconnection. Pulses of GnRH have been detected in the hypophysial portal blood of female monkeys (Carmel et al., 1983) and irrefutable evidence of the dependency of LH release on GnRH secretion was provided by Clarke and Cummins (1982) and Levine et al., (1982). They showed that all the major secretory episodes of LH secretion are preceded by a transient increase in GnRH in the hypophysial portal blood.

ELECTROPHYSIOLOGY OF GnRH RELEASE

Selective identification of neurones is a prerequisite for the study of hypothalamic neurone activity; the neurones have to fulfil a number of electrophysiological criteria that permit identification during an experiment; methods have to be developed to permit physiological activation of these identified cells and lastly there has to be an on-line method of assessment of hormone release. Only in the context of the magnocellular system have these criteria been satisfactorily fulfilled (Poulain and Wakerley, 1982), so our understanding of the neural determinants of GnRH release is poor. The problems have been exacerbated further by the discovery that the release of GnRH and hence LH was disrupted or inhibited by anaesthesia (Everett and Sawyer, 1950; Blake and Sawyer, 1972; Lincoln and Kelly, 1972). Therefore all experiments on the neural determinants of GnRH secretion must either be carried out on unanaesthetized animals or neurologically isolated tissue.

There have been a number of attempts to monitor the multi-unit activity of neurones that project to the median eminence. The recordings represent the activity of a small population of cell bodies close to the tip of the recording electrode and there have been some surprising correlations observed between the multi-unit activity and the plasma LH profiles. Thiery and Pelletier (1981) reported that there was a statistical correlation between the multi-unit activity in the retrochiasmatic region and plasma LH in ovariectomized ewes. Kawakami et al., (1982) using barbiturate anaesthetized rats showed that there were striking increases in multi-unit activity from the mediobasal hypothalamus that preceded LH release. These latter findings are difficult to reconcile with the known localization of the GnRH-cell bodies in the rat and with the evidence that anaesthesia disrupts the release of GnRH and LH. More recently, Wilson et al., (1984) investigated the changes in multi-unit activity in the arcuate region of both anaesthetized and conscious monkeys and showed that there were marked periods of intense neural activity before LH pulses in the periphery. The magnitude of the responses reported in both the rat and the monkey indicates that the change in neural activity must dominate that particular region of the hypothalamus and this is surprising when one consider that the same areas are thought to control the release of a variety of other factors.

RHYTHMIC PATTERNS OF DISCHARGE FROM PREOPTIC NEURONES

We have recently developed a technique for taking long-term recordings from single cells in the hypothalamus of unanaesthetized, unrestrained rats and rabbits (Summerlee et al., 1979; Summerlee, 1981; Summerlee and Paisley, 1982) and have used this technique to investigate the activity of putative GnRH neurones in the conscious rabbit. We chose the rabbit because it is an induced ovulator so the timing of the LH surge can be controlled.

Approximately 20% of the neurones recorded from the preoptic area of the ovariectomized rabbit were antidromically-identified from a stimulating electrode chronically implanted into the region of the median eminence. These neurones displayed a rhythmic or oscillatory pattern of discharge that distinguished them from all other types of cell recorded in the hypothalamus (Paisley and Summerlee, 1985a). In particular, the discharge was different from so-called 'phasic' activity, typical of vasopressinergic neurones (Poulain and Wakerley, 1982). The neurones discharged in a continuous series of short (1-3s duration) bursts of high frequency activity. The interval between the bursts ranged from 10 - 20s but was remarkably constant for a given neurone providing the basis for the rhythmic or oscillatory pattern of activity. The rhythmic discharge was not affected by the animal's level of arousal, even over long periods of recording which distinguishes these neurones from other cells in the hypothalamus; most hypothalamic cells studied show minute changes in their firing pattern in response to level of arousal (Summerlee and Paisley, 1982). The firing pattern of the oscillatory neurones was affected, however, by treatment of the animal with exogenous oestrogen (Summerlee, 1985 & 1986). Oestradiol benzoate (1ug i.m.) caused a slight decrease in the periodicity of the bursts, but within each burst the frequency of firing increased, so the overall effect on firing rate of the neurone was negligible in most cells studied.

The activity of single oscillatory neurones was studied and compared with the multi-unit activity recorded from neighbouring cells on the same electrode. Where groups of neurones with the oscillatory pattern of discharge were recorded simultaneously, the clusters of cells showed short periods of synchronization (Paisley and Summerlee, 1985a & b; Summerlee, 1986). These periods of synchronization lasted for about 3min and always occurred immediately before a rise in peripheral LH. No evidence of synchronization was observed for any other neurones that were recorded in the preoptic area. Furthermore, where two single neurones with the rhythmic pattern of discharge were recorded simultaneously on the same electrode, the activity of the pairs of neurones synchronized immediately before an LH pulse. These data indicate that neurones approximately $100\mu m$ apart can show significant periods of synchronization before an LH pulse and it is suggested that these cells are the putative GnRH neurones (Paisley and Summerlee, 1985b; Summerlee, 1986). More recently we have made recordings from two oscillatory cells recorded on different electrodes in the same animal. These neurones also showed brief period of synchronous firing before an LH pulse (see Synchronization of firing) which suggests that many, if not all, the GnRH-secretory cells are synchronized throughout the hypothalamus to produce a pulse of hormone. Certainly no evidence of recruitment of oscillatory firing has been observed.

There are at least three different patterns of LH release into the plasma; a tonic low-level secretion, a regular pulsatile output and, in the female, an ovulatory surge of hormone and evidence indicates that differnt patterns of GnRH induces the different LH profiles. We studied further the activity of the oscillatory neurones to observe whether there were any changes in either the oscillatory activity or in the degree of synchronization between the oscillatory cells associated with an LH surge. The ovariectomized female rabbits were treated with oestrogen injections 72h apart which induced behavioural oestrus and the rabbits were mated. This caused a surge of LH in the plasma, which represented at least a ten-fold increase in LH levels compared with peak values during an LH pulse. Moreover, LH titres during the surge remained elevated in the peripheral circulation for at least 2h. The surges were always preceded by periods of synchronized activity from either the pairs of oscillatory neurones or from the single oscillatory neurones and multi-unit recordings (Paisley and Summerlee, 1985b; Summerlee, 1986). In all cases the periods of synchronous discharge lasted up to 90min, compared with an average of 3min associated with an LH pulse. There was also a significant relationship between the strength of synchronous firing and the size of the LH surge. Weakly synchronized activity was observed before an LH pulse but moderate to strong forms of synchrony were seen before and associated with an LH surge (Summerlee, 1986).

The data are consistent with the suggestion that the preoptic neurones with the oscillatory pattern of discharge are the putative GnRH cells. Each burst of discharge releases a quantum of hormone but for the majority of the time the activity of all the GnRH neurones is independent so the concentration of GnRH in the portal circulation is not sufficient to evoke the release of a pulse of LH from the gonadotroph, but it may be important in priming the pituitary to GnRH (Clayton, 1982). Periodically, the activity of the GnRH neurones is coupled together for a few minutes so that the bursts of activity occur synchronously: the output of GnRH into the portal circulation will appear as a small pulse, significantly elevating the concentration arriving at the gonadotroph which elicits a pulse of LH. When the activity of the GnRH neurones is synchronized for longer, for example following mating, then the output of GnRH would be significantly elevated over a longer period of time (perhaps up to 90 minutes) which would result in a massive surge of LH released into the plasma.

Barbiturate anaesthesia disrupts the release of GnRH and LH (Everett and Sawyer, 1950) so we investigated the effects of light pentobarbitone anaesthesia on the oscillatory pattern of discharge of the rabbit preoptic neurones. Barbiturate was injected into the ear vein of a conscious animal and there was an almost immediate disruption of the rhythmic pattern of activity (Summerlee, 1986). The bursts became much shorter, often with only 2 or 3 spikes per burst, and there was considerable variation in the interburst intervals. The ultra-short bursts of activity were similar to those described by Weick and Dyer (1982) who recorded the activity of single neurones in neurologically isolated areas of rat hypothalamus <u>in situ</u> which showed changes in firing pattern that could be correlated with changes in plasma LH.

Opioid peptides have been implicated in the control of the release of GnRH for some time, in particular, early work suggested that the opioids might be involved in inhibiting secretion under conditions of stress. Evidence now suggests that they play a more important and subtle role in the control of GnRH secretion and may mediate the negative feedback effects of gonadal steroids in LH secretion. Opiate agonists and β-endorphin inhibit the secretion of LH in acute studies (Sylvester et al., 1982), and the effect is eliminated by gonadectomy (Bhanot and Wilkinson, 1984). Furthermore, in man (Ropert et al., 1981) and in the rat (Bhanot and Wilkinson, 1983), the opiate antagonist naloxone elevates circulating levels of LH, implicating the endogenous opioids in a tonic inhibition of LH release. The effect of naloxone is only seen under conditions of negative feedback such as during the luteal phase of the menstrual cycle (Ropert et al., 1981) and during the breeding cycle of sheep (Lincoln et al., 1985). Naloxone is similarly effective at promoting an increase in pulsatile LH release in the ovariectomized rabbit after appropriate steroid priming (Summerlee, 1985). We investigated the effects of naloxone on the release of LH and also on the patterns of discharge from the oscillatory neurones (Paisley and Summerlee, 1985b) and found that naloxone induces pulses of LH in the plasma approximately every 20min. Each pulse of LH was preceded by a short (average 3min) period of synchronized activity among the pools of neurones with the rhythmic pattern of firing. The findings are consistent with the hypothesis that the oscillatory neurones are the GnRH-secretory cells of the hypothalamus.

SYNCHRONIZATION OF FIRING OF OSCILLATORY NEURONES

The mechanism underlying the synchronization of the putative GnRH neurones is not known. There is considerable evidence implying that GABA and dopaminergic systems

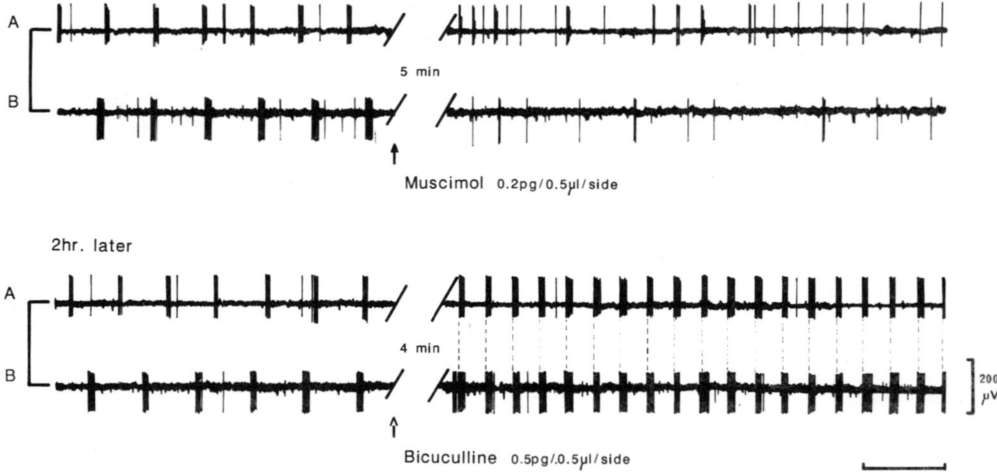

Fig. 1.   Effects of muscimol (GABA agonist) and bicuculline (antagonist) on the activity of rhythmically discharging preoptic neurones in a conscious rabbit. Two neurones (A & B) were recorded simultaneously on different electrodes in the same animal. Note muscimol disrupted, whilst bicuculline synchronized the activity of these neurones.

within the hypothalamus may be involved in modulating the release of LH (Adler and Crowley, 1986; Fuchs et al., 1984; Lamberts et al., 1983) so we set out to investigate whether they might be involved in either promoting or suppressing synchronization among the oscillatory neurones.

Rabbits were fitted with infusion cannulae placed bilaterally into the region of the zona incerta, in addition to the implanted electrodes for recording the activity of oscillatory neurones in the preoptic area of the hypothalamus. Preliminary experiments demonstrated that altering GABA levels in the zona incerta affected LH release and ovulation in intact rabbits. Muscimol, the GABA agonist, inhibited LH release whilst bicuculline, the GABA antagonist, stimulated LH release. Muscimol disrupted the firing pattern of GnRH neurones. The rhythmic pattern of discharge was fragmented and the neurones continued to fire in ultra-short, irregularly-timed bursts (Fig. 1). This pattern of firing was similar to that observed following barbiturate anaesthesia (see previous section). The plasma LH levels in these animals fell to undetectable values. In contrast, bicuculline caused three major changes in the oscillatory discharge: 1) there was a significant increase in the frequency of the rhythmic bursts of activity, 2) the modal interspike interval within each burst increased i.e. there was an increase in frequency of firing with the bursts, and 3) there was an immediate synchronization observed between pairs of oscillatory neurones recorded simultaneously. These changes are illustrated in Figure 1. Plasma LH levels in the rabbits treated with bicuculline infusion were in excess of 30 ng/ml which is higher than at the peak of an LH surge. The nature of the effects of altering GABA in the zona incerta and the relationship between GABA levels, dopamine concentrations, LH release and GnRH neurone activity require further investigation, but they do at least provide a preliminary explanation for the sudden synchronization that is observed among the oscillatory neurones.

Conclusions

These data provide the skeleton of a hypothesis about the neural control of GnRH and LH release. The GnRH neurones are those that exhibit the oscillatory pattern of discharge: it is not the firing rate of these cells _per se_ which is responsible for the pulsatile release of GnRH but the synchronization of pools of oscillatory neurones. The duration and the strength of such synchronization modulates the amount of GnRH released into the portal circulation and hence the size of the LH pulse or surge. The hypothesis, however, must remain skeletal for the essential element of proof is missing: the chronic microwire recording technique that is a pre-requisite for obtaining stable long-term recordings in conscious animals, does not allow us to make a cast-iron identification of the recorded cells. We cannot mark particular cells with immunofluorescent dye and prove that the recorded neurone contained GnRH. The recording sites have been located to be within the areas containing GnRH cell bodies (Summerlee, 1986) but we must await further developments for the final link to be forged in the chain of events that make the neural control of ovulation.

Acknowledgements

I am grateful to B. Delisle Burns, A.J. Emerton, D.W. Lincoln, A.C. Paisley, D.G. Porter, B.T. Pickering and A.C. Webb for their constant help, patience and advice in this work.

References

Adler, G. and Crowley, W.F., 1986, Evidence for τ-aminobutyric acid modulation of ovarian hormonal effects on luteinizing hormone secretion and hypothalamic catecholamine action in the female rat, Endocrinology, 118:91.
Bennet-Clarke, C. and Joseph, S.A., 1982, Immunocytochemical distribution of LHRH neurons and processes in the rat: Hypothalamic and Extrahypothalamic locations, Cell Tissue Research, 221:493.

Bhanot, R. and Wilkinson, M., 1983, Opiatergic control of gonadotrophin secretion during puberty in the rat: a neurochemical basis for the gonadostat? Endocrinology, 113:596.

Bhanot, R. and Wilkinson, M., 1984, The inhibitory effect of opiates on gonadotrophin secretion is dependent upon gonadal steroids, Journal of Endocrinology, 102:133.

Blake, C.A. and Sawyer, C.H., 1972, Ovulation blocking actions of urethane in the rat, Endocrinology, 91:87.

Carmel, P.W., Araki, S. and Ferrin, M., 1976, Pituitary stalk portal blood collection in rhesus monkeys: evidence for pulsatile release of gonadotrophin- releasing hormone (GnRH), Endocrinology, 99:243.

Clarke, I.J. and Cummins, J.J., 1982, The temporal relationship between gonadotrophin-releasing hormone (GnRH) and luteinizing hormone (LH) secretion in ovariectomized ewes, Endocrinology, 111:1737.

Clarke, I.J., Fraser, H.M. and McNeilly, A.S., 1978, Active immunization of ewes against luteinizing hormone, releasing hormone and its effects on ovulation and gonadotrophin, prolactin and ovarian steroid secretion, Journal of Endocrinology, 78:39.

Clarke, I.J., Cummins, J.J., Findlay, J.K., Burman, K.J. and Doughton, D.W., 1984, Effects on plasma luteinizing hormone and follicle-stimulating hormone of varying the frequency and amplitude of gonadotrophin-releasing hormone pulses in ovariectomized ewes with hypothalamic- pituitary disconnection, Neuroendocrinology, 39:214.

Clayton, R.N., 1982, Gonadotrophin-releasing hormone modulation of its own pituitary receptors: evidence for biphasic regulation, Endocrinology, 111:152.

Crowley, W.F., Filicori, M., Spratt, D.I. and Santoro, N.F., 1985, The physiology of gonadotrophin hormone-releasing hormone (GnRH) secretion in men and women, Recent Progress in Hormone Research, 41:473.

Ellis, G.B., Desjardins, C. and Fraser, H.M., 1983, Control of pulsatile LH release in male rats, Neuroendocrinology, 37:177.

Everett, J.W. and Sawyer, C.H., 1950, A 24-hour periodicity in the 'LH-release apparatus' of female rats disclosed by barbiturate sedation, Endocrinology, 47:198.

Fraser, H.M. and McNeilly, A.S., 1983, Differential effects of LH-RH immunoneutralization on LH and FSH secretion in the ewe, Journal of Reproduction and Fertility, 69:569.

Fuchs, E., Mansky, T., Stock, W.-W., Vijayan, E. and Wuttke, W., 1984, Involvement of catecholamines and glutamate in GABAergic mechanism regulatory to luteinizing hormone and prolactin secretion, Neuroendocrinology, 38:484.

Karsch, F.J., Bittman, E.L., Foster, D.L., Goodman, R.L., Legan, S.J. and Robinson, J.E., 1984, Neuroendocrine basis of seasonal breeding, Recent Progress in Hormone Research, 40:185.

Kawakami, M., Uemura, T. and Hayashi, R., 1982, Electrophysiological correlates of pulsatile gonadotrophin release in rats, Neuroendocrinology, 35:63.

King, J.C. and Anthony, E.L.P., 1984, LHRH neurones and their projections in humans and other mammals; species comparisons, Peptides, 5, Suppl.1 195.

Lamberts, R., Vijayan, E., Graf, M., Mansky, T. and Wuttke, W., 1983, Involvement of preoptic-anterior hypothalamic GABA neurons in the regulation of pituitary LH and prolactin secretion, Experimental Brain Research, 52:356.

Levine, J.E., Kwok-Yuen, F.P., Ramirez, V.D. and Jackson, G.L., 1982, Simultaneous measurement of luteinizing hormone-releasing hormone and luteinizing hormone release in unanaesthetized, ovariectomized sheep, Endocrinology, 111:1449.

Lincoln, D.W. and Kelly, W.A., 1972, The influence of urethane on ovulation in the rat, Endocrinology, 90:1594.

Lincoln, D.W., Fraser, H.M., Lincoln, G.A., Martin, G.B. and McNeilly, A.S., 1985, Hypothalamic pulse generators, Recent Progress in Hormone Research, 41:369.

Lincoln, G.A., 1979, Use of a pulsed infusion of luteinizing hormone releasing hormone to mimic seasonally induced endocrine changes in the ram, Journal of Endocrinology, 83:251.

Lincoln, G.A. and Fraser, H.M., 1979, Blockade of episodic secretion of luteinizing hormone in the ram by administration of antibodies to luteinizing hormone-releasing hormone, Biology of Reproduction, 21:1239.

Lincoln, G.A. and Short, R.V., 1980, Seasonal breeding: Nature's Contraceptive, Recent Progress in Hormone Research, 36:1.

Paisley, A.C. and Summerlee, A.J.S., 1985a, Activity of preoptic neurones in conscious rabbits, Journal of Physiology, 364:49P.

Paisley, A.C. and Summerlee, A.J.S., 1985b, Rabbit preoptic neurones and luteinizing hormone release, Journal of Physiology, 364:50P.

Poulain, D.A. and Wakerley, J.B., 1982, Electrophysiology of hypothalamic neurones secreting oxytocin and vasopressin, Neuroscience, 7:773.

Ropert, J.F., Quigley, M.E. and Yen S.C.C., 1981, Endogenous opiates modulate pulsatile luteinizing hormone release in humans, Journal of Clinical Endocrinology and Metabolism, 52:583.

Silverman, A.J., Antunes, J.L., Abrams, G.M., Nilaver, G., Thau, R., Robinson, J.A., Ferrin, M. and Krey, L.C., 1982, The luteinizing hormone-releasing hormone pathways in the rhesus (Macaca mulatta) and pigtailed (Macaca nemestrina) monkeys: new observations on thick unembedded sections, Journal of comparative Neurology, 211:309.

Summerlee, A.J.S., 1981, Extracellular recordings from oxytocin neurones during the expulsive phase of birth in unanaesthetized rats, Journal of Physiology, 321:1.

Summerlee, A.J.S., 1985, Steroids determine the response of the GnRH pulse generator to naloxone in rabbits, Neuroscience Letters, Suppl. 22, S598.

Summerlee, A.J.S., 1986, The neural control of luteinizing hormone release, Progress in Neurobiology, 26:147.

Summerlee, A.J.S. and Paisley, A.C., 1982, The effect of behavioural arousal on the activity of hypothalamic neurones in unanaesthetized, freely moving rats and rabbits, Proceedings of the Royal Society, Series B 214:263.

Summerlee, A.J.S., Lincoln, D.W. and Webb, A.C., 1979, Long-term electrical recordings from single neurosecretory cells in the conscious lactating rat, Journal of Endocrinology, 83:41P.

Sylvester, P.W., van Vugt, D.A. Aylsworth, C.A., Hanson, E.A. and Meites, J., 1982, Effects of morphine and naloxone on inhibition by ovarian hormones of pulsatile release of luteinizing hormone in ovariectomized rats, Neuroendocrinology, 34:269.

Thiery, J.C. and Pelletier, J., 1981, Multiunit activity in the anterior median eminence and adjacent areas of the hypothalamus of the ewe in relation to luteinizing hormone secretion, Neuroendocrinology, 32:217.

Wilson, R.C. Kesner, J.S., Kaufman, J-M., Uemura, T. Akema, T. and Knobil, E., 1984, Central electrophysiological correlates of pulsatile luteinizing hormone secretion in the rhesus monkey, Neuroendocrinology, 39:256.

Weick, R.F. and Dyer, R.G., 1982, Stimulation of single unit activity in the preoptic area and anterior hypothalamus, and secretion of luteinizing hormone, by ovarian steroids in ovariectomized rats with diencephalic islands, Proceedings of the Royal Society, Series B 216:461.

Weindl A. and Sofroniew, M.V., 1981, The morphology of LRH and oxytocin neurones, in: Experimental Brain Research, Suppl 3, 1-7, Springer-Verlag, Berlin, Germany.

Witkin, J.W., Paden, C.M. and Silverman, A.J., 1982, The luteinizing hormone-releasing hormone (LHRH) systems in the rat brain, Neuroendocrinology, 35:429.

ELECTRICAL ATTRIBUTES OF NEUROSECRETORY TERMINALS AND THEIR RELATIONSHIP TO SECRETION

E. L. Stuenkel[1] and I. M. Cooke[2]

[1]Department of Physiology, University of California
San Francisco, CA 94143
[2]Bekesy Laboratory of Neurobiology, 1993 East-West Road
Honolulu, HI 96822

INTRODUCTION

This article briefly reviews the electrical attributes of nerve terminals and of their relationship to secretion for a neurosecretory system in crabs termed the X-organ sinus gland (XOSG). This neurosecretory system is a purely peptidergic secretory system whose hormonal secretions are known to influence or govern nearly all physiologic and homeostatic processes (reviews, Cooke and Sullivan, 1982; Keller, 1983). A portion of the system termed the sinus gland (SG) is a discrete spherical (<u>c</u>. 1.5 mm. diameter) structure composed of an extensive arborization of axon varicosities and terminals together with glial and supportive cells which surround a network of hemolymph sinuses. The cell soma, where axons project to and compose the sinus gland, are located in a cluster named the X-organ. The relatively large size of individual axon terminal dilatations in certain species of crabs (e.g., <u>Cardisoma carnifex, Podophthalamus vigil</u>) is novel among peptidergic neurosecretory systems. The terminals in these species are of sufficient size (up to 30 um) to have allowed extensive investigation with intracellular microelectrodes in both <u>in situ</u> and <u>in vitro</u> preparations (reviews, Cooke and Stuenkel, 1985; Stuenkel and Cooke, 1988). Recently a preparation of dissociated terminals from the sinus gland has been developed and investigated with the various recording configurations of the patch clamp technique (Lemos et al., 1986). A number of electrical and secretory attributes of the peptidergic terminals have been characterized including: 1) the ionic dependence of regenerative activity, 2) 'whole terminal' patch clamp analysis of voltage dependent macroscopic ionic currents and of two novel cation channel types and 3) correlation of terminal electrical behaviour with hormonal secretion monitored simultaneously under different secretory regimes.

INTRACELLULAR RECORDING

In the crab (e.g., <u>Cardisoma carnifex</u>) it is possible to record intracellularly from the soma, axon and terminal regions of the neurosecretory neuron (Cooke, 1977, 1985; Stuenkel, 1985; Nagano and Cooke, 1987). This allows comparison of the electrical attributes from each region with the understanding that intracellular recordings average the contributions from large areas of membrane and will underestimate the extent of regional differences. Average resting membrane potentials (-50 to -65 mV) show little regional differences as do the percentage of soma (64-75%) and terminal (77- 93%) recordings exhibiting spontaneous impulse activity. Clear differences, however, are apparent in the form of the action potentials from each region. Soma impulses show a slower rate of rise and a larger percentage of soma exhibit non-overshooting action potentials than do axons and terminals. These differences may represent variability in the extent of impulse invasion of the soma, as impulses are likely initiated in the proximal axon. Soma impulses in the crayfish <u>Procambarus clarkii</u> have been reported to be consistently overshooting (Iwasaki and Satow, 1971). Other qualitative differences in impulse form include, 1) the presence of hyperpolarizing afterpotentials in both terminal and soma impulses but depolarizing afterpotentials in axons and 2) longer halfamplitude durations in terminal and soma impulses (c. 8 to 10 msec) than in axons (c. 6 msec).

Approximately 30% of terminal recordings that show spontaneous activity exhibit a patterning of their impulses (Cooke and Stuenkel, 1985; Cooke, 1985; Nagano and Cooke, 1987). The patterning can occur in many forms with most containing an underlying depolarizing potential from which the impulses arise. In some cases, subsequent spikes within a 'burst' exhibit an increase in impulse duration as a result of a decrease in the rate of the falling phase of the action potential (Cooke, 1985). The physiological significance of this spike broadening, similar to that of bursting itself, awaits elucidation. Speculation and analogy with facilitation of hormone release from the mammalian neural lobe under bursting stimulus regimes (Dutton and Dyball, 1979; Bicknell et al., 1982) suggests a relationship with enhanced hormone secretion.

Studies of the ionic dependence of impulse activity have led to the conclusion that soma and terminal action potentials include both Na^+ and Ca^{2+} inward currents while those of axons are primarily mediated by Na^+ (Iwasaki and Satow, 1971; Cooke, 1977, 1985; Stuenkel, 1985; Nagano and Cooke, 1987). Reduction of $[Na^+]_o$ or application of the selective Na^+ conductance blocker TTX during terminal or soma recording does not eliminate their responsiveness to intracellularly applied depolarizing current. A decrease in the rate of rise and degree of overshoot is observed. Similarly, application of blockers of Ca^{2+} currents (e.g., Cd^{2+}, Mn^{2+}) alone does not result in blockage of impulse activity. Whole-cell patch clamp techniques on crayfish X-organ somata, however, have found major inward currents to be blocked by Cd^{2+} in axotomized cells (Garcia et al., 1986). TTX sensitive currents are observed in soma recordings where a portion of the axon is present. Therefore, contribution of inward Na^+ current to the soma may result from electrotonic spread from the axon. Failure to show regenerative responses to extracellular stimulation or intracellular depolarizing current under conditions of reduced $[Na^+]_o$ or in the presence of TTX substantiates a conclusion that Na^+ alone accounts for the inward current associated with axonal action potentials. The importance of stimulus induced Ca^{2+} influx to secretion, as initially described by Douglas and coworkers (1963, 1964a,b) for the vertebrate adrenal medulla and neurohypophysis and by Katz and Miledi (1967) for the neuromuscular junction, clearly provides a physiologic relevance to the calcium component of the nerve terminal action potential. Simultaneous monitoring of the terminal region compound electrical responses and secretion, while stimulating the XOSG nerve tract extracellularly, has shown a reduction in secretion in reduced Ca^{2+} salines while the efficacy of stimulation and production of compound action potentials remains unaltered (Cooke et al., 1977). The relevance for inward Ca^{2+} current in the soma is less clear. Ca^{2+} is a well known second messenger involved in the activation of a number of kinases as well as other physiological processes and may be important in such processes as entrainment of secretory protein synthesis to release or in a feedback mechanism whereby responses to synaptic input are modulated. We have attempted to monitor hormone secretion from the region of soma and collaterals in response to depolarizing concentrations of potassium (Stuenkel, 1985). These studies have yielded inconclusive evidence but suggest that secretion from the soma or the neuron's collateral arbor cannot be ruled out.

PATCH CLAMP ANALYSIS

Recently we have developed, in conjunction with J. Lemos and J. J. Nordmann, a preparation of dissociated SG nerve terminals which readily form gigohm seals with patch pipettes (Lemos et al., 1986). The preparation results in hundreds of isolated spherical terminal structures which at the EM level are morphologically indistinguishable from the non-dissociated nerve endings. Observation of macroscopic ionic currents under the "whole-terminal" recording configuration have demonstrated, in response to depolarizing voltage steps, the presence of initial inward followed by outward maintained current. The inward currents exhibit properties consistent with conductances to Na^+ and Ca^{2+} ions. The calcium component is partially blocked by external application of Cd^{2+} and is unaffected by TTX whereas the Na^+ component is susceptible to TTX blockade. Outward currents are strongly inhibited following external application of TEA and completely blocked when K^+ is replaced by Cs^+ in the patch pipette confirming that outward current is carried by K^+. The presence of both Na^+ and Ca^{2+} components to terminal inward currents is as expected from the intracellular recordings which demonstrated Na^+ and Ca^{2+} components to the terminal action potential.

Single channel currents were recorded from the isolated terminals in both the "terminal attached" and "terminal detached" recording configurations. More recent studies have also included recordings from surface terminals of isolated but otherwise intact, sinus glands (Ruben and Cooke, 1987). Two apparently novel types of voltage insensitive cation channels have been observed (Fig. 1a,b). The first type, termed f, is most predominant and occurs in nearly every patch. It characteristically exhibits rapid flickering to an open state which in some cases occurs as bursts interrupted by intervals of seconds. Under conditions of symmetrical KCl (inside out) the channel exhibits a mean slope conductance of 69 pS (Fig. 1c). The reversal potential was observed to shift in the presence of ionic gradients as expect ed for a purely cation selective channel. Substitution of Na_2SO_4 and K_2SO_4 for NaCl or KCl and CsCl for KCl show failure of the channel to pass SO_4^{2-}, Cl^- or Cs^+. The f channel is activated by $[Na^+]_i$ of 40 mM or greater, which extends the length of time spent in the open state.

The second cation channel, the s channel, was observed more infrequently. In symmetrical KCl (310 mM) it exhibits a mean slope conductance of 213 pS (Fig. 1d). Exposure of this channel to similar tests as above for ionic selectivity show it likewise to be purely cation selective in allowing Na^+ and K^+ to pass equally well while failing to pass anions. The s channel, however, shows no rectification of the current voltage relations in the presence of CsCl and thus allows Cs^+ to pass. Unlike the f channel, which is activated by Na^+, the s channel is activated by $[Ca^{2+}]_i$ in the uM range while remaining unaffected by $[Na^+]_i$. In this manner the s channel resembles calcium-activated K^+ channels. The physiologic role played by either the f or s channel is unknown. Their conductance characteristics, however, are not inconsistent with their involvement in either broadening of spikes or in generation of plateau potentials seen to underly bursting in some terminals. Recent application of patch clamp methodology to terminal preparations of the neurohypophysis (Lemos and Nordmann, 1986; Mason and Dyball, 1986), together with the report from one of these preparations (Lemos and Nordmann, 1986) of channels exhibiting similar properties as above, points to the necessity for further study of the properties of these channels and of their physiologic relevance.

ELECTROPHYSIOLOGICAL CORRELATES OF SECRETION

The role of electrical activity in secretion is believed related to the depolarization induced activation of inward calcium current associated with propagated action potentials. The influx of calcium then initiates and participates in a cascade of events leading to the exocytotic secretion of the secretory granule contents. Direct evidence for the role of propagated impulse activity in the release of red pigment concentrating hormone (RPCH) has been presented by Cooke et al., (1977). Secretion was dependent on the presence of extracellular calcium and was reduced to basal secretory levels when axonal propagation of impulses was blocked. These data, together with the above electrophysiological evidence for voltage-dependent channels carrying inward calcium current and for mixed Na^+ and Ca^{2+} dependence of the terminal action potential, strongly substantiate the above secretory hypothesis for the XOSG system.

The description of calcium currents which exhibit substantial voltage- or calcium-dependent inactivation and those which are more resistant to inactivation led us to investigate secretion under conditions of maintained depolarization (i.e. in the presence of elevated $[K^+]_o$) (Cooke and Haylett, 1984; Stuenkel, 1985). Significant differences in the secretory time course were observed between RPCH and the release of ^{3}H-labelled peptides. The labelled peptides primarily represent crustacean hyperglycemic hormones and peptide H and related cleavage fragments (Stuenkel, 1983, 1986). The secretion of RPCH reaches peak secretory rates within 5 min. from which it falls to near basal levels over the next 2 min. In contrast, radiolabel efflux reaches peak levels on a similar time course but remains elevated for an extended period ($t_{1/2}$ = 47 min). The secretion of both RPCH and of ^{3}H- labelled

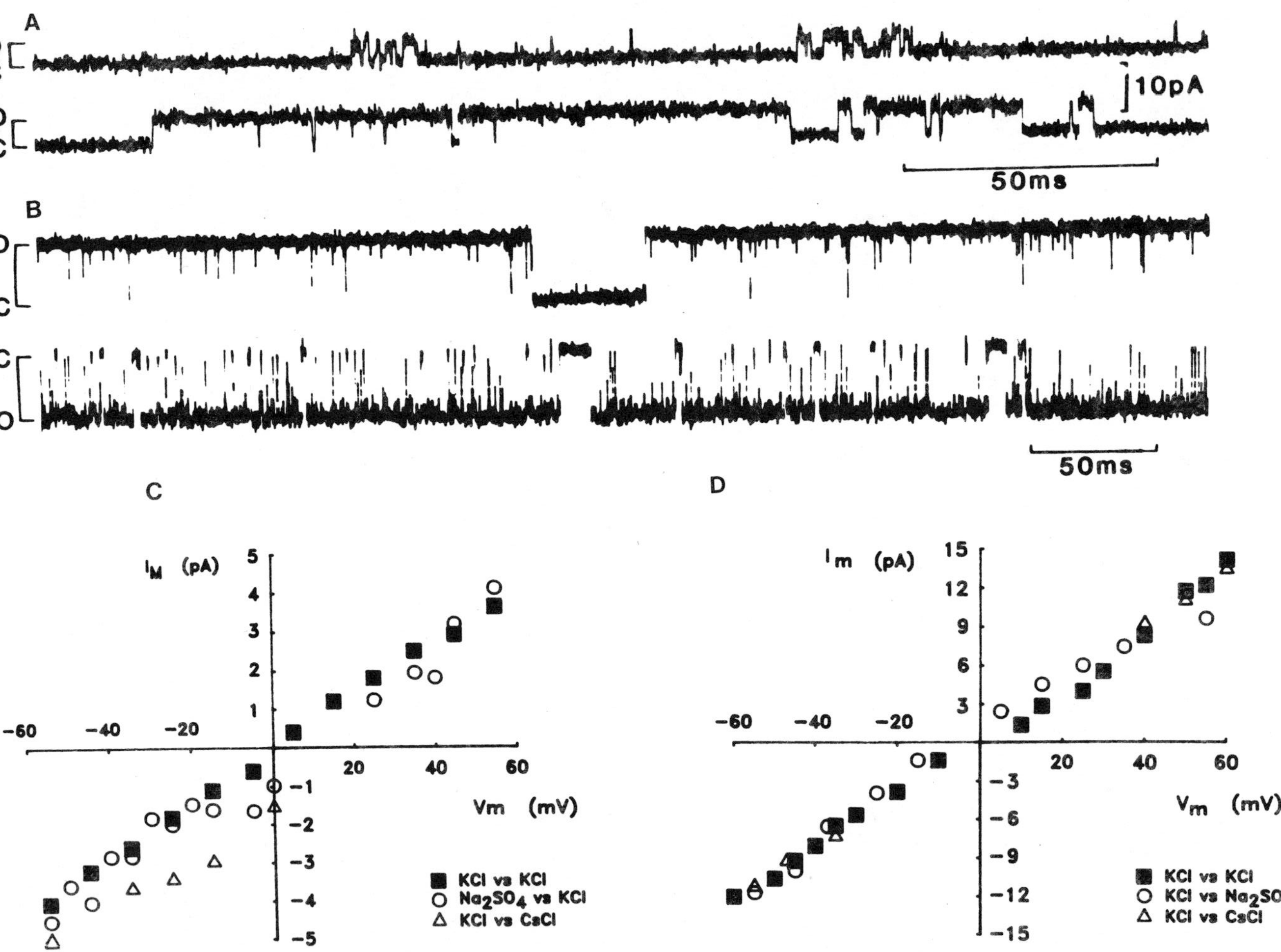

A
O
C
O
C
10pA
50ms
B
O
C
C
O
50ms
C
D
IM (pA)
Vm (mV)
KCl vs KCl
Na2SO4 vs KCl
KCl vs CsCl
Im (pA)
Vm (mV)
KCl vs KCl
KCl vs Na2SO4
KCl vs CsCl

peptides are dependent on extracellular calcium. Results from additional experiments suggest that the difference in the time course reflects a voltage-dependent inactivation of the RPCH release, perhaps attributable to inactivation of Ca^{2+} channels, and a lack of such inactivation in the majority of terminals responsible for ^{3}H-labelled peptide secretion. Several distinct calcium channel types have now been described that show differences in voltage activation, single channel conductance and inactivation properties from a number of preparations (e.g., Nowycky et al., 1985; Matteson and Armstrong, 1986). Clearly, other explanations for the different secretory time courses can be considered, such as depletion of a readily releasable pool, or differences in the secretory cascade beyond calcium entry. A hypothesis that different Ca^{2+} channel types are involved in mediating secretory responses is supported by a study where substance P release from vertebrate dorsal root ganglion neurons was found to be dihydropyridine sensitive while norepinephrine release from sympathetic neurons showed resistance to this drug (Perney et al., 1986). Using the isolated terminal preparation coupled with patch clamp methodology, it is now possible to directly test the hypothesis that multiple Ca^{2+} channel types are present in different neurosecretory terminals and that these are correlated to the different secretory time courses. Subsequent antibody determination of the terminal type should allow correlation to release studies on the isolated intact system.

SUMMARY AND PERSPECTIVES

The electrophysiological characteristics of peptidergic nerve terminals of the XOSG system can be briefly summarized as follows. 1) Spontaneous regenerative activity is present in the majority of recordings with impulses occurring in either tonic, irregular or a grouped discharge pattern. 2) Impulses are both Na^+- and Ca^{2+}-mediated. 3) Voltage-dependent ionic currents include inward Na^+ and Ca^{2+} currents and outward K^+ currents. 4) Two unique cation channel types are found in the terminals. The relationship of these electrical attributes to secretion is clearly manifest through impulse mediated Ca^{2+} entry. The effects of spike broadening, impulse bursting and of cation channels on secretion is less clear. Also, the mechanisms underlying differences in the secretory time course remain unresolved. The development of both antibodies against particular XOSG peptides and of an isolated terminal preparation together with patch clamp methodology should facilitate research towards an understanding of these parameters. Additionally, the development of a pure terminal preparation provides the material necessary to propose and test hypotheses regarding intraterminal biochemical cascades distal to impulse evoked Ca^{2+} entry, which lead to or modulate exocytotic secretion. Further, while research has begun to define the pharmacology of regulatory and modulatory synaptic input to the XOSG system (review, Fingerman, 1985) and of its electrical correlates (Glantz, et al., 1983; Nagano, 1986) this remains an area requiring further investigations.

Fig. 1.   (Opposite page) Single channel currents from inside-out patches showing two types of cation channels in isolated nerve terminals. A, First (f) channel type: Upper, patch held at +55 mV in symmetrical KCl. Upward openings (o) indicate outward current. Lower, Same patch held at +45 mV with 310 mM KCl outside and 210 mM Na_2SO_4 inside. Note the increased duration of the openings. B, Second (s) channel type: Upper, patch held at +55 mV with 310 mM KCl inside and 210 mM Na_2SO_4 outside and with 3 uM $[Ca^{2+}]_i$. Lower, Same patch as Upper but held at -40 mV. C, I-V relationship of the f channel. The calculated regression has a slope conductance of 69 pS (n = 6) in symmetrical 310 mM KCl (■). D, I-V relationship of the s channel. The calculated regression has a slope conductance of 213 pS (n = 4) in symmetrical 310 mM KCl (■). C, D, Other symbols indicate I-V relationship when KCl is replaced by Na_2SO_4 (o) or CsCl (△). (From Lemos, Nordmann, Cooke and Stuenkel, 1986).

REFERENCES

Bicknell, R., Flint, J., Leng, A.P.F., and Sheldrick, E.L., 1982, Phasic pattern of electrical stimulation enhances oxytocin secretion from the isolated neurohypophysis, Neuroscience Letters, 30:47.

Cooke, I.M., 1977, Electrical activity of neurosecretory terminals and control of peptide hormone release, in: "Peptides in Neurobiology," H. Gainer, ed., Plenum, New York.

Cooke, I.M., 1985, Electrophysiological characterization of peptidergic neurosecretory terminals, J. Exp. Biol., 118:1.

Cooke, I.M., and Haylett, B.A., 1984, Ionic dependence of secretory and electrical activity evoked by elevated K+ in a peptidergic neurosecretory system, J. Exp. Biol., 113:289.

Cooke, I.M., Haylett, B.A., and Weatherby, T.M., 1977, Electrically elicited neurosecretory and electrical responses of the isolated crab sinus gland in normal and reduced calcium salines, J. Exp. Biol., 70:125.

Cooke, I.M., and Stuenkel, E.L., 1985, Electrophysiology of invertebrate neurosecretory cells, in: "The Electrophysiology of the Secretory Cell," A.M. Poisner and J. Trifaro, eds., Elsevier, Amsterdam.

Cooke, I.M., and Sullivan, R.E., 1982, Hormones and neurosecretion, in: The Biology of Crustacea," H.Atwood and D.Sandeman, eds., Academic Press, New York.

Douglas, W.W., and Poisner, A.M., 1964a, Stimulus-secretion coupling in a neurosecretory organ: The role of calcium in the release of vasopressin from the neurohypophysis, J. Physiol., 172:1.

Douglas, W.W., and Poisner, A.M., 1964b, Calcium movements in the neurohypophysis of the rat and its relation to the release of vasopressin, J. Physiol., 172:19.

Douglas, W.W., and Rubin, R.P., 1963, The mechanism of catecholamone release from the adrenal medulla and the role of calcium in stimulus-secretion coupling, J. Physiol., 167:288.

Dutton, A., and Dyball, R.E.J., 1979, Phasic firing enhances vasopressin release from the rat neurohypophysis, J. Physiol., 290:433.

Fingerman, M., 1985, The physiology and pharmacology of crustacean chromatophores, Am. Zool., 25:233.

Garcia, U., Onetti, C., and Valdiosera, R.F., 1985, Calcium currents in the secretory neuron soma of the X-organ of the crayfish, Soc. Neurosci. Abstr., 11:520.

Glantz, R.M., Kirk, M.D., and Arechiga, H., 1983, Light input to crustacean neurosecretory cells, Brain Res., 265:307.

Iwasaki, S., and Satow, Y., 1971, Sodium- and calcium-dependent spike potentials in the secretory neuron soma of the X-organ of the crayfish, J. Gen. Physiol., 57:216.

Katz, B., and Miledi, R., 1967, The timing of calcium action during neuromuscular transmission, J. Physiol., 189:535.

Keller, R., 1983, Biochemistry and specificity of the neurohemal hormones in Crustacea, in: "Neurohemal Organs of Arthropods," A.P. Gupta, ed. Thomas, Springfield.

Lemos, J.R., and Nordmann, J.J., 1986, Ionic channels and hormone release from peptidergic nerve terminals, J. Exp. Biol., 124:53.

Lemos, J.R., Nordmann, J.J., Cooke, I.M., and Stuenkel, E.L., 1986, Single channels and ionic currents in peptidergic nerve terminals, Nature, 319:410.

Mason, W., and Dyball, R., 1986, Single ion channel activity in peptidergic nerve terminals of the isolated rat neurohypophysis related to stimulation of neural stalk axons, Brain Res., 383:279.

Matteson, D.R., and Armstrong, C.M., 1986, Properties of two types of calcium channels in clonal pituitary cells, J. Gen. Physiol., 87:161.

Nagano, M., 1986, Regional inhibitory effects of 5HT and GABA on the spontaneous electrical activity of a crab neurosecretory system, Biomedical Res., 7:267.

Nagano, M., and Cooke, I.M., 1987, Comparison of electrical responses of terminals, axons, and somata of a peptidergic neurosecretory system, J. Neurosci., 7:634.

Nowycky, M.C., Fox, A.P., and Tsien, R.W., 1985, Three types of neuronal calcium channels with different calcium agonist sensitivity, Nature, 316:440.

Perney, T.M., Hirning, L.D., Leeman, S.E., and Miller, R.J., 1986, Multiple calcium channels mediate neurotransmitter release from peripheral neurons, Proc. Natl. Acad. Sci. USA, 83:6656.

Ruben, P., and Cooke, I., 1987, Comparison of cation channels in isolated and *in situ* peptidergic neurosecretory terminals, Biophys. J., 51:250a.

Stuenkel, E., 1983, Biosynthesis and axonal transport of proteins and identified peptide hormones in the X-organ sinus gland neurosecretory system, J. Comp. Physiol., B207:191.

Stuenkel, E.L., 1985, Simultaneous monitoring of electrical and secretory activity in peptidergic neurosecretory terminals of the crab, J. Physiol., 359:163.

Stuenkel, E.L., 1986, A common precursor to two major crab neurosecretory peptides, Peptides, 7:397.

Stuenkel, E.L., and Cooke, I.M., 1988, Electrophysiological characteristics of peptidergic nerve terminals correlated with secretion, *in*: "Current Topics in Neuroendocrinology. Vol 9. Stimulus-secretion Coupling as Applied to Neuroendocrine Subjects," D. Gantan and D. Pfaff, eds., Springer-Verlag, New York, in press.

NEUROANATOMICAL AND ELECTROPHYSIOLOGICAL ANALYSIS OF THE BRAIN-SINUS GLAND NEUROSECRETORY SYSTEM IN A CRUSTACEAN

C.G.H. Steel and R.G. Chiang

Department of Biology
York University
North York,
Ontario, M3J 1P3
Canada

INTRODUCTION

In recent years, this laboratory has been exploiting terrestrial isopod crustaceans (woodlice, sowbugs) as model organisms in which to investigate the organization and physiology of neurosecretory systems. The initial attraction of these animals derived from theoretical considerations, but was supported by the existing literature, albeit sparse (see Chiang and Steel, 1984). These animals, being small and terrestrial, are easy to rear in large numbers in the laboratory, where they undergo physiological events controlled by neurohormones (e.g., moulting, reproduction) with repeatable regularity (see Heeley, 1941).

Our initial studies demonstrated that the moulting cycle exhibited a very precise timetable and was repeated every 28 days at $21^{\circ}C$ (Steel, 1980). The rigidity and frequency of this cyclical process implied that the roles of neurohormones in the control of moulting would be more readily elucidated in these crustaceans than in the more extensively studied decapod crustaceans (see Skinner, 1985), in which moulting is often infrequent and unpredictable. Using the isopod <u>Oniscus asellus</u> (L.), we divided the moulting cycle into 15 distinct stages which could be recognized without injury to the animal (Steel, 1982). A specific sensory cue which initiates premoult in normal animals was also identified (Steel, 1980), permitting study of the neuroendocrine changes responsible for premoult initiation. Such changes necessarily precede morphological changes in the animal and had previously been inaccessible to experimentation (see Passano, 1960).

The nervous system of isopods also displays features conducive to the study of neurosecretion. The nervous system is anatomically simple (Walker, 1935) and the main neurohaemal organ, the sinus gland (SG), is so conspicuously white in the head of the living animal that it was assumed by classical authors to be a statocyst (Schobl, 1880). The SG is a bulbous structure protruding from the optic lobe on each side of the brain. It is only about 60 μm in diameter (Gabe, 1952; see also Fig. 1), suggesting it contains rather few neurosecretory terminals whose detailed connections were investigated using histology, electron microscopy and backfilling with cobalt. Although neurosecretory cells (NSCs) have been described in terrestrial isopods using both histology (e.g., Matsumoto, 1959) and electron microscopy (e.g., Martin, 1972), the presumed axonal pathways to the SG have not been examined. It was easy to draw the SG into a suction electrode in restrained animals and hence to monitor on-going electrical activity in the terminals. It was found that each of the five structurally identified types of terminal generated a characteristic type of electrical potential. Changes in this on-going electrical activity were then monitored throughout a complete cycle of moulting in order to elucidate the times of neurohormone release <u>in vivo</u> from the identified NSCs.

HISTOLOGY

The first level of analysis employed the classical selective stain for neurosecretion, paraldehyde fuchsin (PAF), applied as described by Steel (1977) to 5 μm serial sections of the head (Chiang and Steel, 1985a). The SG stained prominently, and was found to contain numerous PAF-positive droplets about 10 μm in diameter as seen originally by Gabe (1952). The droplets were interpreted as putative neurosecretory terminals; these were connected to a tract of similarly staining axons which passed into the optic lobe of the brain (Fig. 1) and which could be traced in serial sections to three anatomically distinct clusters of PAF-positive perikarya in the brain. These clusters were names B, β and τ cells according to the original terminology introduced by Matsumoto (1959) for the isopod <u>Armadillidium</u>, based on their location and appearance. The total number of perikarya is small; 5B, 5β and 15τ cells. Fine branches of the axons were seen in the neuropiles adjacent to the cell bodies, suggestive of dendritic arborizations (see Fig. 2). It was concluded that the SG contained terminals from three distinct clusters of stainable NSC which differed in location and cytology. Moreover, analysis of cyclical cytological changes in these perikarya in heads fixed at various stages of the moult cycle and/or oocyte development, indicated that β cells underwent changes associated with moulting and τ cells underwent changes associated with oocyte development (details in Chiang and Steel, 1985a). Preliminary indications of distinct functions for these separate clusters of cells were therefore obtained from histological studies.

COBALT BACKFILLING

Further details of the neuroanatomy of this neurosecretory system were obtained by backfilling the NSCs from their terminals in the SG. It was found unnecessary to sever the axons in order to fill them with cobalt. Rather, a glass capillary was prepared with an inside diameter of 50-60 μm and filled with 0.5 M $CoCl_2$. The entire bulb of the SG was drawn by suction into this capillary and the cells were allowed to fill for 15-20 h at room temperature. Thus, cobalt was applied extracellularly to the bulb of the SG. These animals were intact save for a small portion of the exoskeleton of the head which was removed in order to insert the capillary. The animals were not anaesthetized, but merely restrained with Plasticene. Following filling with cobalt, precipitation and intensification employed routine methods (details in Chiang and Steel, 1985a).

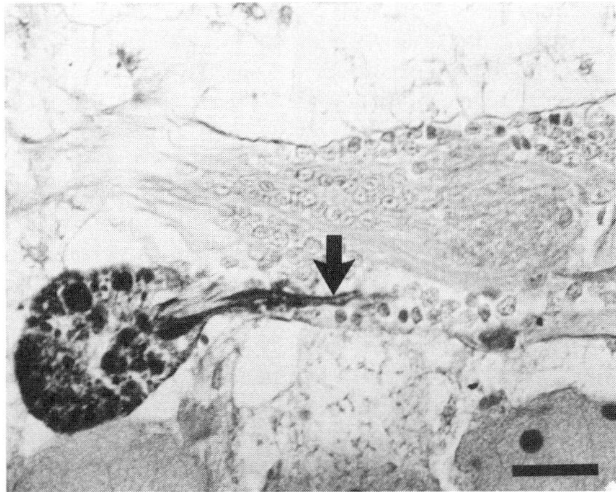

Fig. 1.　　Transverse section of the head of <u>Oniscus asellus</u>, showing the optic lobe (eye to the left) with the sinus gland containing paraldehyde-fuchsin stained neurosecretory terminals and their axons (arrow). Scale bar 20 μm (from Chiang and Steel, 1985a).

The prominent tract of PAF-positive material in the optic lobe filled readily, verifying that this indeed comprised axons ending in the SG. Its prominence was due not to the presence of large numbers of axons but to the presence of conspicuous varicosities along the length of a small number of axons, as shown in Fig. 2. The arborizations of B and β cells were revealed in great detail in the location detected with PAF. An important additional finding was that part of the arborizations extended across the midline of the brain and terminated in the vicinity of the arborizations of the B and β cells from the opposite site of the brain (Fig. 2). This anatomical finding suggests a possible pathway for coupling of the electrical activity of homologous cells from left and right sides of the brain. Such coupling was found in later electrophysiological studies (see below).

The three clusters of PAF-positive cells all filled with cobalt, confirming that these cells are indeed neurons which project to, and terminate in, the SG. The SG therefore contains terminals from a total of 25 NSCs whose perikarya reside in at least three anatomically distinct locations (Fig. 3). In some preparations, a fourth cluster of neurons very close to the SG was also filled (cluster unnamed in Fig. 3); these cells could not be identified with other techniques and consequently their neurosecretory nature is unclear.

ULTRASTRUCTURE

The small size and compact shape of the SG in <u>Oniscus</u> made it possible to examine its entire ultrastructure and to relate the structure of neurosecretory axons and terminals within it to their perikarya of origin identified above by histology and backfilling. Serial sections of the bulb of the SG showed that the distal region comprised exclusively neurosecretory terminals separated by fine glial processes; the size and distribution of terminals was similar to the droplets seen with PAF. The terminals were of three kinds, distinguishable by the size, shape and electron-density of their contained neurosecretory granules (types 1, 2 and 3 in Fig. 4). The axons of these three terminal types could be traced proximally in serial sections; it was found that types 1 and 2 terminals extended to the protocerebrum and originated from β and B cells, whereas type 3 terminals ended in the optic lobe where τ cells are located (Chiang and Steel, 1985b). The appearance of granules in the axons was closely similar to those in the terminals, making it highly unlikely that one NSC might give rise to more than one type of terminal. There is therefore an excellent

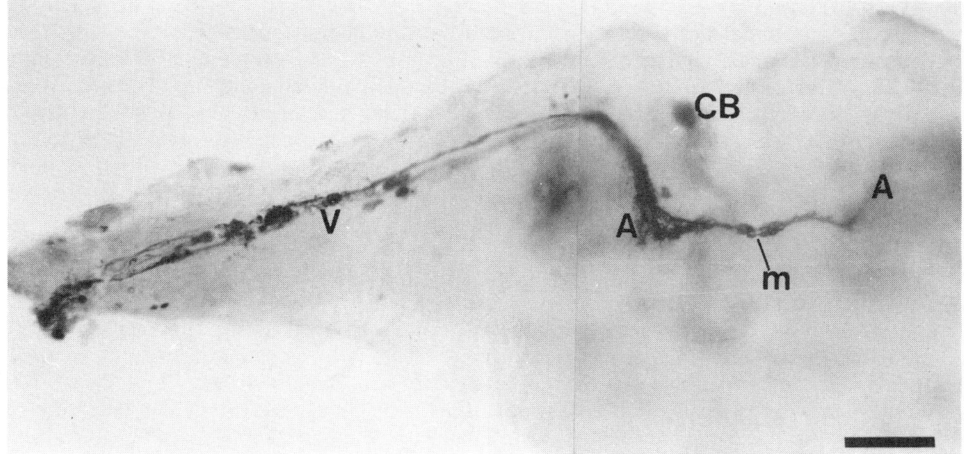

Fig. 2.   Posterior view of the left protocerebrum showing the axon pathways filled with cobalt applied to the sinus gland. Varicosities (V) are seen in the axons in the optic lobe and arborizations (A) are seen in the central protocerebral neuropile. Note that some arborizations traverse the midline (m) and extend into the contralateral protocerebrum. Cell bodies (CB) of these cells are out of the plane of focus. Scale bar 100 μm (from Chiang and Steel, 1985a).

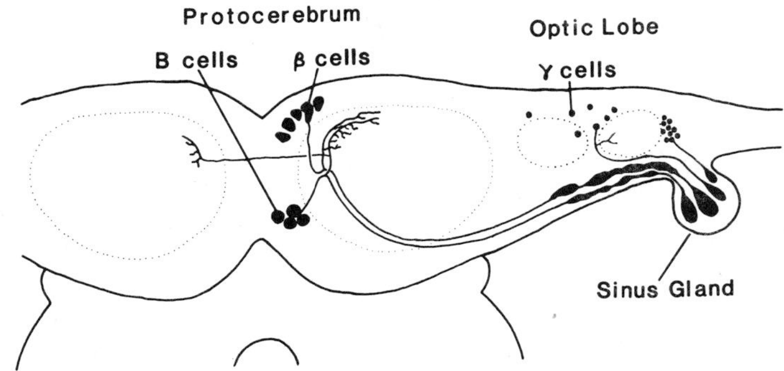

Fig. 3.   Composite diagram of the neurosecretory system as revealed by cobalt backfilling (from Chiang and Steel, 1985a).

correspondence between the histology of the perikarya, the axonal projections revealed by backfilling and the ultrastructure of terminals in the SG.

The stalk of the bulb contains the axons of these three terminal types in its central region, surrounded by glial cell nuclei and processes. The surface of the optic lobe lateral to the stalk of the bulb also contains neurosecretory terminals but of two kinds different from those described above (types 4 and 5 in Fig. 4). This region possesses a thin neural sheath and exocytotic profiles were evident in some terminals; therefore this area is also a neurohaemal site and has been called the 'lateral extension' of the SG (Chiang and Steel, 1985b). The perikarya of origin of these terminals have not been located; but they are clearly different from the other three as they do not stain with PAF. Thus, the bulb of the SG contains the terminals of the three previously identified NSC types, whereas an additional two types of terminal were found in the lateral extension. There is therefore a regional segregation of terminal types between the bulb and the lateral extension of the SG, with the PAF-positive terminals confined to the bulb (Fig. 4). The two regions of the SG contain a total of five ultrastructurally different types of terminal; similar numbers have been reported in various other crustacean SGs (references in Chiang and Steel, 1985b) and it has been speculated that each type may release functionally distinct neurohormones (Cooke, 1977; Kleinholz and Keller, 1979).

ELECTROPHYSIOLOGY

The primary objective of the electrophysiological analysis was to determine the timing of on-going electrical activity of the SG in almost intact animals in relation to known physiological events during a cycle of moulting. We thereby hoped to ascertain the times of neurohormone release from the SG _in situ_. It is well established for NSCs of both vertebrates and invertebrates that action potentials elicit hormone release and that the pattern and frequency of action potentials is correlated with the quantity of hormone released (see Cooke and Sullivan, 1982; Vincent and Dufy, 1982). However, this approach has rarely been used to identify naturally occurring release times in NSCs; among invertebrates, it has been employed in the mollusc _Aplysia_ (Pinsker and Dudek, 1977) and the insect _Rhodnius_ (Orchard and Steel, 1980). There have been many valuable studies of the electrophysiology of the decapod crustacean SG (see Stuenkel, this volume), but these employed isolated eyestalks deprived of their normal synaptic inputs. In the present studies, surgical interference was minimal and animals in numerous, accurately determined stages of moulting

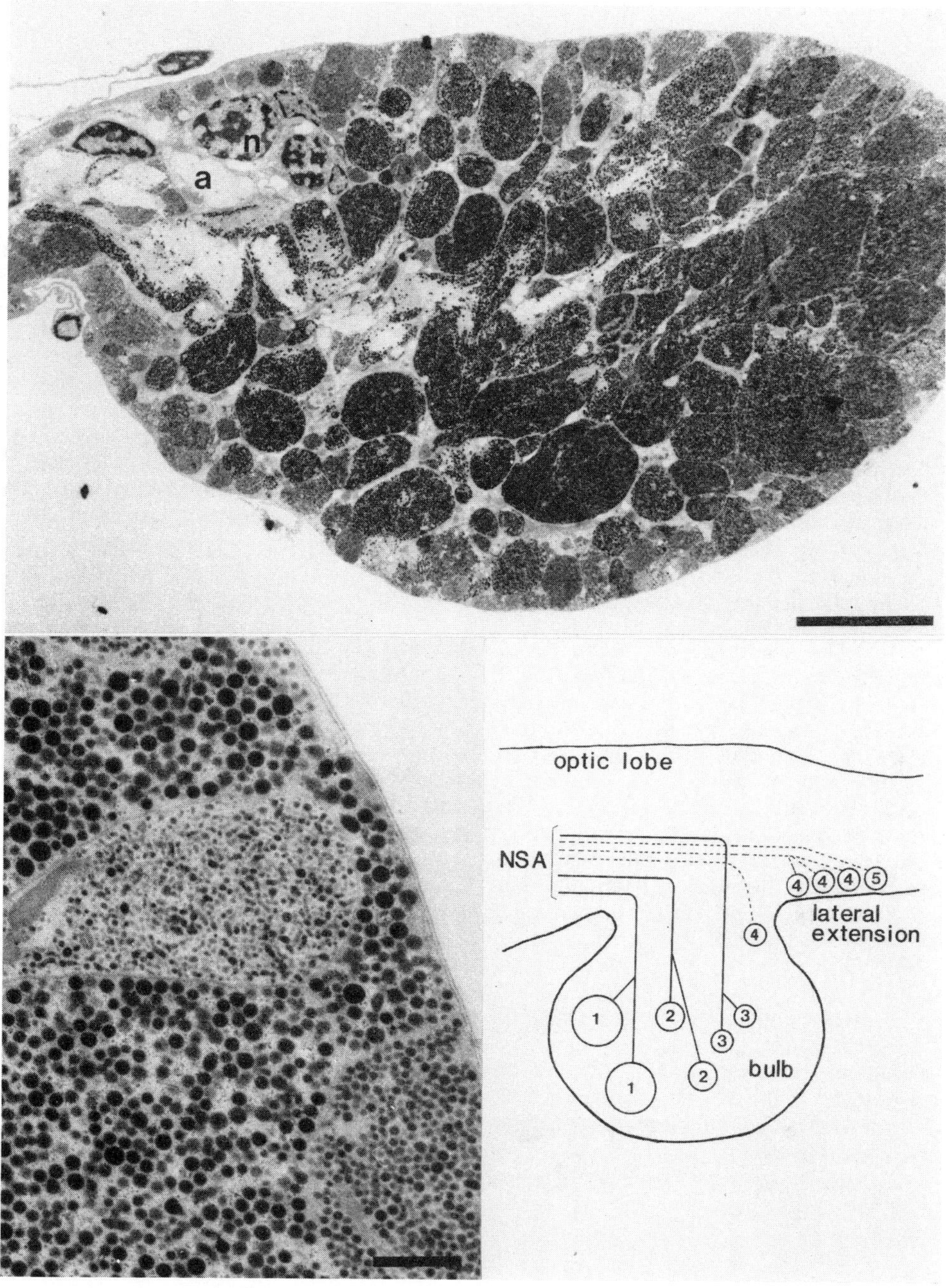

Fig. 4.   Top: the bulb of the sinus gland, showing neurosecretory terminals in the distal
region (right side). The narrow region to the left connects the gland to the optic
lobe and contains neurosecretory axons (a) and nuclei of glial cells (n). Scale
bar, 10 μm. Lower left: three distinct types of terminal occur in the gland.
Scale bar, 1 μm. Lower right: scheme of the neuroanatomical organization of
the sinus gland. Solid lines are neurosecretory axons (NSA) from perikarya in
the brain to the three types of terminal in the bulb, while dashed lines are NSA
to the two additional types of terminal in the lateral extension (from Chiang and
Steel, 1985b).

were selected by the criteria of Steel (1982). Recordings were made from restrained animals from which a small piece of exoskeleton overlying the SG was removed. A suction electrode was inserted and the entire bulb of the SG drawn into it. Such preparations survived for more than 24h.

<u>Types of SG Potentials</u>

Records of SG electrical activity revealed five classes of potential which clearly differed in both amplitude, shape and discharge pattern (for a detailed description see Chiang and Steel, 1986). All showed durations longer than those recorded simultaneously from non-neurosecretory nerve bundles (Chiang and Steel, 1986) indicating that all were neurosecretory. Moving the suction electrode away from the optic lobe without dislodging the bulb of the SG did not alter the size or shape of the three largest potential types (A, B and C) but the size of the smallest two (D and E) decreased progressively as the electrode was moved. Types A, B and C potentials therefore emanate from the bulb of the SG and types D and E from the optic lobe. Types A, B and C were found to display a small upward deflection following the main spike, such as previously seen in terminal potentials of crab SGs (Cooke et al., 1977). These deflections were attenuated in calcium-free saline (see Chiang and Steel, 1986) as expected for neurosecretory terminal potentials (Cooke and Stuenkel, 1985). Types A, B and C potentials are therefore terminal potentials in the bulb of the SG. Such a deflection was absent from D and E potentials; these appear to arise from neurosecretory axons in the lateral extension of the SG on the optic lobe (see Chiang and Steel, 1986).

<u>Structural Origins of Potentials</u>

The simplest interpretation of the finding that there are three Types of potentials from the bulb of the SG and two from the lateral extension is that these arise from the three structural types of endings in the bulb and the two from the lateral extension. This conclusion implies that each structurally distinct NSC type generates a visibly distinct type of electrical potential. Indirect evidence in support of this conclusion is derived from consideration of the relative volumes occupied by the different types of terminal within the electrode (Chiang and Steel, 1986). In essence, the relative amplitude of each potential in the SG should increase in relation to the proportion of the recording area occupied by the structure producing it. The relative areas occupied by different terminal types had been quantified at the ultrastructural level by Chiang and Steel (1985b), enabling us to propose that the large Type A potentials emanated from the large Type 1 terminals (and hence the β cells) while the smallest Type C potentials emanated from Type 3 terminals (and hence the τ cells).

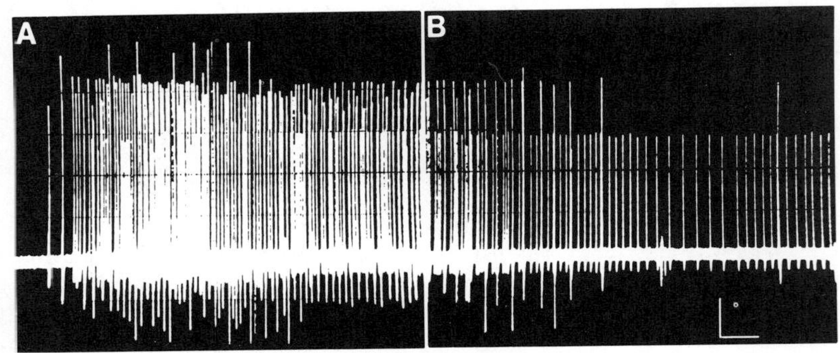

Fig. 5.   Burst of potentials recorded extracellularly from the sinus gland. A: the initial stages of the burst contain potentials of three different characteristic amplitudes. B: the later stages of the same burst contain spikes of only one amplitude. Scales, horizontal 0.5 s, vertical 75 mV (from Chiang and Steel, 1986).

An attempt was made to confirm these relationships directly by cutting through the protocerebrum in a location that severed the B and β cells and their arborizations from the SG, while leaving the τ cells intact (Chiang and Steel, 1986). The result was ambiguous. We did not find one of the three types of potential in the bulb of the SG remained active in these preparations (which could have been attributed to τ cells), but rather that all three types were eliminated. One interpretation of this result is that all three types of potential are generated by the B and β cells, implying that one of these clusters of cells is electrically heterogeneous and that the τ cells are electrically silent. A more conservative interpretation is that electrical activity in the τ cells depends on nervous input from the severed part of the protocerebrum, such that the cells became silent when it is removed. The notion that such input exists is evidenced from the synchronous onset of bursting in the three types of potential described below. It may therefore be that the neural machinery that synchronizes bursting activity between the different clusters of NSC is located in this region of the central protocerebrum.

Correspondence between the anatomy and potential type in the lateral extension is less equivocal; the two potential types (D and E) from this region evidently derive from the two classes of terminal (4 and 5) described therein.

Coupling of Electrical Activity

The on-going electrical activity of the SG consisted of intense bursts of action potentials lasting for up to two minutes, followed by a variable period of electrical silence (Chiang and Steel, 1985c). Such a bursting firing pattern is common among peptidergic neurons (see Poulain, this volume). However, a burst usually comprised several different types of potentials (Fig. 5) which are synchronized with each other. Since we interpret the potentials as due to activity in different NSCs, we infer that the brain contains a mechanism for synchronizing the discharges from anatomically distinct NSCs. Moreover, the pattern of firing of the various potential types within a burst shows a high degree of temporal order; commonly, bursts commence with rapid discharges of A, B and C potentials, but A and B potentials soon disappear from the burst leaving only a steady discharge of type C potentials. Thus, both the onset and termination of firing are synchronized between type A and B potentials, whereas type C potentials are synchronized with A and B with respect to onset but not termination. Potential types D and E behave independently of each other and of types A, B and C; type D occur in infrequent, low frequency bursts and type E occur in low frequencies for long periods with no evident bursting pattern (Chiang and Steel, 1986).

Further evidence of coupling was found when simultaneous recordings were made from both left and right SGs of the same animal (Chiang and Steel, 1985c). The large bursts were closely synchronized between the two sides. This synchrony results from coupling between the NSCs producing potential types A, B and C, which is presumably accomplished via the arborizations which traverse the midline of the brain seen in cobalt preparations. A and B potentials from the two sides frequently discharged with 1:1 correlation, indicating a strong coupling between the cells generating A and B potentials. On the other hand, type C potentials from the two sides, while synchronized in onset and termination, discharged independently from each other. Thus, there is impressive synchrony in the activity of both left and right SGs and of the component potential types within a burst. It would be intriguing to know whether this synchronous electrical activity results in the synchronous release of pulses of several hormones into the blood.

Changes in Electrical Activity During Moulting

Attention was next focused on changes in the on-going electrical activity of the SG during the moulting cycle, with the objective of elucidating the natural release times of neurohormones and their relation to known physiological events. Much of the classical endocrinology of crustacean moulting has employed decapods, in which the anatomy of the neurosecretory system is greatly altered by the presence of eyestalks, which are absent in isopods. We have carefully compared the homologies of the main neuropile regions of the brain in isopods and decapods (Chiang and Steel, 1985a) and concluded that the B and β cells represent the equivalent of the "X-organ", the main source of NSCs ending in the SG of decapods. We therefore focused interest on changes in potential types A, B and C during

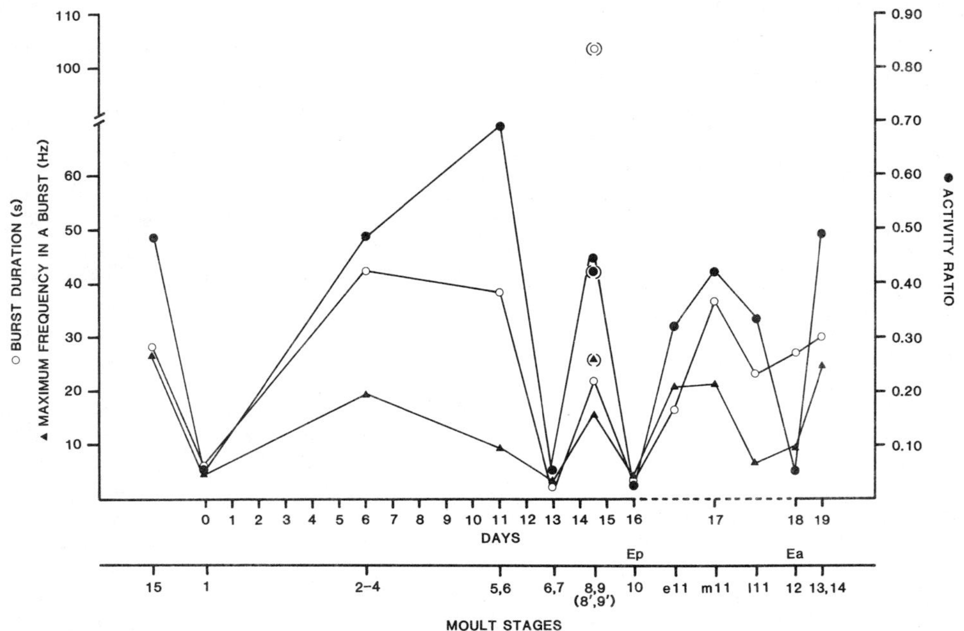

Fig. 6. Changes during the moulting cycle in the bursting firing pattern of the sinus gland. Three parameters of the pattern are displayed versus days after the natural initiation of premoult (Day 0). Morphological stages of the cycle are also shown (stages 1-15 of Steel, 1982). Ep, Ea ecdysis of posterior and anterior halves of the animal (from Chiang and Steel, 1987).

moulting. Potential types D and E were eliminated from the analysis using an amplitude window discriminator when counting potentials. Recordings were made from the SG of 3-6 animals in each moult stage and three parameters of the firing pattern were measured for each (see Chiang and Steel, 1987): average burst duration, average maximum spike frequency within a burst and the proportion of the recording time for which the SG was electrically active (the "activity ratio") (Fig. 6).

Animals in intermoult (stage 15, Fig. 6) showed high values for all these parameters, indicating massive release from the SG during intermoult. Premoult was initiated in these animals by transferring them from $4^{\circ}C$ to $21^{\circ}C$ under 8L:16D; premoult is initiated 10 days after the transfer (see Steel, 1980). At the time of premoult initiation (stage 1, Fig. 6), there is a dramatic reduction in electrical activity. Classical studies had established that moulting in decapods was regulated by a moult inhibiting hormone (MIH) from the SG whose release presumably occurs throughout intermoult. The present result is the first finding of an attenuation of release from the SG associated with the onset of premoult and strongly suggests that MIH release is inhibited at premoult initiation. However, the reduction in electrical activity occurs in all three types of potential, indicating that reduction in the release of other neurohormones also occurs at this time.

Morphological evidence that premoult has begun is not seen until several days after these endocrine changes, by which time electrical activity has begun to recover (stages 2-4, Fig. 6). By this time, the haemolymph level of moulting hormone (MH) has begun to rise significantly (C.G.H. Steel, unpublished), presumably in response to the withdrawal of inhibition by MIH. The time-course of MH increase is apparently not dependent on MIH, but may be regulated by other (neurosecretory ?) factors from the brain (Hopkins, 1983); the resumption of electrical activity during premoult could reflect the release of such factors. At 11-13 days after premoult initiation, the SG exhibits a second rapid attenuation in

214

activity (Fig. 6). All moult-related processes in the animal are proceeding rapidly at this time and it is possible that continued release of SG hormones is not needed to sustain them. It is also possible that release from the SG is actively suppressed by the rapidly rising levels of MH in the blood (C.G.H. Steel, unpublished); suppression of electrical activity and inhibition of NSC by MH have been reported in insects (see Steel, 1975) but not in crustaceans. During the last two days of premoult, the activity of the SG undergoes a brief resurgence (Fig. 6). The material released from the SG at this time could be involved in the final preparations for ecdysis, which include rapid resorption of calcium salts from the old exoskeleton, secretion of exocuticle in the posterior half of the animal and a rapid decrease in MH levels (Steel, 1982 and unpublished).

In isopods, the old exoskeleton of the posterior half of the body is shed a day before the anterior half. The SG of animals caught in the behaviour of shedding either half was quiescent, but both behaviours were each preceded and followed by high levels of activity (Fig. 6). Numerous physiological changes occur rapidly before and after each ecdysis, most notably in the storage and translocation of calcium reserves and their use in the initial calcification of the new exoskeleton (Steel, 1977). Earlier suggestions that these events were controlled by hormones from the head (Steel, 1977) are supported by the increased activity of the SG at these times.

CONCLUSIONS

Expectations were outlined in the Introduction that terrestrial isopods should prove to be good model animals for analysis of the structure and physiology of a neurosecretory system. These initial expectations have been vindicated. But in addition, many of our specific findings provide additional reasons for the use of this system as a model. Some of these are summarized below.

The total number of NSCs innervating the SG is small (about 25) and these are localized in at least three anatomically separate groups of a few cells each; each group appears to be homogeneous in cytology, ultrastructure and electrical properties, suggesting that each may also be homogeneous in function. It may therefore become possible to attribute specific functions to these groups. This possibility is increased by our finding of localized release times _in vivo_ in records of on-going electrical activity (see below).

The number and variety of terminals in the SG is also small, enabling us to attribute each structural type of terminal both to an anatomical group of NSC and to a type of electrical potential. Thus, it has been possible to demonstrate a direct correspondence between anatomical, ultrastructural and electrophysiological parameters of neurosecretion in this system.

We also found an impressive degree of synchrony in the firing of different NSCs of the same SG and between left and right SGs of the same animal. Four of the five potential types discharge naturally in bursts, but the bursts are synchronized between the different types; a burst generator providing common elements of synaptic input to different NSCs is postulated in the protocerebrum. In addition, left and right SGs are synchronized, probably via arborizations which are seen in cobalt fills to traverse the midline and end in the vicinity of the arborizations from the other side.

The on-going electrical activity of the entire SG can be monitored with a single suction electrode in near intact animals, which has enabled us to examine the natural release times during a cycle of moulting. The records contained A, B and C potentials; since these emanate from different cell types, classical reasoning would lead us to expect that these would reflect the release of three hormones. Certainly, there are three major release times during a moulting cycle. The most dramatic of these is during intermoult, which is a state presumed to be maintained by continuous inhibition of secretion of MH by release of MIH from the SG (see Skinner, 1985). We have found for the first time that the natural initiation of premoult is accompanied by a drastic reduction of release from the SG. After this reduction, the blood level of MH begins to rise and premoult becomes apparent. The second major release time accompanies ecdysis and is probably involved in the rapid movements of stored calcium salts into and out of the exoskeleton at this time; it had previously been suggested that these events were controlled by hormones from the head (Steel, 1977). The third release time occurs during the first half of premoult; there is no evidence relating directly to what hormone(s) might be released at this time. Thus, information concerning the release times from specific NSC types _in vivo_ is enabling us to begin to correlate types of cell with physiological changes in the animal and may lead to the elucidation of specific functions of neurohormones released at specific times.

ACKNOWLEDGEMENTS

Research in the authors' laboratory is supported by grant A6669 from the Natural Sciences and Engineering Research Council of Canada. Our thanks to Dr. X. Vafopoulou for escaping Mustafa in the nick of time.

REFERENCES

Chiang, R.G., and Steel, C.G.H., 1984, Neuroendocrinology of growth and moulting in terrestrial isopods, in: "The Biology of Terrestrial Isopods", S.L. Sutton and D.M. Holdrich, eds., Symp. Zool. Soc. Lond., 53:109.

Chiang, R.G., and Steel, C.G.H., 1985a, Structural organization of neurosecretory cells terminating in the sinus gland of the terrestrial isopod, Oniscus asellus, revealed by paraldehyde fuchsin and cobalt backfilling, Can. J. Zool., 63:543.

Chiang, R.G., and Steel, C.G.H., 1985b, Ultrastructure and distribution of identified neurosecretory terminals in the sinus gland of the terrestrial isopod, Oniscus asellus, Tiss. Cell, 17:405.

Chiang, R.G., and Steel, C.G.H., 1985c, Coupling of electrical activity from contralateral sinus glands, Brain Res., 331:142.

Chiang, R.G., and Steel, C.G.H., 1986, Electrical activity of the sinus gland of the terrestrial isopod, Oniscus asellus: characteristics of identified potentials recorded extracellularly from neurosecretory terminals, Brain Res., 377:83.

Chiang, R.G. and Steel, C.G.H., 1987, Changes during the moult cycle in the bursting firing pattern of the electrical activity recorded extracellularly from the sinus gland of the terrestrial isopod, Oniscus asellus, Brain Res., 402:49.

Cooke, I.M., Haylett, B.A., and Weatherby, T.M., 1977, Electrically elicited neurosecretory and electrical responses of the isolated crab isinus gland in normal and reduced calcium salines, J. exp. Biol., 70:125.

Cooke, I.M., and Stuenkel, E.L., 1985, Electrophysiology of invertebrate neurosecretory cells, pp. 115-164, in: "The Electrophysiology of the Secretory Cell", A.M. Poisner and J.M. Trifaro, eds., Elsevier Science Publishers.

Cooke, I.M., and Sullivan, R.E., 1982, Hormones and neurosecretion, pp. 205-391, in: "The Biology of Crustacea", vol. 3, H.L. Atwood and D.C. Sandeman, eds., Academic Press, New York.

Gabe, M., 1952, Sur l'existence d'un cycle secretoire dans la glande du sinus (organe pseudofrontale) chez Onisus asellus L, C. R. Acad. Sci. (Paris), 235:900.

Heeley, W., 1941, Observations on the life histories of some terrestrial isopods, Proc. Zool. Soc. Lond. Ser. B., 111:79.

Hopkins, P.M., 1983, Patterns of serum ecdysteroids during induced and uninduced proecdysis in the fiddler crab, Uca pugilator, Gen. comp. Endocrinol., 52:350.

Kleinholz, L.H., and Keller, R., 1979, Endocrine regulation in crustacea, pp. 159-213, in: "Hormones and Evolution", vol. 1, E.J.W. Barrington, ed., Academic Press, New York.

Martin, G., 1972, Analyse ultrastruturale des cellules neurosecretrices du protocerebron de Porcellio dilatatus (Brandt) (Crustace, Isopode, Oniscoide), C. R. Acad. Sci., (Paris), 274:243.

Matsumoto, K., 1959, Neurosecretory cells of an isopod, Armadillidium vulgare (Latreille), Biol. J. Okayama Univ., 5:43.

Orchard, I., and Steel, C.G.H,. 1980, Electrical activity of neurosecretory axons from the brain of Rhodnius prolixus: relation of changes in the pattern of activity to endocrine events during the moulting cycle, Brain Res., 191:53.

Passano, L.M., 1960, Molting and its control, pp. 473-536, in: "The Physiology of Crustacea", vol. 1, T.H. Waterman, ed., Academic Press, New York.

Pinsker, H.M., and Dudek, E.F., 1977, Bag cell control of egg laying in freely behaving Aplysia, Science, 197:490.

Schobl, J., 1880, Die Fortpflanzung isopoder Crustaceen, Ark. Mikr. Anat. Bonn, 17:125.

Skinner, D.M., 1985, Interacting factors in the control of the crustacean molt cycle, Am. Zool., 25:275.

Steel, C.G.H., 1975, A neuroendocrine feedback mechanism in the insect moulting cycle, Nature, 253:267.

Steel, C.G.H., 1977, Cuticle hardening by calcification and its hormonal control in isopod crustacea, Am. Zool., 17:899.

Steel, C.G.H., 1977, The neurosecretory system in the aphid Megoura viciae with reference to unusual features associated with long distance transport of neurosecretion, Gen. comp. Endocrinol., 31:307.

Steel, C.G.H., 1980, Mechanism of coordination between moulting and reproduction in terrestrial isopod Crustacea, Biol. Bull., 159:206.

Steel, C.G.H., 1982, Stages of the intermoult cycle in the terrestrial isopod Oniscus asellus and their relation to biphasic cuticle secretion, Can. J. Zool., 60:429.

Vincent, J.-D., and Dufy, B., 1982, Electrophysiological correlates of secretion in endocrine cells, pp. 107-145, in: "Cellular Regulation of Secretion and Release", P.M. Conn, ed., Academic Press, New York.

Walker, R., 1935, The central nervous system of Oniscus (Isopoda), J. comp. Neurol., 62:197.

INTRINSIC AND SYNAPTIC FACTORS REGULATING MAMMALIAN
MAGNOCELLULAR NEUROSECRETORY NEURON ACTIVITY

Leo P. Renaud

Neurosciences Unit,
Montreal General Hospital and McGill University
Montreal, Quebec, Canada

INTRODUCTION

The magnocellular oxytocin-synthesizing and vasopressin-synthesizing neurons of the hypothalamic supraoptic, paraventricular and accessory magnocellular nuclei are the classical mammalian neurosecretory neurons. Although research on these special endocrine neurons continues to provide increasing information on their morphology, peptide synthesis and neurosecretory mechanisms in the neurohypophysis, data has also been rapidly accumulating on their intrinsic membrane properties and synaptic inputs. The initial part of this chapter will report intracellular current- and voltage-clamp data which illustrate several intrinsic membrane conductances of supraoptic neurons magnocellular neurosecretory cells (magnocellular neurons); the second part will focus on a few samples of the synaptic inputs that can directly influence their activity.

INTRINSIC MEMBRANE PROPERTIES

Passive Properties

Owing to their relatively small dimensions and proximity to major blood vessels, stable intracellular recordings from magnocellular neurons have necessitated the use of <u>in vitro</u> slice (Mason, 1980; Dudek et al., 1980) explant (Bourque and Renaud, 1983), tissue or organ cultured preparations (Gahwiler and Dreifuss, 1979; Legendre et al., 1982). Magnocellular neurons typically have high input impedances (range 50-300 MΩ) with membrane time constants of 9-15msec. Voltage-current relationships reveal linearity near resting membrane potentials and a small but variable inward rectification at more hyperpolarized levels. On the other hand, depolarizing currents near and above threshold reveal a strong outward rectification that is calcium dependent (Bourque and Renaud, 1985a; Bourque, 1986).

Serial reconstruction of magnocellular neurons after Lucifer yellow injection reveals a simple morphology of one to three main dendrites that have minimal branches but multiple spinous processes, and extend within the long axis of the supraoptic neurons for 30-725 μm (Randle et al., 1986d). The electrotonic voltage response of a magnocellular neurone to a current pulse displays an exponential time course from which only one time constant can be resolved. This would suggest that the neurons behave as single isopotential compartments and can be adequately spaceclamped, an issue that is of major importance in the interpretation of voltage-clamp data (see Bourque, 1987d).

<u>Action potentials</u>

Action potentials are composed of both a low threshold sodium conductance and a high threshold tetrodotoxin-resistant calcium conductance, the latter contributing to a distinct shoulder on the recovery phase of the action potential (Bourque and Renaud, 1985a). Action potential durations can vary substantially depending on firing frequency, from a mean of 1.75ms during relative quiescence (i.e. 0.5 Hz) to a mean of 2.68ms at frequencies of 20 Hz (Bourque and Renaud, 1985b). Moreover, spike duration has a time and pattern dependency. Action potential broadening occurs progressively during the initial 15-20 spikes at the onset of spontaneous or current-induced bursts, and remains in this prolonged state until the end of the burst. This process is accelerated at the onset of the burst by the occurrence of several spikes with short interspike intervals, prior to spike frequency accommodation. The return of action potentials of shorter duration is progressive during a silent period, with a time constant of approximately 5s. Frequency-dependent spike broadening can be reversibly decreased or blocked by depletion of calcium from the media or by cobalt, cadmium or manganese. Thus,the phenomenon may reflect facilitated calcium entry per spike and/or a frequency-dependent modulation of a potassium conductance with subsequent enhanced calcium entry per spike. Interestingly, a frequency-dependent broadening of compound action potentials can also be detected in the terminals of neurosecretory neurons (Gainer et al., 1986). What is not clear is the relationship, if any, of this phenomenon to the facilitation (or fatigue) of hormone release from isolated neural lobe terminals when they are stimulated in a phasic rather than a continuous mode (ses Shaw et al., 1984).

<u>Afterpotentials</u>

The repolarization phase of action potentials merges with a prominent hyperpolarizing afterpotential (Andrew and Dudek, 1984a; Bourque et al., 1985). The features of this afterpotential (resistance to synaptic blockade or intracellular chloride iontophoresis, sensitivity to calcium channel blockers, association with an increase in membrane conductance, reversal potential of approx. -75 mV which shifts with extracellular K^+ concentration) indicate the activation of a calcium-dependent potassium conductance. The time course of this conductance corresponds closely to the "silent period" noted to follow suprathreshold antidromic activation of the neurons, and is likely to contribute to the phenomenon previously ascribed to are current synaptic inhibitory mechanism (Barker et al., 1971; Dreifuss and Kelly, 1972). During depolarizing current pulses, evoked action potential trains demonstrate a progressive spike frequency adaptation associated with potentiation of successive hyperpolarizing afterpotentials, an effect that is blocked by calcium channel blockers (Bourque et al., 1985).

<u>In vivo</u> extracellular recordings from putative oxytocin secreting magnocellular neurons in lactating suckled rats reveal intermittent high frequency (up to 60 Hz) bursts of activation prior to milk ejection (Lincoln and Wakerley, 1974). Typically each burst lasts 1-2s and is followed by a silent interval lasting for several seconds. Similar phenomena follow repetitive antidromic activation (Dreifuss and Kelly, 1972). During intracellular recordings <u>in vitro</u>, current-evoked bursts of action potentials are followed by post-burst protracted afterhyperpolarizations and associated reductions in membrane excitability corresponding to the extracellularly observed silent interval (Andrew and Dudek, 1984a; Bourque et al., 1985). An underlying calcium-activated potassium conductance is also deemed responsible for this event.

Evidently there are at least two different calcium-dependent potassium channels in magnocellular neurons. Bourque (1987d) has recently described the properties of a transient calcium-dependent outward current which is unaffected by tetraethylammonium but is sensitive to agents (eg 4-aminopyridine, dendrotoxin) known to block I_A in other neurons. The kinetics and voltage-dependency of this transient calcium-dependent outward current suggest a role in both spike repolarization and the peak and <u>initial</u> phase of the hyperpolarizing afterpotential. The calcium-dependent potassium conductance which underlies the <u>prolonged</u> afterhyperpolarizations, and spike frequency accommodation during a burst, is markedly sensitive to apamin (a bee venom toxin) and d-tubocurarine (Bourque and Brown, 1987). The actions of apamin to reduce spike afterhyperpolarizations often unmask or enhance a subsequent depolarizing afterpotential which contributes to bursting and regenerative spiking behaviour similar to that detected in the presence of noradrenalin (see below). The supraoptic neurons contains high density binding for apamin (Mourre et al., 1986). Other endogenous apamine-like ligands appear to be available to activate these receptors (see Fossett et al., 1984).

In many magnocellular neurons, their hyperpolarizing afterpotential is followed by a depolarizing after potential (Andrew and Dudek, 1983, 1984b). Subsequent current- and voltage-clamp studies have indicated that this depolarizing afterpotential exhibits a marked voltage- and calcium-dependency (Bourque, 1986, 1987b). Moreover, activation of the current underlying the depolarizing afterpotential imparts a narrow region of negative resistance in their current-voltage relationship between -60 and -65 mV i.e. close to spike threshold. Depolarizing afterpotentials generated by successive action potentials occurring at short interspike intervals can summate and establish a plateau potential that will sustain regenerative firing (Andrew and Dudek, 1983). Under these conditions, regenerative inward current further depolarizes the cell towards threshold and continued activity.

Phasic Bursting

Extracellular _in vivo_ recordings identify putative vasopressin-synthesizing cells by their unique phasic firing pattern (see Renaud et al., 1985 for review). _In vitro_ staining has confirmed that the ability to produce phasic firing is indeed associated with vasopressin rather than oxytocin-synthesizing cells (Yamashita et al., 1983; Cobbett et al., 1986). Intrinsic membrane conductances that involve a voltage and calcium sensitive pacemaker mechanism, rather than patterned synaptic input, underly phasic bursting in magnocellular neurons (Andrew and Dudek, 1983; 1984b; Bourque, 1986, 1987a,b; Andrew, 1987a). This mechanism can, however, be modulated by extrinsic (synaptic or osmotic) influences (Bourque, 1987a; Andrew, 1987b). The actual onset of a phasic burst may depend on any combination of events that facilitates the achievement of threshold for spike generation. For instance, random excitatory postsynaptic potentials may initiate action potentials whose depolarizing afterpotentials summate, establish a plateau potential and thereby sustain regenerative firing (Andrew et al., 1983). Since magnocellular neurons demonstrate voltage dependent, non-synaptic depolarizing potentials near threshold, these may also trigger action potentials (Bourque et al., 1986b). These non-synaptic depolarizing potentials are sensitive to calcium channel blockers and are presumed to represent openings of channels that carry inward current. A less consistent feature of magnocellular neurons is a calcium-dependent slow depolarization that progressively activates during the silent interval (Andrew, 1987a). In any event, once initiated, the burst is sustained by a plateau potential that will sustain firing at 7-15 Hz for a few seconds or for several minutes, depending in part on the external depolarizing drive on the neuron. As mentioned above, both the depolarizing afterpotentials and plateau forming mechanism are calcium- and voltage-dependent. At the termination of a phasic burst, the cell typically stops firing followed by the collapse of the plateau potential, apparently due to the progressive calcium inactivation of the underlying calcium conductance (Bourque et al., 1986a; Bourque, 1987a,b). A silent period may also be initiated by a hyperpolarizing event eg. current pulse, or one or more inhibitory postsynaptic potentials that will tend to remove the membrane potential from the zone of activation.

Osmosensitivity

In the rat, increasing plasma osmotic pressure potently stimulates release of both vasopressin and oxytocin into the circulating plasma. The mechanisms for osmoreception appear to depend largely on several structures located within the so-called anteroventral third ventricle (AV3V) area. Supraoptic neurons display osmotic-induced depolarizations, even in the presence of synaptic blockade (Mason, 1980). Under voltage-clamp, supraoptic neurons reveal a sustained inward current which is accompanied by an increase in membrane conductance (Bourque, 1987c). It therefore appears that neurons possess an endogenous osmosensitivity. However, magnocellular neurons _in vivo_ alter their firing frequency following microinfusions of hyper- or hypo-osmotic media into the AV3V region (Honda et al., 1987), thereby indicating the presence of synaptic components in the central osmoregulatory process.

SYNAPTIC INPUTS TO MAGNOCELLULAR NEURONS

Exposure of magnocellular neurons to an array of neurotransmitters and neuropeptides results in an alteration in their excitability (see Renaud et al., 1985 for review), presumably reflecting the nature of their afferent innervation and corresponding membrane receptors. Many of these inputs can be visualized with retrograde and anterograde tracers and their chemical composition determined with immunohistochemical methods (Swanson and Sawchenko, 1983; Tribollet et al., 1985). The neurophysiological assessment of specific afferent pathways although far from complete has yielded interesting insights into the complex neural circuitry that engages magnocellular neurons in the rat. In most instances, the initial investigations have been conducted _in vivo_ where extracellular recordings from the cells can be verified on the basis of their antidromic activation from the neural lobe, and further distinguished as originating from oxytocin-secreting or vasopressin-secreting cells on the basis of their firing patterns and response to a brief drug-induced rise in mean arterial pressure. Local circuits have been further examined in slice and explant preparations using both extra-and intracellular recordings.

Catecholamines

All four magnocellular nuclei receive dense catecholamine innervation. In part this is due to dopaminergic fibers, possibly arising from neurons in the ventral tegmental region (Buijs et al., 1984) and one _in vitro_ study suggests that dopamine selectively enhances the firing of a portion of the oxytocin cells (Mason, 1983). However, the majority of these catecholamine fibers are noradrenergic and originate mainly from the caudal ventrolateral medulla A1 cell group, with a smaller contribution from the dorsomedial medulla A2 cell group (Swanson and Sawchenko, 1983). Considerable controversy has existed on the role of noradrenaline in regulating vasopressin release, although the current data consistently indicate that this transmitter enhances the excitability of the cells (Day et al., 1985; Randle et al., 1986a) and promotes hormone release both _in vivo_ (Willoughby et al., 1987) and _in vitro_ (Randle et al., 1986c) through an α_1 receptor mechanism. This is in agreement with the observed excitatory responses of the supraoptic neurons that follow electrical stimulation of the A1 and A2 regions (Day and Renaud, 1984; Raby and Renaud, 1987). It is also of developmental interest that A1 inputs are selective for vasopressin cells (Day and Renaud, 1984). Natural stimuli for the activation of these endogenous A1 and A2 inputs to magnocellular neurons include carotid chemoreceptors (Harris, 1979) renal nociceptors (Day and Ciriello, 1987) hypotension and unloading of peripheral baroreceptors (McAllen and Blessing, 1987).

During intracellular recordings in perfused hypothalamic explants, supraoptic cells respond to 10-60 μM noradrenalin with a gradual voltage-dependent membrane depolarization and emergence of bursting and phasic discharges (Randle et al., 1986a). Noradrenaline can be seen to shorten the duration of the hyperpolarizing afterpotential and enhance the magnitude of the depolarizing afterpotential. These features closely resemble the responses of the neurons to apamin (Bourque, 1987e) and strongly argue for the actions of noradrenaline to reduce a calcium-dependent potassium conductance which is somewhat similar in other ways to I_A.

Subfornical organ and AV3V

The magnocellular nuclei are innervated by neurons located in the AV3V area and the subfornical organ, regions known to participate in body fluid homeostasis and blood pressure regulation (Miselis, 1980). Electrical stimulation of the subfornical organ evokes a unique long duration excitation of both vasopressin- and oxytocin-secreting cells (Sgro et al., 1984) and a corresponding increase in plasma levels of these hormones (Ferguson, 1987). Circulating angiotensin II acting at the subfornical organ increases the excitability of magnocellular neurons (Ferguson and Renaud, 1986). Angiotensin II is also detected in subfornical organ pathways to all four nuclei (Lind et al., 1984). Evidence that saralasin blocks the central excitatory actions of subfornical organ on the supraoptic cells (Jhamandas and Renaud, 1987) favours the possibility that angiotensin II has a neurotransmitter role in this pathway.

<u>GABAergic afferents</u>

Two features link the diagonal band of Broca, lateral septum and amygdala with magnocellular neurons: anatomical tracer studies indicate that their neurons send axons to the perinuclear zone (rather than the interior) of the nuclei, and electrical stimulation in these sites more often depresses (rather than excites) the cells (Renaud, 1987). Certain diagonal band of Broca neurons with supraoptic connections are sensitive to baroreceptor activation (Jhamandas and Renaud, 1986a) and this input is particularly directed towards depressing the firing of vasopressin-secreting neurons (Jhamandas and Renaud, 1986b). Both _in vivo_ (Jhamandas and Renaud, 1986b, 1987) and _in vitro_ data (Randle et al., 1986b; Randle and Renaud, 1987) indicate that this depressant action is mediated through GABA-A receptors. These observations, together with recent ultrastructural evidence that more than 50% of the synapses on the magnocellular cells demonstrate GABA-like immunoreactivity (Buijs et al., 1987), attest to a prominent role for this transmitter in regulating excitability.

SUMMARY

Recent _in vitro_ intracellular current and voltage clamp data reveal that magnocellular cells, similar to other central neurons, contain an array of intrinsic membrane conductances that influence action potential shape, duration and activity patterns. Examples of voltage- and ligand (transmitter)-activated conductances are now available, and further features are expected. Several anatomically- and (in certain instances) chemically-defined inputs to magnocellular neurons have been identified. Electrophysiological studies indicate that brainstem catecholamine and subfornical organ inputs enhance excitability, whereas diagonal band and septal-amygdalar inputs engage perinuclear GABAergic neurons to depress excitability. These data are a mere reflection of the exciting events to come as the scope of intrinsic and synaptic factors that influence MNC excitability are elaborated.

I wish to acknowledge the financial assistance of the Canadian MRC and Quebec Heart Foundation, and typographical aid of Gwen Peard.

REFERENCES

Andrew, R.D., 1987a, Endogenous bursting by rat supraoptic neuroendocrine cells is calcium dependent, J. Physiol. Lond., 384:451.

Andrew, R.D., 1987b, Isoperiodic bursting by magnocellular neuroendocrine cells in the rat hypothalamic slice, J. Physiol. Lond., 384:467.

Andrew R.D. and Dudek, F.E., 1983, Burst discharge in mammalian neuroendocrine cells involves an intrinsic regenerative mechanism, Science, 221:1050.

Andrew, R.C. and Dudek, F.E., 1984a, Intrinsic inhibition in magnocellular neuroendocrine cells of rat hypothalamus, J. Physiol. Lond., 353:171.

Andrew, R.C. and Dudek, F.E., 1984b, Analysis of intra-cellularly recorded phasic bursting by mammalian neuroendocrine cells, J. Neurophysiol., 51:552.

Barker, J.L., Crayton, J.W. and Nicoll, R.A., 1971, Anti-dromic and orthodromic responses of paraventricular and supraoptic neurosecretory cells, Brain Res., 33:353.

Bourque, C.W., 1986, Calcium-dependent spike after-current induces burst firing in magnocellular neurosecretory cells, Neurosci. Lett., 70:204.

Bourque, C.W., 1987a, Intrinsic features and control of phasic burst onset in magnocellular neurosecretory cells in: "Organization of the Autonomic Nervous System: central and peripheral mechanisms," J. Ciriello, F.R. Calaresu, L.P. Renaud and C. Polosa, eds., A. Liss, New York.

Bourque, C.W., 1987b, Current- and voltage-clamp studies of transient and pacemaker currents in neurosecretory neurons of the supraoptic nucleus. in: "Inactivation of Hypersensitive Neurons,"M. Chalazonitis and E. Gola, eds., A. Liss, New York.

Bourque, C.W., 1987c, Osmotic induction of burst firing in magnocellular neuroendocrine cells: in vitro analysis using perfused hypothalamic explants, Neurosci. Lett., Suppl. 29:S16.

Bourque, C.W., 1987d, Transient calcium-dependent potassium current in magnocellular neurosecretory cells of the rat supraoptic nucleus, J. Physiol. Lond., in press.

Bourque, C.W., 1987e, Apamin and d-tubocurarine block the after hyperpolarization of rat supraoptic neurosecretory neurons, <u>Neurosci. Lett.</u>, in press.

Bourque, C.W., Brown, D.A. and Renaud, L.P., 1986a, Bariumions induce prolonged plateau depolarizations in neurosecretory neurones of the adult rat supraoptic nucleus, <u>J. Physiol. Lond.</u>, 365:573.

Bourque, C.W., Randle, J.C.R. and Renaud, L.P., 1985, Calcium-dependent potassium conductance in rat supraoptic nucleus neurosecretory neurons, <u>J. Neurophysiol.</u>,. 54:1375.

Bourque, C.W., Randle, J.C.R. and Renaud, L.P., 1986b, Non-synaptic depolarizing potentials in rat supraoptic neurones recorded in vitro., <u>J. Physiol. Lond.</u>, 376:493.

Bourque, C.W. and Renaud, L.P., 1983, A perfused in vitro preparation of hypothalamus for electrophysiological studies on neurosecretory neurons, <u>J. Neurosci. Meth.</u>, 71:203.

Bourque, C.W. and Renaud, L.P., 1985a, Calcium dependent action potentials in rat supraoptic neurosecretory neurones recorded in vitro, <u>J. Physiol. Lond.</u>, 363:419.

Bourque, C.W. and Renaud, L.P., 1985b, Activity dependence of action potential duration in rat supraoptic neurosecretory neurones recorded in vitro, <u>J. Physiol. Lond.</u>, 363:429.

Buijs, R.M., Geffard, M., Pool, C.W. and Hoorneman, E.M.D., 1984, The dopaminergic innervation of the supraoptic and paraventricular nucleus. A light and electron microscopical study, <u>Brain Res.</u>, 323:65.

Buijs, R.M., van Vulpen, E.H.S. and Geffard, M., 1987, Ultrastructural localization of GABA in the supraoptic nucleus and neural lobe, <u>Neuroscience</u>, 20:347.

Cobbett, P., Smithson, K.G. and Hatton, G.I., 1986, Immuno-reactivity to vasopressin- but not oxytocin-associated neurophysin antiserum in phasic neurons of rat hypothalamic paraventricular nucleus, <u>Brain Res.</u>, 362:7.

Day, T.A., and Ciriello, J., 1987, Effects of renal receptor activation on neurosecretory vasopressin cells, <u>Am. J. Physiol.</u>, 253:R234.

Day, T.A., Randle, J.C.R. and Renaud, L.P., 1985, Opposing α- and β-adrenergic mechanisms mediate dose- dependent actions of noradrenaline on supraoptic vasopressin neurones in vivo, <u>Brain Res.</u>, 358:171.

Day, T.A. and Renaud, L.P., 1984, Electrophysiological evidence that noradrenergic afferents selectively facilitate the activity of supraoptic vasopressin neurons, <u>Brain Res.</u>, 303:233.

Dreifuss, J.J. and Kelly, J.S., 1972, Recurrent inhibition of antidromically identified rat supraoptic neurones, <u>J. Physiol. Lond.</u>, 220:87.

Dudek, F.E., Hatton, G.I. and MacVicar, B.A., 1980, Intracellular recordings from the paraventricular nucleus in slices of rat hypothalamus, <u>J. Physiol. Lond.</u>, 301:101.

Ferguson, A.V. 1987, The subfornical organ: a central integrator in the control of neurohypophysial hormone secretion, <u>in</u>: "Organization of the Autonomic Nervous System: central and peripheral mechanisms", J. Ciriello, F.R. Calaresu, L.P. Renaud and C. Polosa eds. A. Liss, New York.

Ferguson, A.V. and Renaud, L.P., 1986, Systemic angiotensin acts at subfornical organ to facilitate activity of neurohypophysial neurons, <u>Am. J. Physiol.</u>, 251:R712.

Fossett, M., Schmid-Antomarchi, H., Hugues, M., Romey, G. and Lazdunski, M., (1984), The presence in pig brain of an endogenous equivalent of apamin, the bee venom peptide that specifically blocks CA^{2+} dependent K^+ channels, <u>Proc. Nat. Acad. Sci. USA.</u>, 81:7228.

Gahwiler, B.H. and Dreifuss, J.J., 1979, Phasically firing neurones in long-term cultures of the rat hypothalamic supraoptic area: pacemaker and follower cells, <u>Brain Res.</u>, 177:95.

Gainer, H., Wolfe, Jr., S.A., Obaid, A.L. and Salzberg, B.M., 1986, Action potentials and frequency-dependent secretion in the mouse neurohypophysis, <u>Neuroendocrinology</u>, 43:557.

Harris, M.C., 1979, The effect of chemoreceptor and baroreceptor stimulation on the discharge of hypothalamic supraoptic neurones in rat, <u>J. Endocrinol.</u>, 82:115.

Honda, K., Negoro, H., Higuchi, T. and Tadokoro, Y. 1987, Activation of neurosecretory cells by osmotic stimulation of anteroventral third ventricle, <u>Am. J. Physiol.</u>, 252:R1039.

Jhamandas, J.H. and Renaud, L.P., 1986a, Diagonal band neurons may mediate arterial baroreceptor input to hypothalamic vasopressin secreting neurons, <u>Neurosci. Lett.</u>, 65:214.

Jhamandas, J.H. and Renaud, L.P., 1986b, A τ-aminobutyric-acid-mediated baroreceptor input to supraoptic vasopressin neurones in the rat, J. Physiol. Lond., 381:595.

Jhamandas, J.H. and Renaud, L.P., 1987a, Saralasin diminishes subfornical organ-evoked excitation of hypothalamic supraoptic neurosecretory neurons, Can. J. Physiol. Pharm,. 65:Axvii.

Jhamandas, J.H. and Renaud, L.P., 1987, Neurophysiology of a central baroreceptor pathway projecting to hypothalamic vasopressin neurons, Can. J. Neurol. Sci., 14:17.

Legendre, P., Cooke, I.M. and Vincent, J.D., 1982, Regenerative responses of long duration recorded intracellularly from dispersed cell cultures of fetal mouse hypothalamus, J. Neurophysiol., 48:1121.

Lincoln, D.W. and Wakerley, J.B., 1974., Electrophysiological evidence for the activation of supraoptic neurones during the release of oxytocin. J. Physiol. Lond., 242:533.

Lind, R.W., Swanson, L.W. and Ganten, D., 1984, Angiotensin II immunoreactivity in the neural afferents and efferents of the subfornical organ of the rat, Brain Res., 321:209.

Mason, W.T., 1980, Supraoptic neurones of rat hypothalamus are osmosensitive, Nature, 287:154.

Mason, W.T., 1983, Excitation by dopamine of putative oxytocinergic neurones in the rat supraoptic nucleus in vitro: evidence for two classes of continuously firing neurones, Brain Res., 267:113.

McAllen, R.M. and Blessing, W.W., 1987, Neurons (presumably A_1 cells) projecting from the caudal ventrolateral medulla to the region of the supraoptic nucleus respond to baroreceptor inputs in the rabbit, Neurosci. Lett., 73:247.

Miselis, R., 1981., The efferent projections of the subfornical organ of the rat: a circumventricular organ with a neural network subserving water balance, Brain Res., 230:1.

Mourre, C., Hugues, M. and Lazdunski, M., 1986, Quantitative autoradiographic mapping in rat brain of the receptor of apamin, a polypeptide toxin specific for one class of Ca^{2+}-dependent K^+ channels, Brain Res., 382:239.

Raby, W. and Renaud, L.P., 1987, Characterization of a norepinephrine pathway from dorsomedial medulla (A2) to hypothalamic supraoptic nucleus in the rat, Can. J. Physiol. Pharm., 65:Axxviii.

Randle, J.C.R., Bourque, C.W. and Renaud, L.P., 1986a, α_1 adrenergic receptor activation depolarizes rat supraoptic neurosecretory neurons in vitro. Am. J. Physiol., 251:R569.

Randle, J.C.R., Bourque, C.W. and Renaud, L.P. 1986b, Characterization of spontaneous and evoked inhibitory postsynaptic potentials in rat supraoptic neurosecretory neurons in vitro, J. Neurophysiol., 56:1703.

Randle, J.C.R., Mazurek, M., Kneifel, D., Dufresne, J. and Renaud, L.P., 1986c, α_1 adrenergic receptor activation releases vasopressin and oxytocin from perfused hypothalamic explants, Neurosci. Lett., 65:219.

Randle, J.C.R., Bourque, C.W. and Renaud, L.P., 1986d, Serial reconstruction of Lucifer yellow-labelled supraoptic nucleus neurons in perfused rat hypothalamic explants, Neuroscience, 17:453.

Randle, J.C.R. and Renaud, L.P., 1987, Actions of τ-aminobutyric acid on rat supraoptic nucleus neurosecretory neurones in vitro, J. Physiol. Lond.,387:629.

Renaud, L.P., 1987, Magnocellular neuroendocrine neurons: update on intrinsic properties, synaptic inputs and neuropharmacology, TINS, in press.

Renaud, L.P., Bourque, C.W., Day, T.A., Ferguson, A.V. and Randle, J.C.R., 1985, Electrophysiology of mammalian hypothalamic supraoptic and paraventricular neurosecretory cells, in: "The Electrophysiology of the Secretory Cell," A. Poisner and J. Trifaro, eds., Elsevier, Amsterdam.

Sgro, S., Ferguson, A.V. and Renaud, L.P., 1984, Subfornical organ-supraoptic nucleus connections: An electro-physiological study in the rat, Brain Res., 303:7.

Shaw, F.D., Bicknell, R.J. and Dyball, R.E.J., 1984, Facilitation of vasopressin release from the neurohypophysis by application of electrical stimuli in bursts, Neuroendocrinology, 39:271.

Swanson, L.W. and Sawchenko, P.E., 1983, Hypothalamic Integration: organization of the paraventricular and supraoptic nuclei, A. Rev. Neurosci., 6:269.

Tribollet, E., Armstrong, W.E., Dubois-Dauphin, M. and Dreifuss, J.J., 1985, Extra-hypothalamic afferent inputs to the supraoptic nucleus area of the rat as determined by retrograde and anterograde tracer techniques, Neuroscience, 15:135.

Willoughby, J.O., Jervois, P.M., Menadue, M.F. and Blessing, W.W., 1987, Noradrenaline, by activation of alpha-1 adrenoreceptors in the region of the supraoptic nucleus, causes secretion of vasopressin in the unanaesthetized rat, Neuroendocrinology, 45:219.

Yamashita, H., Inenaga, K., Kawata, M. and Sano, Y., 1983, Phasically firing neurons in the supraoptic nucleus of the rat hypothalamus: immunocytochemical and electrophysiological studies, Neurosci. Lett., 37:87.

INTRINSIC ELECTROPHYSIOLOGICAL REGULATION OF FIRING PATTERNS OF BURSTING NEURONS IN APLYSIA

Robert S. Zucker

Physiology-Anatomy Department
University of California
Berkeley
CA 94720 U.S.A.

INTRODUCTION

Endogenous generation of action potential bursts occurs commonly among peptidergic neurosecretory cells (Barker and Gainer, 1975; Gähwiler and Dreifuss, 1977; Cooke, 1982; Andrew and Dudek, 1984). Burst generation has also been observed in more typical central neurons (Miller and Selverston, 1982; Grace and Bunney, 1984; Legendre et al., 1985), indicating that it is a property of many kinds of neurons. It is easily imagined that burst generation functions to enhance transmitter release, which by synaptic facilitation may be particularly sensitive to spike frequency, but it has been difficult to prove this point (Dutton and Dyball, 1979). Bursting neurons have also been implicated in the organization of rhythmic motor patterns (Miller and Selverston, 1982).

MEMBRANE CURRENTS REGULATING BURSTING

The origin of repeated spike bursts in a neuron isolated from all synaptic input has intrigued neurobiologists for a long time. The most extensive studies of this process have been on molluscan neurons, particularly cells R15 and L2-L6 in the abdominal ganglion of Aplysia californica. Many voltage-clamp studies (reviewed in Adams and Benson, 1985) have revealed a complex array of membrane currents which are involved in generating spike bursts. These currents flow through different channels from the usual voltage-dependent sodium and potassium channels involved in shaping each action potential. I shall summarize the most important of these currents involved in burst generation in Aplysia, and the evidence for our interpretation of the responsible ionic mechanisms.

The Pacemaker Current

Bursting neurons are endowed with an inward current that is partly activated at rest (i.e., at potentials prevalent in the interval between bursts). Although not a strong current, it is nevertheless the dominant current at potentials near -40 mV, leading to a negative slope in the steady-state current-voltage relation of bursting neurons. It is often called a "negative slope resistance current." As the dominant current leading to a weak net inward current, it slowly depolarizes cells until they reach spike threshold and begin to fire a burst. It is often referred to as the "slow inward current," not because its kinetics are slow, but because its effect on charging the membrane capacitance and depolarizing the neuron accumulates slowly.

The most comprehensive and coherent study of this current's role in burst generation is that of Gorman et al., (1982). We have repeated and expanded many of their observations (Kramer and Zucker, 1985b). The pacemaker current activates rapidly with depolarization, and deactivates rapidly with hyperpolarization. It is partially activated (and therefore incompletely inactivated) at "resting potentials." In addition to rapidly activating components (Chad et al., 1984), components with slower kinetics have been observed (Eckert and Lux, 1976; Kramer and Zucker, 1985b), suggesting the involvement of more than one type of channel (Chesnoy-Marchais, 1985)

The current through these channels is blocked by calcium antagonists (divalent ions such as Co^{2+} and Mn^{2+}), but not by sodium blockers such as tetrodotoxin. The current is little affected by short exposures to low sodium medium, but disappears when external calcium is chelated. It persists at potentials above the sodium equilibrium potential, and has an extrapolated reversal potential not far from that of calcium. It is apparently predominantly (and perhaps exclusively) a calcium current.

The pacemaker current is blocked by intracellular calcium injection and enhanced by EGTA injection. It is therefore not activated by internal calcium. Since reducing internal calcium should not directly affect the driving force or the net influx of current (from the Goldman single-ion equation), these effects of internal calcium and EGTA probably reflect modulation by calcium-dependent inactivation (Eckert and Chad, 1984).

Prolonged treatment with low sodium reduces the pacemaker current (reviewed in Adams and Benson, 1985), but this effect is prevented when EGTA is injected (Gorman et al., 1982). Sodium removal prevents Na/Ca exchange, which probably leads to elevated intracellular calcium and partial inactivation of the calcium pacemaker current. Low sodium also blocks a second depolarizing current (see below), which in early studies was not distinguished from the pacemaker current.

Calcium-Activated Potassium Current ($I_{K(Ca)}$)

Each spike in a burst is followed by a hyperpolarizing afterpotential lasting roughly 50-100 ms. This is slow compared to spike duration, but faster than subsequent afterpotentials discussed below. It participates in spike termination. This afterpotential is generated by an outward current of similar duration following a depolarization, which we have called the phase I tail current (Kramer and Zucker, 1985a,b). This current reverses at the potassium equilibrium potential, even when the latter is moved by altering potassium

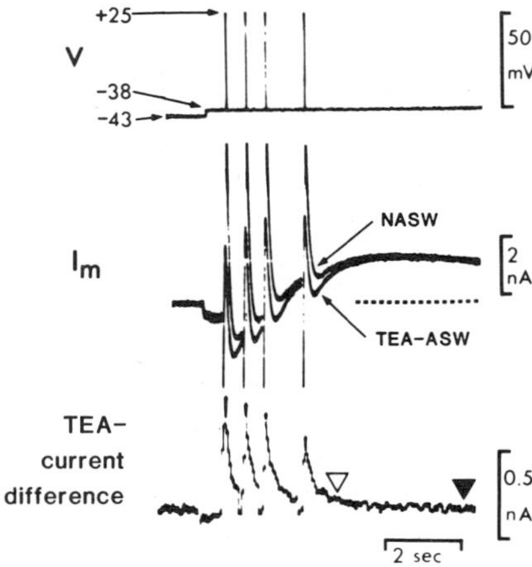

Fig. 1.   Time course of $I_{K(Ca)}$, determined as the difference (bottom trace) between currents in the presence and absence of 50 m$\underline{M}$ TEA (middle traces), in response to a simulated spike burst (top trace) under voltage clamp.

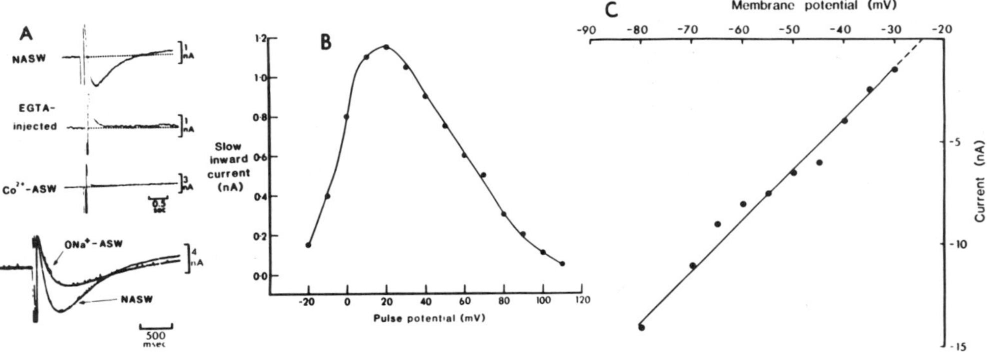

Fig. 2.   Properties of $I_{NS(Ca)}$. A) The slow inward current is blocked by EGTA injection and calcium antagonists, and reduced in sodium-free sea water. B) The inward tail current magnitude has a bell-shaped dependence on the potential of the pulse used to evoke it. C) The current depends linearly upon membrane potential, and reverses at about -25 mV.

concentration. It is blocked by millimolar concentrations of external tetraethylammonium (TEA), and is clearly a potassium current (Fig. 1). Its magnitude shows a bell-shaped dependence on the potential _during_ the pulse used to evoke it, with a peak at the same potential at which calcium influx is maximum during the pulse. Larger pulses, further from the potassium equilibrium potential, elicit smaller phase I tail currents. Its magnitude also depends strongly and nonlinearly on the potential at which it is measured _after_ the pulse, being large at depolarized potentials and very small at potentials below the potassium equilibrium potential, so the underlying conductance is voltage-dependent. It is blocked by EGTA injection and calcium antagonists, so it is a calcium-activated potassium current.

Calcium-Activated Nonspecific Cation Current ($I_{NS(Ca)}$)

Each spike in a burst is followed by a one-second-long depolarizing afterpotential which triggers the next spike in the burst. This afterpotential is mirrored in a second tail current which we have called phase II, a slow inward current which follows each spike and accumulates in a burst (Lewis, 1984; Adams, 1985). This leads to acceleration of spike frequency until a slower outward current begins to predominate (see below). The afterpotential and underlying current are sensitive to both external sodium and calcium concentrations (Thompson and Smith, 1976; Kramer and Zucker, 1985a; Smith and Thompson, 1987), and in R15, the current is partially triggered by the depolarization from axon spikes (Adams and Levitan, 1985). When axon spikes are prevented, this current has a bell-shaped dependence upon the voltage of the pulse used to evoke it; it is mimiced by calcium injection when $I_{K(Ca)}$ is blocked with TEA; and the current is blocked by EGTA injection; so it is a calcium-activated current (Fig. 2). When isolated from $I_{K(Ca)}$, it has a reversal potential at about -25 mV in Aplysia, and is normally carried by sodium, calcium and potassium ions. Changes in concentrations of these ions shift the reversal potential and affect the conductance magnitude. Large cations like TEA, tris, choline, glucosamine, and tetramethylammonium can substitute, at least partially, for sodium (Swandulla and Lux, 1985). Barium and lithium also penetrate the responsible channel, although cobalt, nickel, manganese, hydrogen and chloride do not. Preventing potassium accumulation by reducing potassium current does not affect either depolarizing afterpotential or slow inward current, nor do treatments such as lithium ringer or ouabain which block Na/Ca and Na/K exchange. Unlike $I_{K(Ca)}$, this current is linearly dependent upon the membrane potential during its measurement, indicating a conductance independent of voltage. We therefore have a relatively nonspecific cation current activated by intracellular calcium helping to boost the burst.

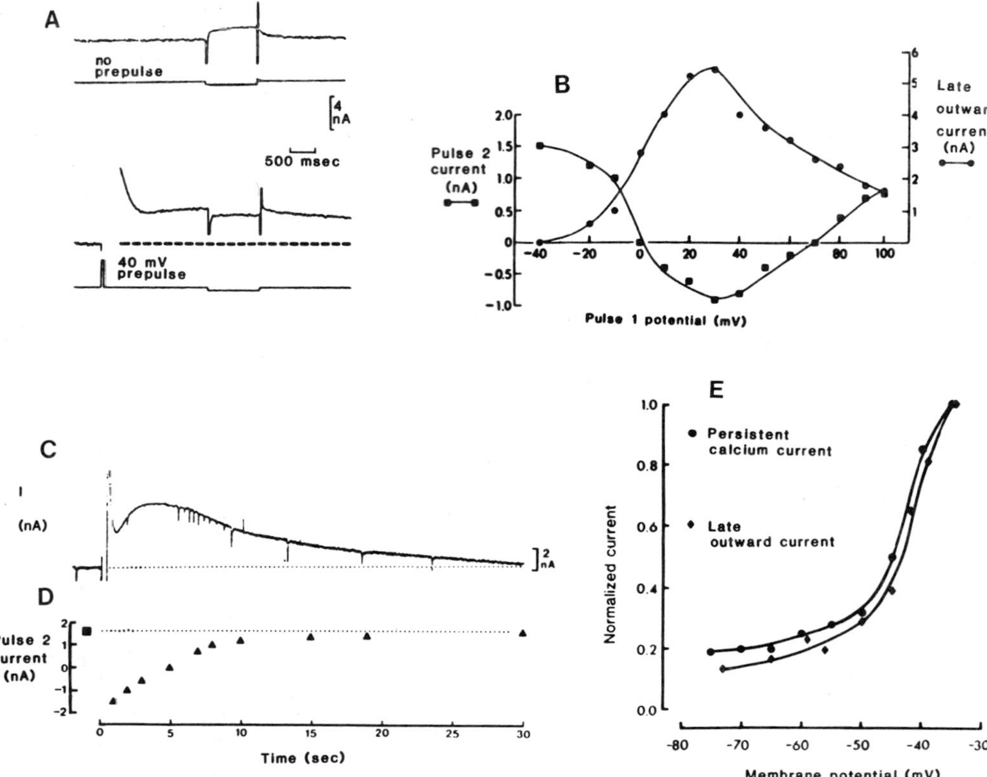

Fig. 3.   Properties of the late outward current. A) A small hyperpolarization normally elicits an outward current at -40 mV, due to the negative resistance of the membrane at this potential. A larger prepulse which elicits the late outward current blocks the negative resistance. B) The negative resistance (pulse 2 outward current) is blocked with the same bell-shaped dependence on pulse amplitude as the late outward current. C) Decay of late outward current after a pulse. D) Decay of negative resistance after an identical pulse. E) Voltage dependence of persistent calcium pacemaker current and late outward current.

The Late Outward Current

Bursts are terminated by a deep hyperpolarization, due to a slow outward current that accumulates gradually during each burst, and only comes to match the other slow currents late in the burst (Lewis, 1984; Adams, 1985). This current correlates with the intracellular accumulation of calcium measured spectrophotometrically with the dye arsenazo III (Gorman and Thomas, 1978). It is blocked by calcium antagonists and EGTA injection, and shows a bell-shaped dependence on pulse potential (Gorman et al., 1982; Kramer and Zucker, 1985b), so it is evidently another calcium-dependent current. Originally thought to be a calcium-activated potassium current, it is distinguished from $I_{K(Ca)}$ at room temperature by its insensitivity to TEA and charybdotoxin and changes in potassium concentration and its failure to reverse at the potassium equilibrium potential (Adams and Levitan, 1985; Kramer and Zucker, 1985; A. Hermann, personal communication). Only at low temperatures and in some dorid bursting neurons does a slow component of $I_{K(Ca)}$, insensitive to TEA, appear to contribute significantly to the late outward current (Deitmer and Eckert, 1985; Thompson et al., 1986). A TEA-insensitive component of outward current in response to calcium injection can also be detected. This current is unaffected by blocking Na/Ca and Na/K transport and replacing chloride in the ringer.

If not by affecting anion flow, electrogenic transport, or potassium conductance, how can calcium cause an increased outward current? A reduced inward current would resemble an increased outward current. Both would appear as an increase in steady-state conductance to voltage or current pulses (Junge and Stephens, 1973; Barker and Gainer, 1975). The persistent inward current most eligible to be reduced by calcium is the calcium pacemaker current, since it is subject to calcium-dependent inactivation (Eckert and Chad, 1984). If the late outward current is a reduction in pacemaker current, they should show the same sensitivity to voltage <u>during</u> the current. They do (Kramer and Zucker, 1985b). The negative resistance characteristic that serves to mark the presence of pacemaker current is reduced during the late outward current. As the latter decays, the former recovers. Both effects show the same bell-shaped dependence upon activating pulse potential <u>preceding</u> the current (Fig. 3). Apparently, bursts are terminated as calcium accumulating during a burst shuts off the pacemaker current.

KINETICS OF CALCIUM-DEPENDENT CURRENTS

At this point, we see that bursting is initiated by a calcium current activated at potentials subthreshold for action potentials, and boosted by a calcium-activated nonspecific cation current triggered by spikes in the burst. A more rapidly decaying calcium-activated potassium current acts mainly to assist spike termination, while a slower late outward current reflecting calcium-dependent inactivation of the pacemaker current terminates bursts. The slow decay of this block of pacemaker current sets the interburst interval, until the pacemaker current recovers and initiates a new burst (Fig.4)

Calcium entering the neuron during a burst has three effects: activation of $I_{K(Ca)}$, activation of $I_{NS(Ca)}$, and block of the pacemaker current. If all three effects are due to the activity of submembrane intracellular calcium, why do they decay at different rates? I initially thought $I_{K(Ca)}$ decayed faster than $I_{NS(Ca)}$ because the former was activated cooperatively by calcium, and followed some power greater than one of calcium concentration. However, in recent experiments (Landò and Zucker, in preparation), we have studied the calcium dependence of $I_{K(Ca)}$ and $I_{NS(Ca)}$ in bursting neurons by activating these conductances directly by intracellular calcium release from the photodynamic chelator nitr-5 (Adams et. al., 1986; Tsien and Zucker, 1986). We found both currents to be activated linearly and without saturation by cytoplasmic calcium increments similar to those occurring during bursting (Fig. 5). One calcium ion appears to activate each current at all voltages tested.

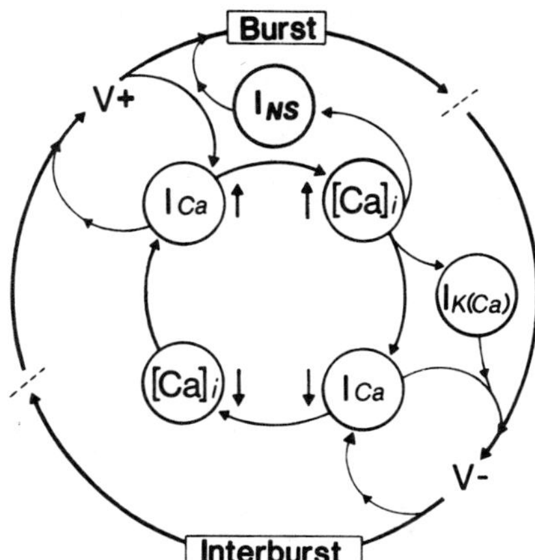

Fig. 4.   Schematic model of the bursting mechanism. Cycles of membrane potential are on the outer circle, and cycles of calcium current and internal calcium levels are on the inner circle. Calcium-activated currents are also shown. The drop in I_{Ca} corresponds to the late outward current.

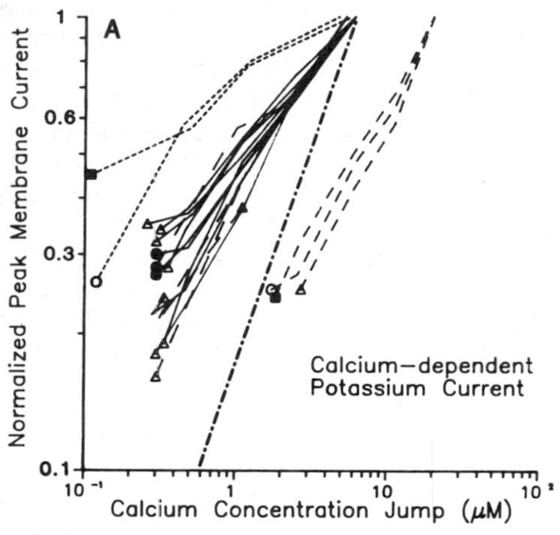

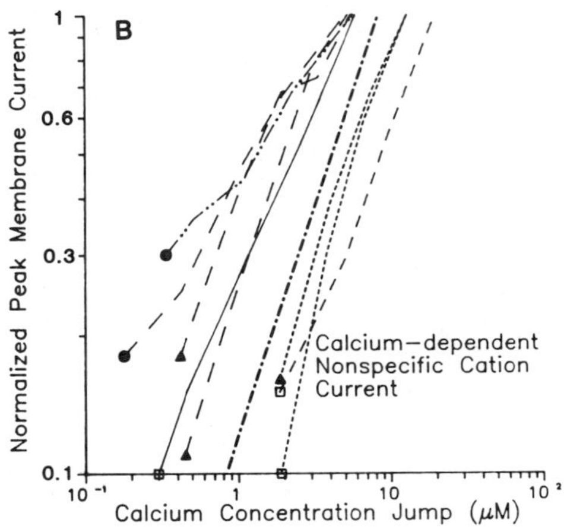

Fig. 5.   Dependence of $I_{K(Ca)}$ (in A) and of $I_{NS(Ca)}$ (in B) upon increments in intracellular calcium concentration caused by flash photolysis of nitr-5 or nitr-7. Each line is from a different experiment. The dot-dash line has unity slope.

Membrane currents activated by flash photolysis of nitr-5 decay after the flash, mostly due to diffusional equilibration of calcium (released mainly at the surface facing the light) and the buffer constituents. These currents, measured at a <u>constant</u> voltage, decay at the same rate in any single cell, again indicating an equal degree of calcium cooperativity in activating both currents. However, as indicated above, $I_{NS(Ca)}$ channels are voltage-insensitive while $I_{K(Ca)}$ are strongly affected by voltage. After a depolarization, $I_{K(Ca)}$ will show a voltage-dependent relaxation, while $I_{NS(Ca)}$ will only decay as submembrane calcium drops. We have measured these voltage-dependent relaxations of $I_{K(Ca)}$, and found that they, combined with the expected fall in submembrane calcium following a burst, are sufficient to explain the more rapid decay rate of $I_{K(Ca)}$ compared to that of $I_{NS(Ca)}$.

The very slow time course of pacemaker current inactivation and its recovery are still a matter of speculation. A very high calcium sensitivity might explain its long persistence, but not its slow build-up during a burst. However, in preliminary experiments we have been unable to block calcium currents during depolarizing pulses, or the negative resistance characteristic in the subthreshold range, by releasing calcium photolytically from nitr-5 injected intracellularly into bursting pacemaker neurons. This suggests that calcium-inactivation of the pacemaker current is the calcium-dependent process least sensitive to calcium accumulation. A similar conclusion was reached by considering the high local activity of calcium in the small "domains" surrounding calcium channel mouths, where calcium presumably acts to shut down the channels (Chad and Eckert, 1984). Perhaps the slow decay of late outward current reflects the time course of protein phosphorylation thought to be involved in recovery from calcium-dependent inactivation of calcium current (Eckert and Chad, 1984).

ACKNOWLEDGMENTS

I have been fortunate to collaborate with several exceptional scientists who contributed significantly to these studies: Drs. Richard Kramer, Luca Landò, Stephen Smith, and Roger Tsien. Russell English provided continuous technical support, and NIH funded the work generously (Grant NS 15114).

REFERENCES

Adams, S.R., Kao, J.P.Y., and Tsien, R.Y., 1986, Photolabile chelators that "cage" calcium with improved speed of release and pre-photolysis affinity, <u>J. Gen. Physiol.</u>, 88:9.

Adams, W.B., 1985, Slow depolarizing and hyperpolarizing currents which mediate bursting in Aplysia neurone R15, J. Physiol. (Lond.), 360:51.

Adams, W.B. and Benson, J.A., 1985, The generation and modulation of endogenous rhythmicity in the Aplysia bursting pacemaker neurone R15, Prog. Biophys. Molec. Biol., 46:1.

Adams, W.B. and Levitan, I.B., 1985, Voltage and ion dependences of the slow currents which mediate bursting in Aplysia neurone R15, J. Physiol. (Lond.), 360:69.

Andrew, R.D. and Dudek, F.E., 1984, Analysis of intracellularly recorded phasic bursting by mammalian neuroendocrine cells, J Neurophysiol., 51:552.

Barker, J.L. and Gainer, H., 1975, Studies on bursting pacemaker potential activity in molluscan neurons. I. Membrane properties and ionic contributions, Brain Res., 84:461.

Chad, J.E. and Eckert, R., 1984, Calcium domains associated with individual channels can account for anomalous voltage relations of Ca-dependent responses, Biophys. J., 45:993.

Chad, J., Eckert, R., and Ewald, D., 1984, Kinetics of calcium-dependent inactivation of calcium current in voltage-clamped neurones of Aplysia californica., J. Physiol. (Lond.), 347:279.

Chesnoy-Marchais, D., 1985, Kinetic properties and selectivity of calcium-permeable single channels in Aplysia neurones, J. Physiol. (Lond.), 367:457.

Cooke, I.M., 1982, Electrical activity in relation to hormone secretion in the X-organ-sinus gland system of the crab, in "Neurosecretion: Molecules, Cells, Systems", ed. Farner, D.S. and Lederis, K., Plenum Press, New York.

Deitmer, J.W. and Eckert, R., 1985, Two components of Ca-dependent potassium current in identified neurones of Aplysia californica, Pflugers Arch., 403:353.

Dutton, A. and Dyball, R.E.J., 1979, Phasic firing enhances vasopressin release from the rat neurohypophysis, J. Physiol. (Lond.), 290:433.

Eckert, R. and Chad, J.E., 1984, Inactivation of Ca channels, Prog. Biophys. Molec. Biol., 44:215.

Eckert, R. and Lux, H.D., 1976, A voltage-sensitive persistent calcium conductance in neuronal somata of Helix, J. Physiol. (Lond.), 254:129.

Gähwiler, B.H. and Dreifuss, J.J., 1979, Phasically firing neurons in longterm cultures of the rat hypothalamic supraoptic area: pacemaker and follower cells. Brain Res., 177:95.

Gorman, A.L.F., Hermann, A., and Thomas, M.V., 1982, Ionic requirements for membrane oscillations and their dependence on the calcium concentration in a molluscan bursting pace-maker neurone, J. Physiol. (Lond.), 327:185.

Gorman, A.L.F. and Thomas, M.V., 1978, Changes in the intracellular concentration of free calcium ions in a pace-maker neurone, measured with the metallochromic indicator dye arsenazo III, J. Physiol. (Lond.), 275:357.

Grace, A.A. and Bunney, B.S., 1984, The control of firing pattern in nigral dopamine neurons: burst firing, J. Neurosci., 4:2877.

Junge, D. and Stephens, C.L., 1973, Cyclic variation of potassium conductance in a burst-generating neurone in Aplysia, J. Physiol. (Lond.), 235:155.

Kramer, R.H. and Zucker, R.S., 1985, Calcium-dependent inward current in Aplysia bursting pacemaker neurones, J. Physiol. (Lond.), 362:107.

Kramer, R.H. and Zucker, R.S., 1985, Calcium-induced inactivation of cal|cium current causes the inter-burst hyperpolarization of Aplysia bursting neurones, J. Physiol. (Lond.), 362:131.

Legendre, P., McKenzie, J.S., Dupouy, B. and Vincent, J.D., 1985, Evidence for bursting pacemaker neurones in cultured spinal cord cells, Neuroscience, 16:753.

Lewis, D.V., 1984, Spike aftercurrents in R15 of Aplysia: their relationship to slow inward current and calcium influx, J. Neurophysiol., 51:387.

Miller, J.P. and Selverston, A.I., 1982, Mechanisms underlying pattern generation in lobster stomatogastric ganglion as determined by selective inactivation of identified neurons. II. Oscillatory properties of pyloric neurons, J. Neurophysiol., 48:1378.

Smith, S.J. and Thompson, S.H., 1987, Slow membrane currents in bursting pacemaker neurones of Tritonia, J. Physiol. (Lond.) 382:425.

Swandulla, D. and Lux, H.D., 1985, Activation of a nonspecific cation conductance by intracellular Ca^{2+} elevation in bursting pacemaker neurons of Helix pomatia, J. Neurophysiol., 54:1430.

Thompson, S.H. and Smith, S.J., 1976, Depolarizing afterpotentials and burst production in molluscan pacemaker neurons, J. Neurophysiol., 39:153.

Thompson, S., Smith, S.J., and Johnson, J.W., 1986, Slow outward tail currents in molluscan bursting pacemaker neurons: two components differing in temperature sensitivity, J. Neurosci., 6:3169.

Tsien, R. and Zucker, R.S., 1986, Control of cytoplasmic calcium with photolabile 2-nitrobenzhydrol tetracarboxylate chelators, Biophys. J., 50:843

THE NEONATAL RAT SUPRAOPTIC NEURONE IN CULTURE:DEVELOPMENT OF A MODEL FOR CONTROL OF PEPTIDERGIC SECRETION

P. Cobbett, K. Inenaga, P. Legendre and W.T. Mason

Department of Neuroendocrinology
AFRC Institute of Animal Physiology and Genetics Research
Babraham, Cambridge CB2 4AT
United Kingdom

INTRODUCTION

The magnocellular neurones of the hypothalamo-neurohypophysial system have in recent years proved a valuable model in which to examine the complex control of peptidergic secretion, in particular oxytocin and vasopressin. The unique properties of the nuclei containing these neurones - principally homogeneity and high density of the neuronal population - have meant that a variety of _in vivo_ and _in vitro_ studies have been possible.

A central aim of our laboratory has been to develop _in vitro_ systems from which detailed electrophysiological data may be obtained regarding both the extrinsic and intrinsic factors which control the activity of neurones of the supraoptic nuclei (SON), and hence the amount and efficiency of hormone release from their terminals in the neural lobe. Our rapidly expanding knowledge of the supraoptic neurones has demonstrated the very complex nature of afferent input from a number of brain areas, mediated by both neurotransmitters and neuropeptides. Furthermore, intracellular voltage recordings from the hypothalamic slice and explant preparations used in this and other laboratories has suggested strongly that properties intrinsic to the neurones themselves are important in modulation of firing rate and pattern (see Dudek & Andrew, 1985).

Inevitably, the available information about control of oxytocin and vasopressin secretion has been limited by the resolution of the techniques used for their study. The history of our knowledge about the activity of neurones of the SON ranges from measurements of the hormones in blood to single unit recordings of spiking activity _in vivo_. More recently, the _in vitro_ slice and explant preparations have enabled single neurones to be impaled on high resistance micropipettes and dynamic recordings of cell voltage to be made, albeit with the limitations imposed by removal of some synaptic drive. Data from the slice and explant preparations have, and are, providing important information about control of firing, yet are limited by the fact that membrane voltage recordings can provide very little information about flow of membrane current. Likewise, the recording technology employed limits the ideal approach being used: that of voltage clamp recording.

To overcome some of these problems, we have recently attempted to develop a culture system of isolated, single supraoptic neurones from the neonatal rat, and to examine these neurones using both current and voltage clamp recording techniques. Success of such a system has many potential rewards, including the ability to record membrane current flow during application of neurotransmitters and peptides, osmotic stimuli, and, importantly, of voltage itself to examine voltage activated membrane currents. We have begun to examine the characterstics of this system, and in this paper, we are able to report some early data and indicate some pointers for the future.

METHODS

Neonatal rat supraoptic neurones (2-4 days old) are obtained by papain dispersion of dissected nuclei, and cultured for 7-21 days prior to use in Dulbecco's modified Eagle's medium supplemented with 15% newborn calf serum. A detailed description of the procedures employed will be found in Cobbett & Mason (1988). Immunocytochemical staining was performed on paraformaldehyde-fixed neurones using antisera to oxytocin, vasopressin or neurophysin, with detection using peroxidase- antiperoxidase reagents.

Whole cell clamp techniques (Hamill et al., 1981) have generally been used to study these cultured neurones. This approach enables dynamic perfusion of the intracellular milieu with media of defined ionic composition or containing the dye Lucifer Yellow CH, while simultaneously recording either membrane voltage under current clamp, or membrane current under voltage clamp. A slight variation, the outside-out patch, enables recording of single ion channels from the membrane of the same cell. Substances are applied to the neurones either in the petri dish or from multi-barrelled pipettes by pressure ejection.

RESULTS

Cultures of neonatal supraoptic neurones have been found to contain a number of phase-bright, multipolar neurones (Fig. 1A) which stain with neurophysin antisera and are further characterised by either oxytocin- or vasopressin-like immunoreactivity (Fig. 1B). The relatively simple morphology seen with phase contrast optics is deceptive: filling of the neurones with Lucifer Yellow and visualisation under fluorescence optics reveals a complex morphology, comprising the main dendrite-like processes with numerous branching elements (Fig. 1C).

We have initially made recordings of membrane voltage which have shown these neurones to possess firing patterns similar to those recorded in both the intact animal and the slice preparation. Figure 2A shows a typical neurone which fired in a continuous manner and Figure 2B shows a different neurone with a distinct pattern of "phasic" firing typical of vasopressin-containing neurones. In contrast to previous intracellular recordings, cells

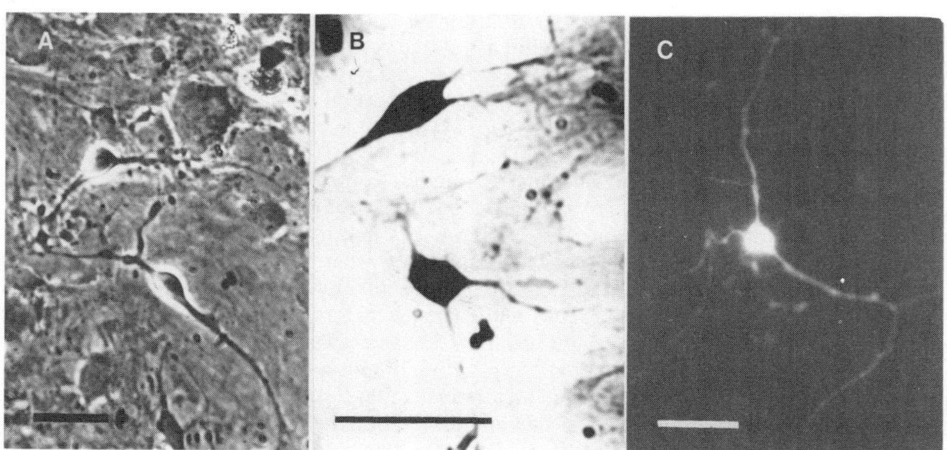

Fig. 1.   Photomicrographs of cultured neurones of the neonatal supraoptic nucleus. (A) Viewed with phase contrast optics, neurones have a bright soma with one or more processes. (B) Neurones stained immunocytochemically viewed in brightfield after incubation with an antiserum against vasopressin. (C) A single neurone seen with fluorescence optics seen after injection with Lucifer Yellow demonstrates the complex outgrowth of processes which occurs in culture. Scale bars represent 50μm.

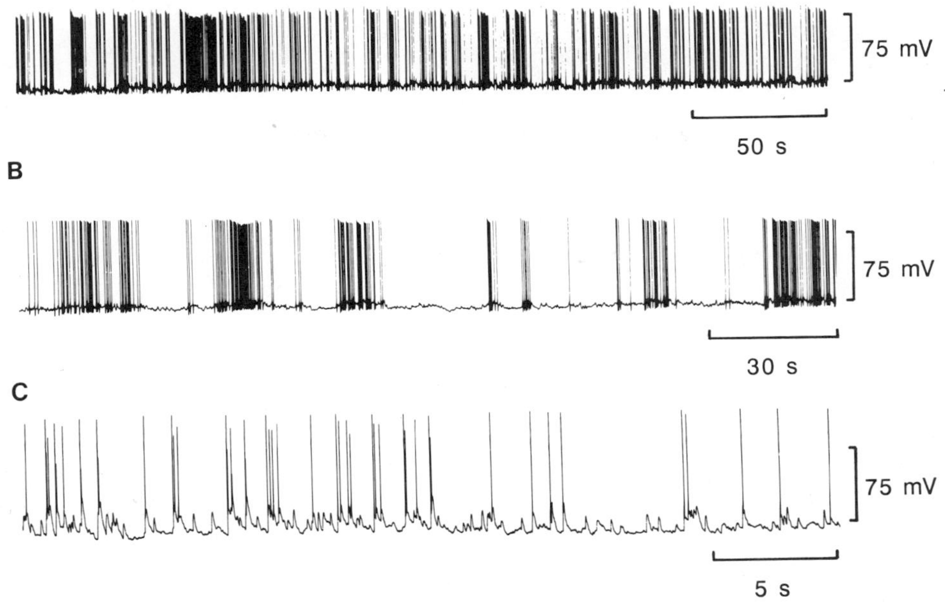

Fig. 2.   Voltage recordings from cultured neurones obtained with the whole cell
technique. Some neurones fired in a continuous manner (A), but in others phasic
bursts of action potentials could be recorded. Depolarising fluctuations of the
membrane potential recorded in some neurones (C) appeared to be postsynaptic
potentials and could initiate action potentials.

recorded with the patch technique have very high membrane input resistances, on the order
of 500MΩ to 10GΩ, with resting membrane potentials of -40 to -90mV.

Significantly, we have also observed large depolarising fluctuations (2-20mV) in
membrane voltage which in many cases exceed the threshold for action potential firing and
are clearly responsible for spike initiation (Fig. 2C). Much of this subthreshold activity is
abolished by 2 mM Co^{2+}, suggesting it to be of synaptic origin.

Aside from these results, perhaps the most important advance we have been able to
make with this preparation is the ability to record membrane current flow in response to a
step change in voltage. In most neurones, we found it possible to achieve reasonably good
spatial clamp, as evidenced by low series resistances and constancy of the latency of current
responses. With physiological salt solutions both inside and out, supraoptic neurones showed
an expected complex membrane current flow in response to depolarising voltage clamp steps.
Consequently, we have used a combination of ionic substitution and pharmacological
blockade to dissect the current flow through specific ion channels (Cobbett & Mason, 1987;
Cobbett et al., 1987).

The results of such experiments are shown in Figure 3. Supraoptic neurones possess
voltage-activated inward flowing Na^+ currents which are specifically blocked by 1μM TTX
(Fig. 3A). These are activated around -55mV, and rapidly inactivate within 2ms of the onset
of activation. Interestingly, other studies not shown here have demonstrated these currents to
be characterised by conventional Hodgkin-Huxley kinetics. These neurones also possess three
different classes of voltage activated K^+ channels, mediating outward current flow.
Interestingly, one of the K^+ currents requires both Ca ions and voltage to operate (not shown
here).

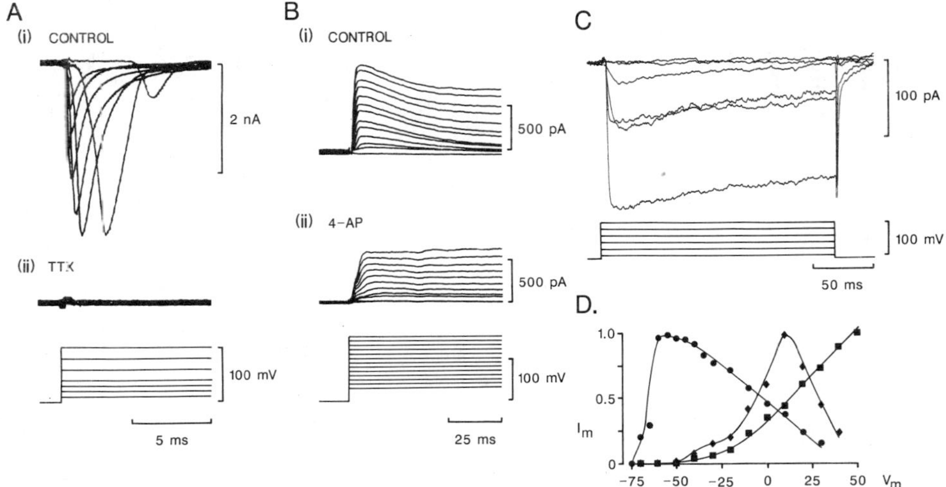

Fig. 3.   Isolated, voltage activated currents recorded with the whole cell technique. (A) With Na^+ as the only permeant ion, positive voltage steps (as indicated in lowest records) from -80mV elicited transient inward currents (i) which were blocked by $1\mu M$ tetrodotoxin (TTX; ii). (B) With K^+ as the permeant ion, similar volatge steps from -100mV elicited two component currents (i) the inactivating component being blocked by 2mM 4-aminopyrdine (4-AP; ii). (C) Using Ba^{2+} as the permeant species, partially inactivating inward currents were recorded during voltage steps from -80mV were recorded. (D) Membrane current (I_m) - voltage (V_m) relationships for Na^+ (circles), K^+ (squares) and Ba^{2+} (diamonds) currents shown in A, B, and C (currents are normalised to the maximum current recorded in that experiment).

The other two K^+ currents have very different potential dependent characteristics of activation and inactivation and may also be separated pharmacologically. Positive voltage steps from -100mV activate a current of two components. One component rapidly activates but shows marked inactivation over 10-50ms and is blocked by 4-aminopyridine (4-AP), the other is insensitive to 4-AP (but is blocked by tetraethylammonium -TEA- ions), is less rapidly activating and does not inactivate (Fig. 3B). The TEA sensitive component but not the 4-AP sensitive component may be activated during voltage steps from -60mV. All three currents must clearly be seen as of potential importance in modulation of firing frequency and interspike interval, as well as of potential significance in the process of phasic plateau potential generation.

Supraoptic neurones also possess at least two classes of voltage activated Ca channels, through which either Ca^{2+} or Ba^{2+} will flow in response to a depolarising step (Fig. 3C) and which are blocked by Co ions. The currents are differentiated by slow activation and in the case of one class of current, sustained activation with little evidence of inactivation. Such a finding could provide an ionic substrate for the plateau potential observed during phasic firing.

An important question originally addressed by the slice preparation is the osmotic sensitivity of the supraoptic neurones. We have found good evidence that SON neurones in culture retain this sensitivity, responding to small changes in osmolarity by Na^+ or mannitol elevation with a depolarisation and increase of firing rate, in some cases accompanied by transition to phasic plateau generation and thus phasic firing (Fig. 4). Whereas we believe the osmoreceptor to be a neuronal complex, one advantage of the current preparation may prove to be that with the cultured cells only one component of the complex is present: when stimulated with increased osmolarity, primary depolarisation of the neurone is observed, but significant changes in synaptic 'noise' are not recorded, unlike the features of osmotic stimulation observed clearly in the slice (Mason, 1980).

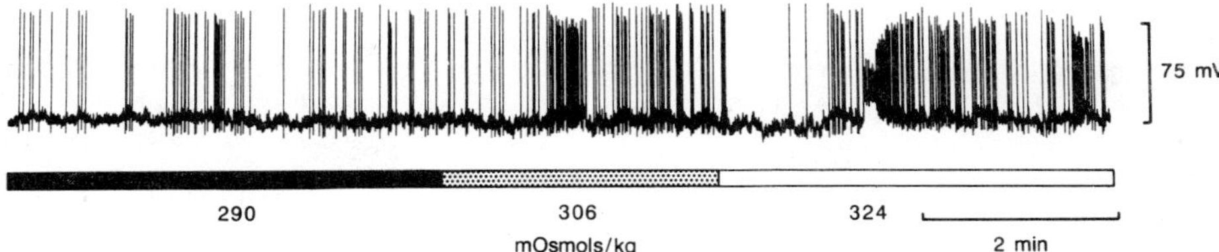

Fig. 4.   Effect of osmotic pressure on the activity of cultured neurones. Increases in the osmotic pressure (by addition of NaCl) produce increases in the mean firing rate and also induce plateau potentials.

Functionality of the neurones in culture also requires they possess or develop receptors for neurotransmitters. We and others (Mason et al, 1987; Randle & Renaud, 1987) have shown that intact SON neurones possess receptors for the putative inhibitory transmitter τ-aminobutyric acid (GABA). Our experiments with cultured SON neurones have confirmed that GABA inhibits firing in these cells, through selective activation of chloride channels (Fig. 5; Inenaga et al., 1987). Manipulation of the chloride equilibrium potential and application of 10-50μM GABA changes the membrane potential or current flow consistent with a 'pure' action on a chloride channel, and examination of single Cl^- channels in outside-out patches revealed a GABA-linked chloride channel which has a single open state of about 20pS conductance, which remains open on average for about 5 ms before closing, and which has a probability of opening dependent on the GABA concentration.

We have also found evidence for functional adrenergic and opioid receptors on these neurones and these receptors appear to be linked to voltage activated channels. Local application of the α-adrenergic agonist phenylephrine or the k-opioid agonist U50 can reduce the amplitude of Ba^{2+} currents through voltage activated Ca^{2+} channels in a reversible manner (Fig. 6). There is also evidence (not shown here) that the activity of voltage activated K^+ channels may be modulated through an α-adrenergic receptor.

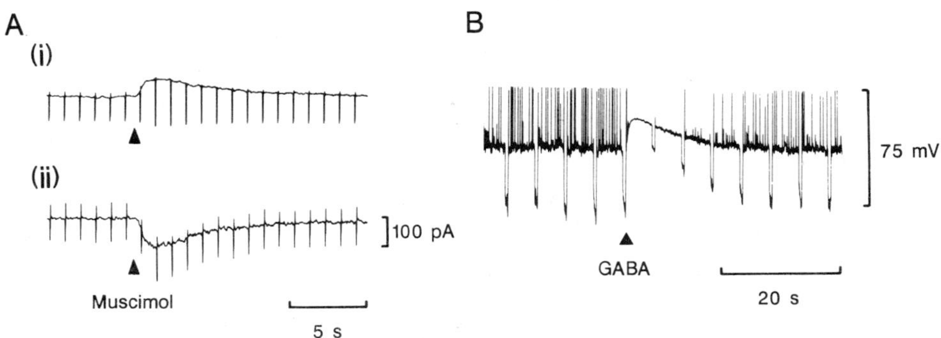

Fig. 5.   Activation of GABA receptors opens chloride channels in cultured supraoptic neurones. (A) Voltage clamp recordings with Cl^- as the only permeant ion show that muscimol, a GABA analog, transiently increases the membrane conductance (an increase in the current response to repeated voltage steps) and alters the steady membrane current at (i) 40mV and (ii) -60mV. The polarity of the steady current response reversed at 0mV, the Cl^- equilibrium potential. (B) Current clamp recordings show that application of GABA produces a transient decrease of the input resistance (a decrease in the voltage response to constant current pulses), a reduced firing frequency and a depolarisation of the membrane (the Cl^- equilibrium potential was 0mV). Muscimol and GABA were applied from micropipettes by pressure ejection.

DISCUSSION

The experiments reported here provide good evidence for the usefulness of the cultured rat SON neurone as a model to study modulation of electrical activity in a peptidergic secretory system. The neurones appear not only to contain significant quantities of the relevant peptide hormones and their associated carrier molecule, neurophysin, but they also develop complex dendritic-like morphology under culture conditions. Our electrophysiological data also suggests the neurones have significant, as yet largely uncharacterised, synaptic input which can result in spike initiation. Experiments reported here with GABA, and with α-adrenergic receptor and k-opioid receptor agonists confirm that these neurones express functional receptors even at this early stage of development, and that such receptors are closely linked to ion channels which can modulate the electrical activity of the SON neurones. Other studies are currently in progress to explore the presence and nature of other receptor-linked ion channels, and preliminary evidence suggests that, amongst others, dopamine may play functional roles in mediation of membrane currents.

Important work remains to be done with this preparation. Having established the neurones are osmosensitive, the ability to record membrane current flow under voltage clamp will prove essential to understand both how the osmosensitive mechanism is linked to changes in ion channel activity and whether osmosensitivity is a specific or general phenomenon in these cells. Likewise, we should be able to make rapid strides in understanding the ionic currents underlying generation of the plateau potential. In these and other respects, we still have little knowledge of how the complex intracellular biochemical machinery of the supraoptic neurone controls ionic channel activity and current

A

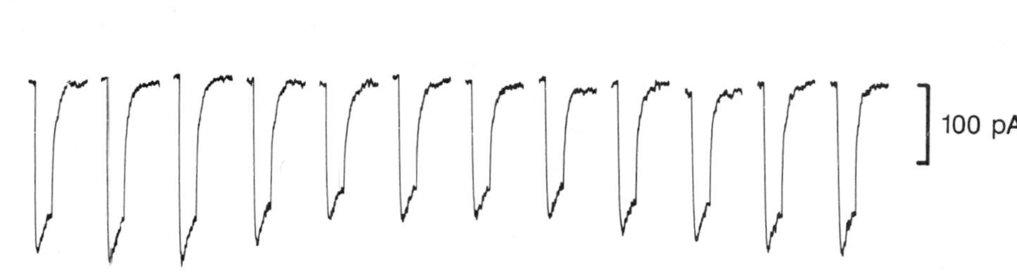

B

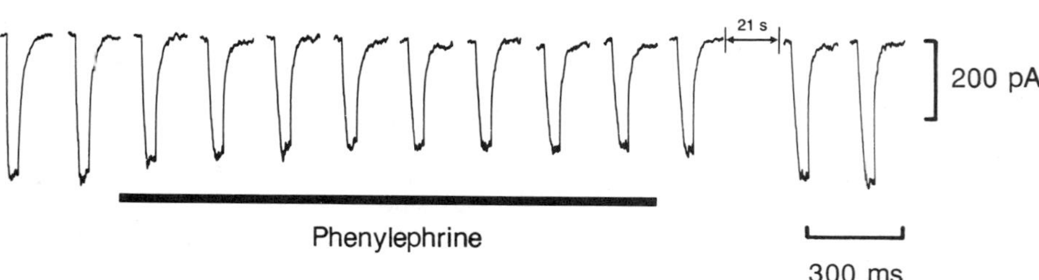

Fig. 6.   Effects on isolated, voltage activated Ba currents of opioid receptor and adrenergic receptor activation. (A) Pressure application of the 3k-opioid agonist U50 reversibly reduced the amplitude of inward Ba current which was elicited by step voltage changes from -80mV to 0mV every 3s. (B) In another neurone, U50 was inactive, but the α-adrenergic agonist phenylephrine reduced the Ba current elicited by the same voltage step protocol.

flow, and it is here that perhaps the best and most exciting discoveries will be made. The patch recording methodology permits manipulation of the intracellular composition. Already, this has shown us that the voltage dependent Ca^{2+} channels in these cells (unlike Na^+ and K^+ channels) must be phosphorylated before they can open, i.e. a source of ATP is required. It is also likely that other key events such as modulation of phasic firing through variation of the resting potential will be found to depend on intrinsic mechanisms, and the approach described here stands a good chance of enabling definition of these important pathways.

ACKNOWLEDGEMENTS

Our work has been generously supported by a Beit Medical Fellowship to P.C., a British Council Scholarship to K.I. and a Nuffield Foundation Small Grant Award to P.L.

REFERENCES

Cobbett, P., and Mason, W.T., 1987, Whole cell voltage clamp recordings from cultured neurons of the supraoptic area of neonatal rat hypothalamus, Brain Res., 409:175.

Cobbett, P., and Mason, W.T., 1988, Potential recordings from neurons of the supraoptic area of the neonatal rat hypothalamus in culture, (Submitted for publication.)

Cobbett, P., Legendre, P., and Mason, W.T., 1987, Three types of K^+ currents recorded from cultured neurones of the region of the supraoptic nucleus of the rat, J. Physiol., 384:20P.

Dudek, F.E., and Andrew, R.D., 1985, Electrophysiological characteristics of paraventricular and supraoptic neurons. in: Vasopressin. Editor: R.W. Schrier. Raven Press New York.

Hamill, O.P., Marty, A., Neher, E., Sakman, B., and Sigworth, E.J., 1981, Improved patch clamp techniques for high resolution current recording from cells and cell-free membrane patches, Pflugers Archiv., 391:85.

Inenaga, K., Legendre, P., and Mason, W.T., 1987, GABA activated chloride channels in cultured neurones from the supraoptic area of neonatal rats, J. Physiol., 388:12P.

Mason, W.T., 1980, Supraoptic neurones of rat hypothalamus are osmosensitive, Nature, 287:154.

Mason, W.T., Cobbett, P., and Poulain, D.A., 1987 τ-aminobutyric acid as an inhibitory neurotransmitter in the rat supraoptic nucleus: intracellular recordings in hypothalamic slices, Neurosci. Letters, 73:259.

Randle, J.C.R., Renaud, L.P., 1987, Actions of τ-aminobutyric acid on rat supraoptic nucleus neurosecretory neurones in vitro, J. Physiol., 387:629.

EFFECTS OF ATRIAL NATRIURETIC POLYPEPTIDE AND ANGIOTENSIN II ON THE SUPRAOPTIC NEURONS IN VITRO

Hiroshi Yamashita and Kiyotoshi Inenaga

Department of Physiology
University of Occupational and Environmental Health
School of Medicine, Japan
Kitakyushu 807, Japan

INTRODUCTION

Atrial natriuretic polypeptide (ANP) was isolated from atrial tissue and exhibits powerful natriuretic, diuretic and vasodilatory activities. Its sequence has been identified and ANP-immunoreactivity has been localized not only in the atria but also in plasma and within rat hypothalamus (Tanaka et al., 1984) with dense accumulations of immunocytochemical staining in regions of the brain that are important in the control of fluid balance and cardiovascular functions (Kawata et al., 1985). In particular, ANP-immunoreactivity has been located in the anterio-ventral region of the third ventricle (AV3V). The AV3V region and the supraoptic nuclei (SON) contain numerous ANP-binding sites and there may be a functional role for ANP in the control of vasopressin release. Peripheral or central administration of atrial natriuretic factor inhibits the release of arginine vasopressin (AVP) caused by dehydration, haemorrhage and changes in plasma osmolarity (Poole et al., 1987). It is possible that ANP might antagonise the effects of AVP, so the first set of experiments were designed to investigate whether or not ANP could influence directly the activity of neurosecretory cells in the SON. The experiments were carried out in trimmed hypothalamus slice preparations and data compared with the effects of ANP on neurons in the AV3V region (Okuya & Yamashita, 1987).

In contrast, intracerebroventricular administration of angiotensin II (AII) increases the release of arginine vasopressin and oxytocin and increases circulating plasma levels of both hormones (Keil et al., 1975; Lang et al., 1981). It has also been shown to excite both oxytocin and vasopressin cells (Akaishi et al., 1981). We have shown that AII can activate neurons in the SON and AV3V area directly and these responses appear to be mediated by specific receptors for AII because the responses were partially or completely blocked by saralasin, an AII antagonist. In addition, AII is considered to have central effects promoting drinking behaviour and a rise in blood pressure (Phillips, 1978). At the same time, there have been several reports that ANP may modulate the effectiveness of AII: ANP appears to inhibit AII-induced vasoconstriction of the aorta (Kleinert et al., 1984), AII-induced drinking (Nakamura et al., 1985) and AII-induced AVP secretion (Yamada et al., 1986). These findings imply that ANP may have an interaction with AII so the second part of this chapter concerns data on the AII and ANP effects on SON neuronal activity in slice tissue.

MATERIALS AND METHODS

Using conventional techniques (Inenaga et al., 1986; Inenaga and Yamashita, 1986), coronal brain slices containing the SON and AV3V (350 μm-450 μm thick) were prepared from the brains of adult male Wistar rats weighing 150-300 g using a vibrating brain slicer. Immediately after sectioning the slices were placed in an incubation medium oxygenated with

95% O_2 and 5% CO_2 at 35°C and left for at least 1 h when they were transferred to a recording chamber. The incubation medium was a modified Yamamoto's solution (pH 7.3-7.5), which contained the following mM concentrations of: NaCl 124, KCl 5, KH_2PO_4 1.24, $MgSO_4$ 1.3, $CaCl_2$ 2.1, $NaHCO_3$ 20, and glucose 10. In the perfusing medium, however the calcium concentration was always 0.75 mM to increase spontaneous activity of neurons in the SON as well as in the AV3V. When required, a low Ca^{2+} (0.5 mM) and high Mg^{2+} (9 mM) solution was used to block synaptic transmission.

Before the experiments started, each slice was carefully trimmed to be an island of the SON or the AV3V. The AV3V contained the OVLT (the organum vasculosum of the lamina terminalis), preoptic suprachiasmatic nucleus, periventricular preoptic nucleus and medial preoptic nucleus. The trimmed slice was placed on a sylgard mat glued to the bottom of the recording chamber which had a volume of 0.8 ml and was held in place with a nylon net and platinum weights. The temperature of the perfusing medium (which was also oxygenated with 95% O_2 and 5% CO_2) was kept at 35 ± 0.5°C. The flow rate of perfusing medium was adjusted to 1-2 ml/min. The perfusing medium in the chamber could be completely exchanged within 1 min. Peptides were applied to the slice by perfusing from a separate storage bottle containing medium to which they had been added.

Extracellular recordings were obtained using glass micropipettes with DC resistance of 20-35 MΩ, filled with 0.5 M sodium acetate containing 2% Pontamine Sky Blue. The micropipette was introduced into the slice under microscopic control. At the end of recording, a cathodal current of 5 μA was passed through the electrode for 3-5 min (tip negative) to deposit a blue spot from which the recorded sites could be determined histologically. Recorded action potentials were displayed, using conventional recording methods, on a storage oscilloscope and stored on magnetic tape for further analysis. To ensure exact recording of the spike shape, DC- or AC-coupled (time constant of 0.1 s) mode was used. A window discriminator and an integrator were used for continuous observation of the firing patterns of the neurons.

The peptides used in these experiments were angiotensin II (AII), [Sar[1], Ala[8]] angiotensin II (saralasin), arginine-vasopressin (AVP) and atrial natriuretic polypeptide (ANP: Rat 1-28) and were supplied by the Peptide Institute (Minoh, Japan).

RESULTS

EFFECTS OF ATRIAL NATRIURETIC POLYPEPTIDE ON SUPRAOPTIC NUCLEUS AND ANTEROVENTRAL THIRD VENTRICLE NEURONS (Okuya and Yamashita, 1987)

Extracellular recordings were made from 15 spontaneously firing cells in the SON and 73 spontaneously firing cells in the AV3V. The SON neurons generally showed a slow and irregular firing pattern, while most of the AV3V neurons showed a continuous or irregular firing pattern. The mean firing rate (mean ± SE of mean) of the spontaneously firing cells in the SON and AV3V was 2.8 ± 0.4 spikes/s and 5.0 ± 0.4 spikes/s respectively. The amplitude of spikes recorded in this study ranged from 0.6 to 5.0 mV.

After making a stable recording from a single cell for at least 10 min, ANP was applied to the slice at a concentration of 10^{-7} M. None of the 15 SON cells tested were inhibited (Fig.1A and B) but 30 (41%) of 73 AV3V cells showed inhibitory responses to application of ANP (Fig.1C). The proportion of cells inhibited by ANP was statistically significant in comparison of two regions (X^2=7.61, P<0.01 at df=1). The size of the response did not depend on initial firing rate. After washing out the ANP with the control medium, the firing rate recovered to the pretreatment values in 5 - 15 min.

To investigate the dose-response relationship of ANP effects, some neurons in the AV3V were tested at different concentrations ranging from 10^{-12} to 10^{-6} M. Fig. 2 shows an example of the effects of ANP at concentrations from 10^{-10} to 10^{-6} M. As ANP concentration increased, the firing rates of all 5 cells tested generally decreased. The threshold ANP concentration to evoke inhibitory responses of the AV3V was approximately

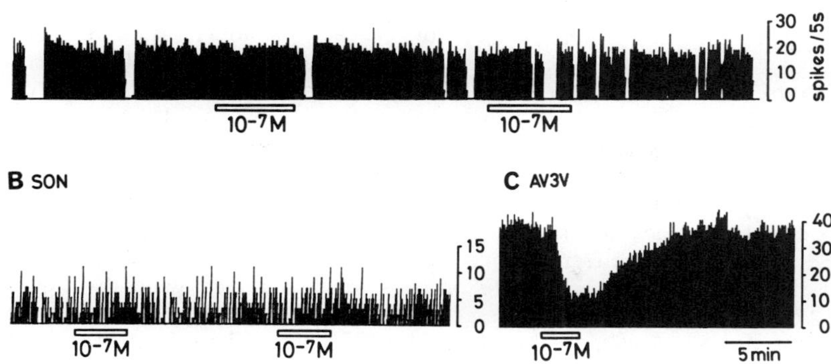

Fig. 1.   The effects of ANP in the SON and AV3V (from Okuya & Yamashita, 1987). A and B show a phasic firing and a slow irregular firing SON cell respectively. Neither neuron was responsive to application of ANP 10^{-7} M (indicated by open bar). C shows an inhibitory response of neuron following application of ANP 10^{-7} M in the AV3V.

10^{-11} M. The inhibitory effects of ANP were reversible and tachyphylaxis was seldom observed with repeated application of the peptide.

These results demonstrate that ANP can influence the AV3V neurons in the physiologically low concentrations while SON neurons are not affected by ANP. Moreover they suggest that ANP may be an active substance functioning as a possible neurotransmitter or modulator in the central nervous system.

EFFECTS OF ANGIOTENSIN II ON SON AND AV3V NEURONS (Okuya et al., 1987)

Extracellular recordings of 106 spontaneously active SON neurons and 86 AV3V

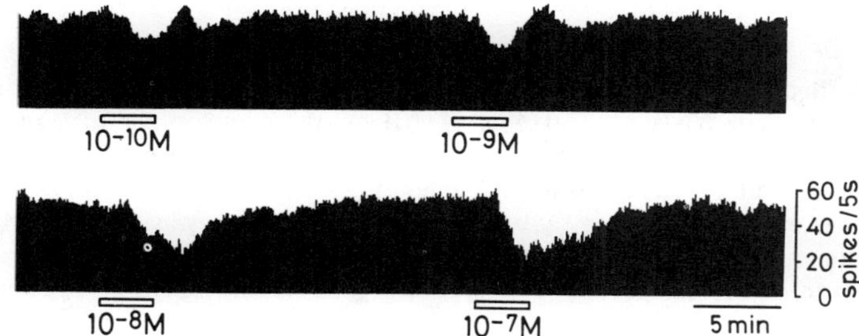

Fig. 2.   A continuous record to show graded responses of a AV3V neuron to different concentrations of ANP (from Okuya & Yamashita 1987). This rate-meter record shows successive clear responses of a neuron to increasing concentrations of ANP (10^{-10}-10^{-7} M) in the AV3V.

neurons were made from trimmed slice preparations. On the basis of their firing patterns, 45 (42%) of 106 SON neurons were classified as phasic firing neurons, characterized by alternating high frequency discharge and quiescence. The burst duration and the silence duration of phasic firing cells fell within the range of 4-300 s and 4-180 s respectively. The mean intraburst firing rates ranged from 3 to 15 spikes/s. The remaining 61 (58%) cells were classified as non-phasic firing neurons. Application of AII (10^{-7} M) (Fig. 3), caused 47 (44%) of 106 SON neurons and 28 (33%) of 86 AV3V neurons to be excited. Only one neuron in the SON was inhibited by AII.

To investigate the dose-response relationships of the AII effects, some neurons in each region were tested at different concentrations ranging from 10^{-10} to 10^{-6} M. As AII concentration increased, the firing rate of cells in each region generally increased. The threshold AII concentration to evoke responses of the SON neurons was approximately 10^{-9} M and that of the AV3V neurons was less than 10^{-10} M. The concentration of AII in the cerebrospinal fluid (CSF) and blood of rats is known to be 3 X 10^{-10} -10^{-9} M^{11}, respectively. Therefore it is reasonable to consider that the threshold concentrations obtained in this study lie in a range which is physiologically meaningful.

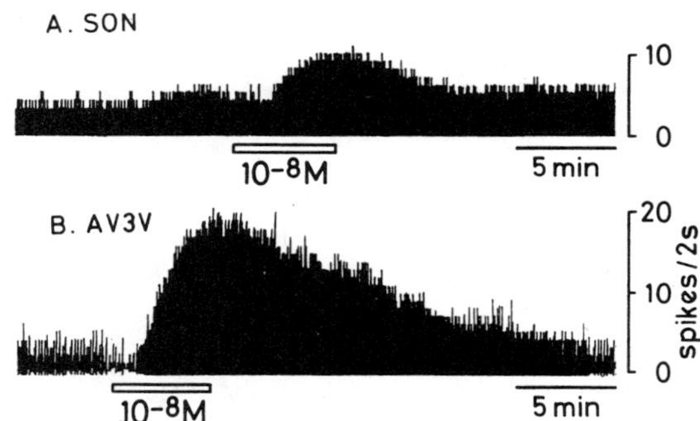

Fig. 3.   Ratemeter records of responses of SON and AV3V neurons to application of AII (from Okuya et al., 1987). A and B show excitatory responses of neurons following application of AII (open bar) in the SON and AV3V respectively.

The recorded SON neurons were divided into two groups, the phasic and non-phasic firing neurons and a comparison of effects of AII on these two types of neurons was made. Of 45 phasic firing cells, 18 (40%) showed an overall increase in firing rate in response to AII. This was achieved by an increase in the length of the bursts of discharge but there was little effect on the intraburst firing rate (Fig 4A). One neuron was inhibited by AII. Of 61 non-phasic firing neurons, 29 (48%) also showed an overall increase in firing rate (Fig. 4B). There was no significant difference between the proportions of each cell type which was excited by AII (x^2=2.30, P>0.1 at df=1).

The effects of saralasin specific AII antagonist: (Palovcik and Phillips, 1984) were tested on AII-induced excitation (Fig. 5). Five neurons in both SON and AV3V were excited by application of AII (10^{-7} M) and saralasin (10^{-6} M) completely blocked the excitatory effects of AII in each case. It did not alter the spontaneous discharge of any of

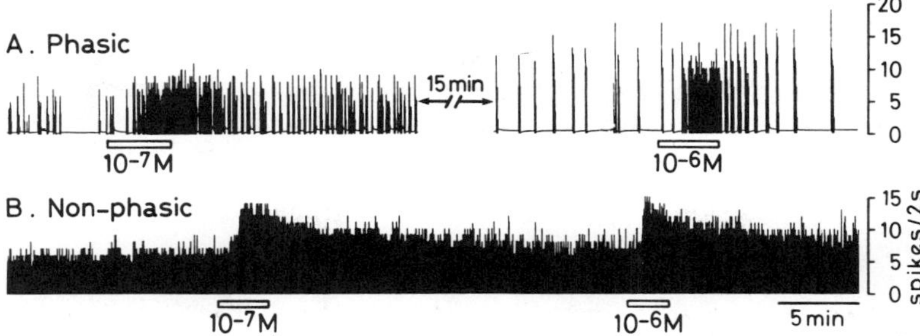

Fig. 4.   Responses of phasic and non-phasic firing neurons in the SON to application of AII (from Okuya et al., 1987). AII increased the duration of bursts in the phasic firing neuron (A) and increased the firing rate of the non-phasic firing neuron (B).

the cells. After washing out saralasin with the control perfusing medium, application of AII induced similar but reduced excitatory responses suggesting long or partly irreversible effects of the antagonist on the SON neurons. These results may provide an evidence for the specificity of the responses of the neurons to AII.

These experiments clearly demonstrate the responses of SON and AV3V neurons to AII. It is not clear whether or not the responses observed were direct or indirect effects of AII, the indirect effects perhaps operating via adjacent neurons in the slice. To determine which alternative was correct, synaptic transmission was blocked with a low Ca^{2+} (0.5mM) and high Mg^{2+} (9mM) medium (Inenaga et al., 1986; Inenaga and Yamashita, 1986). Recordings were taken from 6 SON neurons and 4 AV3V cells. The neurons were first

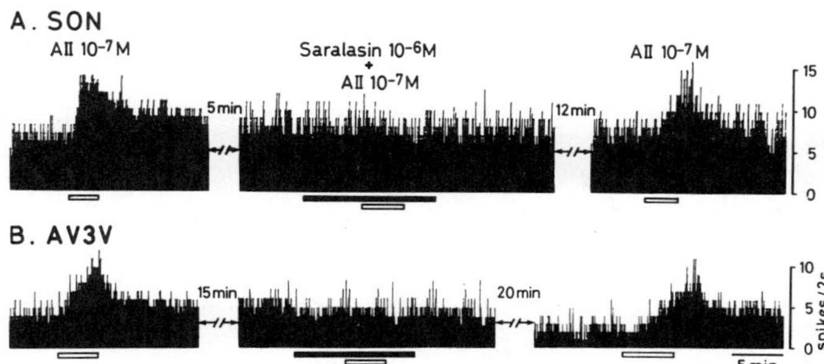

Fig. 5.   Ratemeter records to show the effects of an AII antagonist on the AII-induced activity of neurons in the SON and AV3V (from Okuya et al., 1987). A and B show that the neurons were excited by AII 10^{-7} M (open bar) in the SON and AV3V respectively. Saralasin (AII antagonist) 10^{-6} M when applied (solid bar) completely blocked the effects of AII on each neuron.

excited by AII administration then, 15 min later, the control perfusing medium was replaced by the low Ca/high Mg medium and treatment with AII repeated. All bar one neuron (5/6 SON; 4/4 AV3V) which had shown excitatory responses to AII in the control medium were also excited after blockade of synaptic transmission, implying that the SON and AV3V neurons are directly sensitive to AII.

DISCUSSION

The present studies show that ANP does not affect the activity of SON neurons whereas AII causes excitation of both the phasic and the non-phasic firing neurons. Moreover, AII is active at concentrations that are within physiological ranges found in the csf. Phasic discharge is considered to be the prerogative of vasopressin neurons (Yamashita et al., 1983), so the data imply that AII can activate both oxytocin and vasopressin cells and be act as a neuromodulator or transmitter for the release of the magnocellular peptides.

In contrast to the lack of effect demonstrated on SON neurons, ANP in concentrations within the physiological range in the csf activated neurons in the AV3V area. These results are consistent with the immunocytochemical data that there is a dense accumulation of ANP-immunopositive fibres in the AV3V whilst there are relatively few positive fibres in the SON (Kawata et al., 1985). However, the data conflicts with the finding that there is a significant concentration of ANP binding sites in both the SON and AV3V (Kurihara et al., 1987). This discrepancy may be the result of differential activity expressed by ANP of atrial origin and ANP of central neuronal origin.

Work in this chapter has focussed attention on the role of ANP in controlling the release of vasopressin and oxytocin and raises three important questions: 1) what is/are the origins of ANP that might affect the release of the neurohypophysial hormones? 2) how and 3) where could ANP influence neurosecretion ? It has been shown that both central (Poole et al., 1987; Yamada et al., 1986) and peripheral (Samson, 1985) administration of ANP inhibits vasopressin release, so it possible that both humoral and central ANP could be important in the control of neurohypophysial hormone secretion. It is also possible that the two sources of ANP could give rise to differential effects. Data concerning the mechanisms of ANP action on oxytocin and vasopressin release is less clear. In several experimental paradigms ANP has been shown to inhibit the release of both magnocellular peptides e.g. in response to AII administration (Yamada et al., 1986), hyperosmolarity (Poole et al., 1987) and haemorrhage (Samson, 1985). These findings argue that there may be several sites for ANP action perhaps in AII-sensitive, osmosensitive and haemorrhage-sensitive areas of the brain. However, the magnocellular neurons themselves are clearly sensitive to AII (Okyua et al., 1987) and osmosensitive (Abe & Ogata, 1982; Mason, 1980), therefore it is reasonable to expect that the SON neurons should have responded to ANP. However, it is possible that the SON neurons within the slice are subject to multiple effects from the AII and osmosensitive inputs and this may have affected the results. Last, it is plausible that the ANP receptor-sites may not be present in the SON. AII receptor sites and osmoregulatory sites are located in the AV3V area and in the subfornical organ and these circumventricular organs have direct neural inputs to the SON. Thus the ANP effect may be mediated via these organs.

In summary, these data indicate that there is role for AII and ANP as central modulators or neurotransmitters. The modulatory effect of ANP on vasopressin release in particular requires further investigation.

REFERENCES

Abe, H. and Ogata, N., 1982, Ionic mechanism for the osmotically-induced depolarization in neurones of the guinea-pig supraoptic nucleus in vitro, J. Physiol., 327:157.

Akaishi, T., Negoro, H. and Kobayashi, S., 1981, Electrophysiological evidence for multiple sites of actions of angiotensin II for stimulating paraventricular neurosecretory cells in the rat, Brain Res., 220:386.

Ganten, D., Hutchinson, J.S. and Schelling, P., 1975, The intrinsic brain iso-renin-angiotensin system in the rat: its possible role in central mechanisms of blodd pressure regulation, Clin. Sci. Mol. Med., 48:265s.

Inenaga, K., Dyball, R.E.J., Okuya, S. and Yamashita, H., 1986, Characterization of hypothalamic noradrenaline receptors in the supraoptic nucleus and periventricular region of the paraventricular nucleus of mice in vitro, Brain Res., 369:37.

Inenaga, K. and Yamashita, H., 1986, Excitation of neurones in the rat paraventricular nucleus in vitro by vasopressin and oxytocin, J. Physiol., 370:165.

Kawata, M., Nakao, K., Morii, N., Kiso, Y., Yamashita, H., Imura, H. and Sano, Y., 1985, Atrial natriuretic polypeptide: topographical distribution in the rat brain by radioimmunoassay and immunohistochemistry, Neuroscience, 16:521.

Keil, L.C., Summy-Long, J. and Severs, W.B., 1975, Release of vasopressin by angiotensin II, Endocrinology, 96:1063.

Kleinert, H.D., Maack, T., Atlas, S.A., Januszewicz, A., Sealey, J.E. and Laragh, J.H., 1984, Atrial natriuretic factor inhibits angiotensin-, norepinephrine-, and potassium-induced vascular contractility, Hypertension, 6:Suppl.1, 1143.

Kurihara, M., Saavedra, J.M. and Shigematsu, K., 1987, Localization and characterization of atrial natriuretic peptide binding sites in discrete areas of rat brain and pituitary gland by quantitative autoradiography, Brain Res., 408:31.

Lang, R.E., Rascher, W., Heil, J., Unger, T., Wiedemann, G. and Ganten, D., 1981, Angiotensin stimulates oxytocin release, Life Sci., 29:1425.

Mann, J.F.E., 1982, Brain angiotensin II receptors: possible physiological implications, Exp. Brain Res., Suppl.4:242.

Mason, W.T., 1980, Supraoptic neurones of rat hypothalamus are osmosensitive, Nature, Lond., 287:154.

Nakamura, M., Katsuura, G., Nakao, K. and Imura, H., 1985, Antidipsogenic action of -human atrial natriuretic polypeptide administered intracerebroventricularly in rats, Neurosci. Lett., 58:1.

Okuya, S., Inenaga, K., Kaneko, T. and Yamashita, H., 1987, Angiotensin II sensitive neurons in the supraoptic nucleus, subfornical organ and anteroventral third ventricle of rats in vitro, Brain Res., 402:58.

Okuya, S. and Yamashita, H., 1987, Effects of atrial natriuretic polypeptide on the rat hypothalamic neurones in vitro, J. Physiol., 717:389.

Palovcik, R.A. and Phillips, M.I., 1984, Saralasin increases activity of hippocampal neurons inhibited by angiotensin II, Brain Res., 323:345.

Phillips, M.I., 1978, Angiotensin in the brain, Neuroendocrinology, 25:354.

Poole, C.J.M., Carter D.A., Vallejo, M. and Lightmann, S.L., 1987, Atrial natriuretic factor inhibits the stimulated in vivo and in vitro release of vasopressin and oxytocin in the rat, J. Endocr., 112:97.

Samson, W.K., 1985, Atrial natriuretic factor inhibits dehydration and hemorrhage-induced vasopressin release, Neuroendocrinology, 40:277.

Tanaka, I., Misono, K.S. and Inagami, T., 1984, Atrial natriuretic factor in rat hypothalamus, atria and plasma: determination by specific radioimmunoassay, Biochem. Biophysical Res. Commun., 124:663.

Yamada, T., Nakao, K., Morii, N., Itoh, H., Shiono, S., Sakamoto, M. Sugawara, A., Saito, Y., Ohno, H., Kanai, A., Katsuura, G., Eigyo, M., Matsushita, A., and Imura, H., 1986, Central effect of atrial natriuretic polypeptide on angiotension II-stimulated vasopressin secretion in concious rats, European J. Pharmcol., 125:453.

Yamashita, H., Inenaga, K., Kawata, M. and Sano, Y., 1983, Phasically firing neurons in the supraoptic nucleus of the rat hypothalamus: immunocytochemical and electrophysiological studies, Neurosci. Lett., 37:87.

Molecular Biology and Biosynthesis

Colchicine inhibits intraneuronal transport of neurosecretory material prior to axonal transport.
G. ALONSO.

Molecular forms of locust neuroparsins.
D. BOUREME, A. GIRARDIE & J. GIRARDIE.

POMC expression in foetal rat pituitary gland in organotypic culture.
A. CARR, M.E. BAILEY & L.W. HAYNES.

Proopiomelanocortin (POMC) gene expression in the developing rat spinal cord.
A. CARR, S. BERRY & L.W. HAYNES.

Melanin concentrating hormone: Structure activity studies of a new hormone.
M.E. HADLEY, A.M.L. CASTRUCCI & V.J. HRUBY.

Age-related changes in peptide and amine neurons in the rat hypothalamus.
Y. IBATA, N. MORIMOTO, J. ABE, K. UDA, M. HASEGAWA, I. NAGATSU & N. YANAIHARA.

A schistosome parasite inhibits female reproductive activity of its snail host by (inducing) production of schistosomin(s) which block the action of various female reproductive neuro-hormones.
J. JOOSSE, R. VAN ELK, R.H.M. EBBERINK, H.H. BOER, M. DE JONG-BRINK, W.J.A.G. DICTUS, M. ELSAADANY.

In situ hybridization histochemistry for the analysis of gene expression in the oxytocin and vasopressin neuron system with radiolabelled synthetic oligo-nucleotides.
M. KAWATA, J.T. McCABE, D.W. PFAFF & Y. SANO.

The presence of immunoreactive aMSH in the fish brain.
M. KISHIDA & B.I. BAKER.

Turnover of the vasopressin precursor in salt loaded rats.
L.J. LAWSON & B.T. PICKERING.

In search of the natriuretic hormone : Sodium transport inhibitors from hypothalamus.
G.J. MOORE, P.J. COGGINS & P.M. KEANE.

Processing of neurophysins in the rat neural lobe.
R.W. NEWCOMB & J.J. NORDMANN.

Environmental influences on the production and urophysial content of urotensins in the goby, Gillichthys mirabilis.
D. PEARSON, Dj. ATMANI & D. PULLIAM.

Co-existence of ir MCH and ir MSH in the carp neurohypophysis.
K.A. POWELL & B.I. BAKER.

Insulin-like peptides are encoded on genes expressed in body growth controlling
neuroendocrine cells of the mollusc lymnaea stagnalis.
A.B. SMIT, R. DIRCKS & J. JOOSSE.

Glucocorticoid induced changes in a-N-acetyl endorphin processing in the sheep anterior
pituitary are centrally mediated.
A.I. SMITH, C.A. WALLACE, I.J. CLARKE & J.W. FUNDER.

Does the pig testis contain two vasopressin-like peptides?
A.J. SMITH, H.D. NICHOLSON & B.T. PICKERING.

Two related prohormone genes are expressed in egg-laying behaviour controlling
neuroendocrine cells of the mollusc lymnaea stagnalis.
E. VREUGDENHIL, W. VAN HEUMEN & W.P.M. GERAERTS.

Cell Biology of Neuropeptide Secretion

Suppression of neurohypophysial hormone release by opioids occurs without apparent change
in pituicyte morphology.
G. CLARKE & P.F. HEAP.

Role of connective tissue and glial cells in magnocellular neurosecretory axon regeneration.
H.-D. DELLMANN, L.-F. LUE & S.I. BELLIN.

Changes in the three dimensional morphology of neurosecretory terminal arborisations during
stimulation.
R.E.J. DYBALL & F.D. SHAW.

Innervation of the hypophysial pars intermedia in the newt triturus cristatus and control of
alpha-MSH release.
A. FASOLO, M.F. FRANZONI, V. MAZZI, I. PERROTEAU, J.M. DANGER & H.
VAUDRY.

Increased release of vasopressin and oxytocin from isolated rat neurohypophysis by gaba
receptor stimulation.
B. FJALLAND, J.D. CHRISTENSEN & S. GRELL.

Release of adipokinetic and hypertrehalosaemic neuropeptides from the corpora cardiaca of
locusts, cockroaches and stick insects.
G. GADE.

Ultrastructural localisation of a vasopressin immunoreactive precursor in the homozygous
Brattleboro rat.
S.E.F. GULDENAAR & B.T. PICKERING.

On separate roles of two types of gonadotrophin releasing hormone (GNRH) in the bird.
S. ISHII, A. HATTORI, K. KUBOKAWA, S. MIKAMI, S. YAMADA & M. WADA.

Two functional and structural hemal compartments in the organum vasculosum laminae
terminalis of the rat.
B. KRISCH.

Receptors for atrial natriuretic peptide on pituicyte-like cells in culture.
S.M. LUCKMAN & R.J. BICKNELL.

Effect of propranolol on immobilization stress-induced changes in oxytocin in the pituitary, hypothalamus and spinal cord.
C. MIASKOWSKI, G.L. ONG & J. HALDAR.

Relaxin potentiates vasopressin release in rats _in vivo_.
L.P. PARRY & A.J.S. SUMMERLEE.

The central delivery of neuropeptide hormones following neural transplantation in the endocrine hypothalamus of the brattleboro rat.
D.E. SCOTT.

Hormone release from isolated neurohypophyseal nerve endings and swellings _in vitro_.
F.D. SHAW & J.F. MORRIS.

Release of antidiuretic hormone from specific axons in the corpus cardiacum of the house cricket, _acheta domesticus_, in response to water stress.
J.H. SPRING.

Characterisation of an enriched fraction of nerve terminals (neurosecretosomes) from the bovine neurohypophysis.
E.C. TOESCU, F.D. SHAW & J.F. MORRIS.

A morphological study of neurosecretory axons and endings in the rat neurohypophysis.
C.D. TWEEDLE, K. SMITHSON & G.I. HATTON.

Effects of adrenalectomy on spinal cord vasopressin and oxytocin levels.
J. VRBA, D. LUKIC & J. HALDAR.

Neurohypophysial actions of opioids involve K-opioid receptors localised on secretory terminals of magnocellular neurones.
B-G. ZHAO, C. CHAPMAN & R.J. BICKNELL.

Morphological Techniques in the Study of Neuropeptides

Combined immunoperoxidase and radioimmunocytochemistry: a tool for double immunocytochemical labelling of peptidergic neurons.
G. ALONSO.

Interconnections between vasopressinergic neurons in the suprachiasmatic nucleus.
M. CASTEL, N. FEINSTEIN & S. COHEN.

Neurophysin-containing innervation of the mouse forebrain.
M. CASTEL & J.F. MORRIS.

Immunocytochemical studies on the hypothalamo-hypophysial system of reptiles.
P. FERNANDEZ-LLEBREZ, J. PEREZ, J.M. PEREZ-FIGARES & E.M. RODRIGUEZ.

Photoperiod induces changes in the immunostaining of the LHRH system of the Japanese quail.
R.G. FOSTER, G.C. PANZICA, D.M. PARRY & C. VIGLIETTI-PANZICA.

Neuropeptides and serotonin in the nervous system of _schistosoma mansoni_.
M.K.S. GUSTAFSSON.

Prolactin-like immunoreactivity in the brain and affiliated neuroendocrine structures of the insect _leucophaea maderae_.
G.N. HANSEN, B.L. HANSEN & B. SCHARRER.

Oxytocin and the morphology of 9-day old rat testes.
J. HUMPHRYS & R.T.S. WORLEY.

Cytochemical asymmetry of 5'-nucleotidase in the retinal muller (glial) plasma membrane: A possible role of edenosine as a neuromodulator.
S.T. HUSSAIN & G.W. KREUTZBERG.

Development of neurosecretory nerve endings in the hypothalamus of bufo japonicus deprived of the pituitary primordium.
K. KAWAMURA, S. KIKUYAMA & T. SHIBASAKI.

Development of preoptic recess organ in toad tadpoles.
S. KIKUYAMA, K. KAWAMURA, M. MIYAKAWA, Y. ARAI & I. NAGATSU.

A system of liquor-contacting neurons, containing immunoreactive serotonin and somatostatin, in the hypothalamus of elasmobranchs.
P. MEURLING, E.M. RODRIGUEZ & C.R. YULIS.

Fine structure of avian vasotocin-immunoreacting neurons.
G.C. PANZICA, G.C. ANSELMETTI, G. CORVETTI & C. VIGLIETTI-PANZICA.

Substance P-like immunoreactivity demonstrated by the pap method in hydra attenuata.
P. PIEROBON & M. KEMALI.

Morphometric analysis of the postnatal development of magnocellular neurones in the supraoptic nucleus of brattleboro and long evans rats.
D.V. POW & J.F. MORRIS.

Differential distribution of acetylcholinesterase activity among immunocytochemically identified magnocellular neurosecretory neurones.
D.V. POW & J.F. MORRIS.

Morphological and immunocytochemical evidence for exocytotic release of granules from multiple anatomically distinct compartments in the neural lobe of long evans and homozygous brattleboro rats.
D.V. POW & J.F. MORRIS.

Histomorphological changes in the neurosecretory cells of an indian earthworm metaphire posthuma (vaillant) during development.
O. PRASAD & PAWAN KUMAR.

Subsets of neurons in platyhelminths visualized by 5-HT, RF-amide and FMRF-amide antisera.
M. REUTER.

Aminergic and peptidergic systems in the suprachiasmatic nucleus of an hibernator, Spermophilus richardsonii.
C.M. SCHINDLER, A. KRIETE & F. NURNBERGER.

Peptidergic neurons in the Colorado potato beetle identified with polyclonal and monoclonal antibodies to regulatory peptides.
H. SCHOONEVELD & J.A. VEENSTRA.

Detection of locust neuroparsin immuno-reactivity in invertebrates and vertebrates.
M. TAMARELLE & J. GIRARDIE.

Urotensin-like immunoreactivity in the nervous system of flatworms - whole mount immunocytochemistry.
M. WIKGREN.

Towards oxytocin function in the testes.
R.T.S. WORLEY, J.R. CHALLIS, V.J. AYAD & M.A.C. APPLETON.

Electrical Activity of Peptidergic Neurones

Electrophysiological studies of cholecystokinin action on neurones in the ventromedial
nucleus of rat hypothalamic slices.
P. BODEN & R.G. HILL.

Neurohypophysial hormone release evoked by intraventricular hypertoxic sodium chloride:
The significance of the subfornical organ.
G. CLARKE & S.J. MacMILLAN.

Antagonists of 5-hydroxytryptamine reduce the effect of opioids on the milk-ejection reflex.
G. CLARKE & D.M. WRIGHT.

The bed nucleus of the stria terminalis (BNST): A possible site for mediating the oxytocin-
induced facilitation of milk ejection.
K.L. CUTLER, C.D. INGRAM & J.B. WAKERLEY.

Investigation of the facilitatory effect of stimulation of the hypothalamo-neurohypophysial
system on milk ejection.
I. DAVISON, C.D. INGRAM & J.B. WAKERLEY.

Autoradiographic mapping of different types of neurohypophysial hormone-binding sites in
the long evans rat forebrain and pituitary gland.
M.J. FREUND-MERCIER, M.M. DIETL, M.E. STOECKEL, J.M. PALACIOS & PH.
RICHARD.

Reduced oxytocin response to stress, hyperosmolality and electrical stimulation of the
hypothalamo-neurohypophysial system in lactating rats.
T. HIGUCHI, K. HONDA, S. TAKANO & H. NEGORO.

Activation of the paraventricular neurosecretory cells in osmotic stimulation of the
anteroventral third ventricle in the rat.
K. HONDA, H. NEGORO, T. HIGUCHI & S. TAKANO.

Patch clamp study of potassium channels in teleost fish rostral pars distalis cells.
C.A. LORETZ & C.R. FOURTNER.

Functional interactions between oxytocin cells of the four magnocellular hypothalamic nuclei
in anaesthetized suckled rats.
F. MOOS, G. KUNTZELMAN & PH. RICHARD.

Specific release of oxytocin (OT) inside the hypothalamic magnocellular nuclei during the
milk ejection reflex in rats.
F. MOOS, D. POULAIN, Y. GUERNE, F. RODRIGUEZ, J.D. VINCENT & PH.
RICHARD.

Frequency-dependent effects of different potassium channel blockers on the
neurohypophysial release of vasopressin evoked by electrical stimulation.
K. RACKE, D. JOST & H.P. HOBBACH.

Activation of centrally-projecting oxytocin neurones by naloxone in rats treated chronically
with intracerebroventricular (i.c.v.) morphine.
I.C.A.F. ROBINSON, J.E. COOMBES & J.A. RUSSELL.

Activity of putative vasopressinergic neurones in male and female rabbits.
A.J.S. SUMMERLEE.

Relaxin effects the activity of oxytocin neurones in the conscious rat.
A.J.S. SUMMERLEE, S.A. JONES & K.T. O'BYRNE.

Dynorphin-related peptides act at kappa opiate receptors to inhibit secretion of oxytocin but not vasopressin from the magnocellular system during dehydration.
J.Y. SUMMY-LONG, L.M. ROSELLA-DAMPMAN & G. McLEMORE.

Electrical and secretory activities of nerve terminals visualized by digital imaging.
S. TERAKAWA & M. NAGANO.

Frequency facilitation of hormone release from the neural lobe may be mediated partly by stimulus-evoked increase in extracellular (K+).
S.A. WAY, K. SHIBUKI & G. LENG.

PARTICIPANTS

G. ALONSO	Laboratoire de Neuroendocrinologie, Université de Montpellier 11, Pl. E. Baraillon 3G060, Montpellier, Cedex, France.
B.I. BAKER	School of Biological Sciences, Bath University, Claverton Down, Bath, BA2 7AY, UK.
R.J. BICKNELL	A.F.R.C. Institute of Animal Physiology & Genetics Research, Babraham, Cambridge, CB2 4AT, UK.
G. BILBE	ZMBH, 1m Neuenheimer Feld 282, D-6900 Heidelberg 1, West Germany.
S.D. BIRKETT	Department of Anatomy, The Medical School, Bristol, BS8 1TD, UK.
P. BODEN	Parke-Davis Research Unit, Addenbrookes Hospital Site, Cambridge, UK.
D. BOUREME	Université de Bordeaux 1, Laboratoire de Neuroendocrinologie, UA CNRA 1138, Avenue des Facultes, 33405 Talence Cedex, France.
K. BROWN-GRANT	Memorial University, Faculty of Medicine, St. John's, Newfoundland, Canada, A1B 3V6.
B.J. CANNY	Medical Research Centre, Prince Henry's Hospital, St. Kilda Road, Melbourne, Australia 3004.
M. CASTEL	Department Experimental Zoology, Life Sciences, Hebrew University of Jerusalem, Israel.
V. CHAN-PALAY	Neurology Institute, University Hospital of Zurich, Zurich, CH-8091, Switzerland.
H.M. CHARLTON	Department of Human Anatomy, South Parks Road, Oxford, OX1 3QX, UK.
J.D. CHRISTENSEN	Royal Danish School of Pharmacy, Department of Biology, Universitetsparken 2, 2100 Copenhagen 0, Denmark.
G. CLARKE	Department of Anatomy, The Medical School, Bristol, BS8 1TD, UK.
I.J. CLARKE	Medical Research Centre, Prince Henry's Hospital, St. Kilda Road, Melbourne, Australia 3004.

P. COBBETT

A.F.R.C., Institute of Animal Physiology & Genetics Research, Babraham, Cambridge, CB2 4AT, UK.

D. COLIN

Centre de Neurochimie, 5 rue Blaise Pascal, 67084 Strasbourg Cedex, France.

B.A. CROSS

Institute of Animal Physiology & Genetics Research, Babraham, Cambridge, CB2 4AT, UK.

H.D. DELLMANN

Department of Veterinary Anatomy, Iowa State University, Ames, IA 50011, USA.

P. DENNING-KENDALL

Department of Anatomy, The Medical School, Bristol, BS8 1TD, UK.

J.J. DREIFUSS

Départment de Physiologie, Centre Medical Universitaire, 9 av de Champel, 1211 Genève 4, Switzerland.

H. DUVE

School of Biological Sciences, Queen Mary College, London University, Mile End Road, London, E1 4NS, UK.

R.E.J. DYBALL

University of Cambridge, Department of Anatomy, Downing Street, Cambridge, CB2 3DY, UK.

R.G. DYER

A.F.R.C., Institute of Animal Physiology & Genetics Research, Brabraham, Cambridge, CB2 4AT, UK.

A. FASOLO

Departimento Biologia Animale, Via Accademia Albertino, 17, 10123 Torino, Italy.

P. FERNANDEZ-LLEBREZ

Department of Animal Physiology, Faculty of Sciences, Departamento de Fisiologia Animal, Facultad de Ciencias, 29071 Malaga, Spain.

B. FJALLAND

Department of Biology, Royal Danish School of Pharmacy, Universitetsparken 2, 2100 Copenhagen 0, Denmark.

B.K. FOLLETT

Department of Zoology, University of Bristol, Woodland Road, Bristol, BS8 1UG, UK.

G. GÄDE

Institut für Zoologie IV, der Universität Düsseldorf, Universitsstr. 1, D-4000 Düsseldorf 1, F.R.G.

H. GAINER

National Institute of Child Health, NIH, Building 36, Room 2A21, Bethesda, Maryland 20205, USA.

W.P.M. GERAERTS

Free University of Amsterdam, Biological Laboratory, de Boelelaan 1007, 1007 MC Amsterdam, The Netherlands.

C.L. GILBERT

Department of Anatomy, The Medical School, Bristol, BS8 1TD, UK.

A. GIRARDIE

Université de Bordeaux 1, Laboratoire de Neuroendocrinologie, Avenue des Facultés, 33405 Talence Cedex, France.

J. GIRARDIE

Université de Bordeaux 1, Laboratoire de Neuroendocrinologie, Avenue des Facultés, 33405 Talence Cedex, France.

D.W. GOLDING — Department of Zoology, The University, Newcastle upon Tyne, NE1 7RU, UK.

S.E.F. GULDENAAR — Department of Anatomy, The Medical School, Bristol, BS8 1TD, UK.

M. GUSTAFSSON — Department of Biology, Abo Akademi, Porthansgatan 3, SF-20500 Abo, Finland.

M.E. HADLEY — Department of Anatomy, University of Arizona, Tucson, AZ 85724, USA.

J. HALDAR — St. John's University, Department of Biological Sciences, Grand Central and Utopia Parkways, Jamaica, New York 11439, USA.

M. HAMAMURA — Department of Physiology, Jichi Medical School, Minamikawachi-Machi, Tochigi, Japan 329-04.

B.L. HANSEN — Institute of Medical Microbiology, University of Copenhagen, 22 Juliane Maries Vej, DK 2100 Copenhagen 0, Denmark.

G.N. HANSEN — Institute of Cell Biology and Anatomy, University of Copenhagen, 15 Universitetsparken, DK-2100 Copenhagen 0, Denmark.

A.J. HARMAR — MRC Brain Metabolism Unit, 153 Morningside Drive, Edinburgh, Scotland, EH10 5LG, UK.

G.I. HATTON — Neuroscience Program, Michigan State University, East Lansing, MI 48824-1117, USA.

L.W. HAYNES — Department of Zoology, University of Bristol, Woodland Road, Bristol, BS8 1UG, UK.

P.M. HEADLEY — Department of Physiology, The Medical School, Bristol, BS8 1TD, UK.

P.F. HEAP — Department of Anatomy, The Medical School, Bristol, BS8 1TD, UK.

T. HIGUCHI — Fukui Medical School, Department of Physiology, Matsuoka, Fukui 910-11, Japan.

R.G. HILL — Parke-Davis Research Unit, Addenbrookes Hospital Site, Cambridge, UK.

S. HOLLINGSWORTH — A.F.R.C., Institute of Animal Physiology & Genetics Research, Babraham, Cambridge, CB2 4AT, UK.

K. HONDA — Department of Physiology, Fukui Medical School, Matsuoka, Fukui, 910-11, Japan.

J. HUMPHRYS — Department of Anatomy, The Medical School, Bristol, BS8 1TD, UK.

S.T. HUSSAIN — Department of Biological Sciences, Yarmouk University, Irbid, Jordan.

Y. IBATA

Department of Anatomy, Kyoto Prefectural University of Medicine, Kawaramachi Hirokoji, Kamikyoku Kyoto, Japan 602.

T. ICHIKAWA

Tokyo Metropolitan Institute for Neurosciences, 2-6 Musashidai, Fuchu-City, Tokyo, Japan.

C.D. INGRAM

Department of Anatomy, The Medical School, Bristol, BS8 1TD, UK.

S. ISHII

Department of Biology, Waseda University, Nishi-Waseda 1-6-1, Shiniuku-Ku, Tokyo 160, Japan.

R. IVELL

Institute for Hormone and Fertility Research, Grandweg 64, 2000 Hamburg 54, West Germany.

J. JOOSSE

Vrije Universiteit, Biologisch Laboratorium, Postbus 7161, 1007 MC Amsterdam, The Netherlands.

H. KANNAN

Department of Physiology, University of Occupational and Environmental Health, Japan School of Medicine, 1-1 Iseigaoka, Yahatanishi-ku, Kitakyushu, 807 Japan.

M. KASAI

Department of Physiology, University of Occupational and Environmental Health, Japan School of Medicine, 1-1 Iseigaoka, Yahatanishi-ku, Kitakyushu, 807 Japan.

K. KAWAMURA

Department of Biology, School of Education, Waseda University Nishiwaseda 1, Shinjuku-ku, Tokyo 160, Japan.

M. KAWATA

Department of Anatomy, Kyoto Prefectural University of Medicine, Kawaramachi-Hirokoji, Kamikyo-ku, Kyoto 602, Japan.

P.M. KEEN

Department of Pharmacology, The Medical School, Bristol, BS8 1TD, UK.

M. KEMALI

1st Cibernetica C.N.R., Via Toiano 6, 80072 Arco Felice, Naples, Italy.

S. KIKUYAMA

Department of Biology, School of Education, Waseda University, Nishiwaseda 1, Shinjuku-ku, Tokyo 160, Japan.

M. KISHIDA

University of Bath, Claverton Down, Bath, BA2 7AY, UK.

H. KOBAYASHI

Research Laboratory, Zenyaku Kogyo Co. Ltd., Ohizumi-cho 2-33-7, Nerima-ku, Tokyo 177, Japan.

B. KRISCH

Anatomisches Institut der Universität Kiel, Olshausenstrabe 40, D-2300 Kiel, Germany.

K. KUBOKAWA

Department of Biology, Waseda University, Nishi-Waseda 1-6-1, Shinjuku-Ku, Tokyo 160, Japan.

K. KUROSUMI

Department of Morphology, Institute of Endocrinology, Gunma University, Showa-Machi 3-39-15, Maebashi 371, Japan.

L.J. LAWSON

Department of Anatomy, The Medical School, Bristol, BS8 1TD, UK.

G. LENG — A.F.R.C., Institute of Animal Physiology & Genetics Research, Babraham, Cambridge, CB2 4AT, UK.

D.W. LINCOLN — MRC Centre for Reproductive Biology, 37 Chalmers Street, Edinburgh, Scotland, EH3 9EW, UK.

A. LIVINGSTON — Department of Pharmacology, The Medical School, Bristol, BS8 1TD, UK.

C.A. LORETZ — State University of New York at Buffalo, Department of Biological Sciences, 109 Cooke Hall, Buffalo, New York 14260, USA.

S.M. LUCKMAN — A.F.R.C. Instutute of Animal Physiology, Babraham, Cambridge, CB2 4AT, UK.

S.J.A. MacMILLAN — Department of Anatomy, The Medical School, Bristol, BS8 1TD, UK.

S. MARIVOET — Zoological Institute, Catholic University of Leuven, Naamsestraat 59, B-300 Leuven, Belgium.

T. MARTINEZ — Department of Biological Organic Chemistry, Centro de Investigacion y Desarrollo, Jorge Girona Salgado 18-26, Barcelona - 08034, Spain.

W.T. MASON — A.F.R.C. Institute of Animal Physiology, Babraham, Cambridge, CB2 4AT, UK.

P.B. MEURLING — Department of Zoology, University of Lund, Helgonavägen 3, S-223 62, Lund, Sweden.

C. MIASKOVSKI — St. John's University, c/o Department of Biological Sciences, Jamaica, New York 11439, USA.

G.J. MOORE — University of Calgary, Department Medical Biochemistry, 3330 Hospital Dr. N.W. Calgary, Alberta, T2N 4N1, Canada.

F. MOOS — Laboratoire de Physiologie, Université Louis Pasteur, UA 309 CNRS, 21 rue René Descartes, F67084, Strasbourg Cedex, France.

J.F. MORRIS — Department of Human Anatomy, The University, South Parks Road, Oxford, OX1 3QX, UK.

H.D. NICHOLSON — Department of Anatomy, The Medical School, Bristol, BS8 1TD, UK.

J.J. NORDMANN — INSERM, Unité de Neurochimie Normale et Pathologique, 5 rue Blaise Pascal, 67084, Strasbourg Cedex, France.

F. NURNBERGER — Department of Anatomy & Cytobiology, Aulweb 123, D-6300 Giessen, Federal Republic of Germany.

M. O'SHEA — Laboratoire de Neurobiologie, Université de Genève, 20 rue de l'Ecole de Médicine, CH-1211, Genève 4, Switzerland.

G.C. PANZICA — Department Human Anatomy & Physiology, c.so M. D'Azeglio 52, 1-10126 Torino, Italy.

L. PARRY — Department of Anatomy, The Veterinary School, Park Row, Bristol, BS1 5LS, UK.

D. PEARSON — California State University, 5151 State University Drive, Los Angeles, CA 90032, USA.

J-M. PEREZ-FIGARES — Faculty of Sciences, University of Malaga, Department Biologia Celular Facultad, Ciencias Universidad de Malaga, 29071 Malaga, Spain.

B.T. PICKERING — Department of Anatomy, The Medical School, Bristol, BS8 1TD, UK.

P. PIEROBON — 1st Cibernetica C.N.R., Via Toiano 6, 80072 Arco Felice, Naples, Italy.

J. POLAK — Royal Postgraduate Medical School, Department of Histochemistry, Hammersmith Hospital, Du Cane Road, London, W12 OHS, UK.

D. POULAIN — Unité de Recherches de Neurobiologie des Compartements, INSERM U 176, 33077 Bordeaux Cedex, France.

D.V. POW — Department of Human Anatomy, University of Oxford, South Parks Road, Oxford, OX1 3QX, UK.

O. PRASAD — The University of Allahabad, Faculty of Science, Department of Zoology, Allahabad 211002, India.

K. RACKE — Pharmakologisches Institut der Universität Mainz, 6500 Mainz, Obere Zahlbacher Strabe 67, FRG.

A. RANDALL — Department of Biochemistry, The Medical School, Bristol, BS8 1TD, UK.

S. RAWLINGS — A.F.R.C., Institute of Animal Physiology & Genetics Research, Babraham, Cambridge, CB2 4AT, UK.

L. RENAUD — Division of Neurology, Montreal General Hospital, 1650 Cedar Avenue, Montreal, H3G 1A4, Canada.

M. REUTER — Department Biology, Abo Akademi, Porthansgatan 3, SF-20500 ABO 50, Finland.

PH. RICHARD — Laboratoire de Physiologie, Université Louis Pasteur, UA 309 CNRS, 21 rue René Descartes, F67084, Strasbourg Cedex, France.

J.L. ROBERTS — The Mount Sinai Medical Center, One Gustave L Levy Place, New York, NY 10029, USA.

I.C.A.F. ROBINSON — National Institute for Medical Research, The Ridgeway, Mill Hill, London, NW7 1AA, UK.

E.M. RODRIGUEZ — Instituto de Histologia y Patologia, Facultad de Medicina, Universidad Austral de Chile, Casilla 567, Valdivia, Chile.

J. ROSSIER — CNRS, Départment de Neurophysiologie Appliquée, 91190, Gif Sur Yvettes, France.

E.W. ROUBOS — Vrije Universiteit, Biologisch Laboratorium, Postbus 7161, 1007 MC Amsterdam, The Netherlands.

J.A. RUSSELL — Department of Physiology, University of Edinburgh, University Medical School, Teviot Place, Edinburgh, EH8 9AG, UK.

Y. SANO — Department of Anatomy, Kyoto Prefectural University of Medicine, Kawaramachi-Hirokoji, Kamikyo-ku, Kyoto 602, Japan.

B. SCHARRER — Albert Einstein College of Medicine, 1300 Morris Park Avenue, Bronx, New York 10461, USA.

R. SCHELLER — Department of Biological Sciences, Stanford University, Stanford, California 94305-5020, USA.

C.M. SCHINDLER — Department of Anatomy and Cytobiology, Aulweg 123, D-6300 Giessen, Federal Republic of Germany.

H. SCHOONEVELD — Department of Entomology, Agricultural University, Binnenhaven 7, 6709 PD Wageningen, The Netherlands.

F.D. SHAW — Department of Anatomy, Downing Street, Cambridge, CB2 3DY, UK.

A.B. SMIT — Free University of Amsterdam, Biologisch Laboratorium, Postbus 7161, 1007 MC Amsterdam, The Netherlands.

A.I. SMITH — Medical Research Centre, Prince Henry's Hospital, St. Kilda Road, Melbourne, Victoria 3004, Australia.

J.H. SPRING — Department of Biology, University of Southwestern Louisiana, Lafayette, Louisiana 70504, USA.

C. STEEL — Department of Biology, York University, 4700 Keele Street, Downsview, Ontario M3J 1P3, Canada.

R.J. STERLING — Institute of Animal Physiology & Genetics Research, Roslin, Midlothian, EH25 9PS, Scotland, UK.

E.L. STEUNKEL — Department of Physiology, University of California, San Francisco, CA 94143, USA.

A.J.S. SUMMERLEE — Department of Anatomy, The Veterinary School, Park Row, Bristol, BS1 5LS, UK.

J.Y. SUMMY-LONG — The Pennsylvania State University, Department of Pharmacology, College of Medicine, Hershey, PA 17033, USA.

R.W. SWANN — School of Biological Sciences, University of Bath, Claverton Down, Bath, BA2 7AY, UK.

P.H. TAGHERT — Department of Anatomy & Neurobiology, Washington University School of Medicine, 660 S. Euclid Avenue, St. Louis, MO 63110, USA.

M. TAMARELLE — Laboratoire de Neuroendocrinologie, Avenue Des Facultés, Talence 33405, France.

S. TERAKAWA — National Institute for Physiological Sciences, Myodaiji, Okazaki, 444 Japan.

D. THEODOSIS — Unité de Recherches de Neurobiologie des Compartements, INSERM U 176, 33077, Bordeaux Cedex, France.

E. THOMSEN — Institute of Cell Biology and Anatomy, The Zoological Institutes, University of Copenhagen, 15 Universitetsparken, DK-2100 Copenhagen 0, Denmark.

A. THORPE — School of Biological Sciences, Queen Mary College, London University, Mile End Road, London, E1 4NS, UK.

E.C. TOESCU — Department of Human Anatomy, Oxford University, South Parks Road, Oxford, OX1 3QX, UK.

E. TRIBOLLET — Département de Physiologie, Centre Médical Universitaire, 9 av de Champel, 1211 Genève 4, Switzerland.

C. TWEEDLE — Department of Anatomy, Michigan State University, E. Lansing, MI 48824, USA.

A. URANO — Saitama University, Faculty of Science, Urawa 338, Japan.

X. VAFOPOULOU — York University, Faculty of Science, 4700 Keele Street, North York, Ontario M3J 1P3, Canada.

F. VANDESANDE — Zoological Institute, Catholic University of Leuven, Naamsestraat 59, B-3000 Leuven, Belgium.

J.A. VEENSTRA — Department of Biological Organic Chemistry, Centro de Investigation y Desarrollo, Jorge Girona Salgade, 18-26, 08034 Barcelona, Spain.

C. VIGLIETTI-PANZICA — Department Human Anatomy & Physiology, C.so M. D'Azeglio 52, 1-10126 Torino, Italy.

J.B. WAKERLEY — Department of Anatomy, The Medical School, Bristol, BS8 1TD, UK.

S.A. WAY — A.F.R.C., Institute of Animal Physiology & Genetics Research, Babraham, Cambridge, CB2 4AT, UK.

B. WEATHERHEAD — Department of Anatomy, University of Hong Kong, 5 Sassoon Road, Hong Kong.

M. WIKGREN — Department of Biology, Abo Akademi, Porthansgatan 3, SF-20500 Abo, Finland.

R.T.S. WORLEY — Department of Anatomy, The Medical School, Bristol, BS8 1TD, UK.

D.M. WRIGHT — Upjohn Limited, Fleming Way, Crawley, West Sussex, RH10 2NJ, UK.

H. YAMASHITA — University of Occupational and Environmental Health, Department of Physiology, School of Medicine, Yahatanishi-ku, Kitakyushu 807, Japan.

Z. YARON — Department of Zoology, Tel Aviv University, Tel Aviv 69978, Israel.

W.S.B. YEUNG — Department of Anatomy, The Medical School, Bristol, BS8 1TD, UK.

B. ZHAO — A.F.R.C., Institute of Animal Physiology & Genetics Research, Babraham, Cambridge, CB2 4AT, UK.

R.S. ZUCKER — Department Physiology & Anatomy, University of California, Berkeley, California 4720, USA.

Precursor proteins, 5, 11, 17, 124, 125, 127, 129
Preoptic neurones, 191-195
Preprotachykinin, 31, 32
    gene, 29
Presynaptic terminals, 137
Proctolin, 101, 102, 103
Proinsulin, 69
Prolactin, 164
Pulsatile secretion, 170

Radioimmunoassay, 6, 54, 107
Rana catesbeiana, 48
Recombinant DNA, 7, 124
Recordings
    from brain slices containing AV3V region and
        supraoptic nucleus, 244
    from magnocellular neurones, 167-168, 171
    from preoptic neurones in conscious rabbits,
        192
Red pigment concentrating hormone, 210
Relaxin, 152
RNA splicing, 29, 31, 32
Salt loading, 4, 48, 114
Saralasin, 246-47
Sea anemone, 12
Secretory granules (see also Neurosecretory
        granule), 61, 129, 130, 132, 137, 144
Secretory terminals, 175
Scanning microscopy, 114
Schistocerca gregaria, 100
Sex steroids, 163
Sexual receptivity, 85
Shared synapses, 161-164
Sinus gland
    electrophysiology of cells in, 152, 199-203, 207-
        216
    histology of, in isopod crustaceans, 208
    ultrastructure of neurosecretory cells in, 209
Somatostatin, 30, 35, 89, 90, 92
Spike broadening, 200-201
Spike duration, 173
Subcommissural organ, 72
Subfornical organ, 173, 222
Substance P, 31, 90
Suckling
    afferent pathways transmitting, 170
    and magnocellular neuronal activity, 163, 168,
        171, 183
Suction electrode, 207, 212
Suprachiasmatic nucleus, 4
Supraoptic neurones
    in cultures for electrophysiological recording,
        235-241
    effect of angiotensin II, 245-248
    effect of atrial natriuretic polypeptide, 243-245
Supraoptic nucleus, 4, 49, 50, 75
    electrophysiological recordings from, 168, 171,
        173, 219-221
    inputs to, 222-223
    structural reorganization during lactation, 157-
        164

Stimulus secretion coupling, 147-153, 174, 181, 185
Synaptic plasticity, 157-164
Synaptic terminals, 137-138
Synaptic vesicles, (see also Microvesicles), 137-138,
        141, 144
Synaptoid vesicles, 138, 141
Synaptophysin, 62
Synchronization
    of firing in sinus gland neurosecretory cells, 213
    of GnRH neurones, 193
    of oxytocin cell firing, 164, 170-171, 184

Tannic acid, 130, 138, 141, 143, 149
TATA box, 2
Terminal potentials, 212
Testis, 5
Tetrodotoxin, 181
Thymus, 5
Transverse nerve (TN), 19, 20, 21

Urophysin, 35
Urophysis
    caudal neurosecretary system, 35, 37
Urotensin, 35, 37, 40
    co-expression, 35, 38, 40
    precursor, 35, 36

Varicosities, 144
Vasopressin, 71, 81
    associated glycopeptide, 2,3,4,71,76
    gene, 1, 2, 3, 5, 6, 43, 48, 49
    immunoreactive neurons, 46, 50, 74, 118
    neurophysin, 71, 77, 116
    precursor, 75, 76, 77
    mRNA, 4, 43, 50
    receptors, 84, 86
Vasopressin cells
    electrical activity of, 168-170, 172, 181-188
    effects of atrial natriuretic polypeptide and
        angiotensin II on, 243-248
    intrinsic and synaptic factors regulating, 219-223
Vasotocin
    gene, 48
    immunoreactive neurons, 46
    mRNA, 46, 48, 50
    precursor, 48
Veratridine, 151
VIP, 90
VIP-immunoreactive cells, 64
Voltage-clamping, 235-240
Voltage-dependent channels, 151

X-organ, 199-203
Xenopus laevis, 48

Zebra finch, 48
Zona incerta, 195